KℰE Series on Knots and Everything — Vol. 1

KNOTS
AND
PHYSICS

Louis H. Kauffman

Department of Mathematics,
Statistics and Computer Science
University of Illinois at Chicago

World Scientific

Singapore • *New Jersey* • *London* • *Hong Kong*

.4309613

Published by

World Scientific Publishing Co. Pte. Ltd.
P O Box 128, Farrer Road, Singapore 9128
USA office: 687 Hartwell Street, Teaneck, NJ 07666
UK office: 73 Lynton Mead, Totteridge, London N20 8DH

Library of Congress Cataloging-in-Publication data is available.

KNOTS AND PHYSICS

ISBN 981-02-0343-8
        981-02-0344-6 (pbk)

Printed in Singapore by Loi Printing Pte. Ltd.

KNOTS AND PHYSICS

Preface

This book has its origins in two short courses given by the author in Bologna and Torino, Italy during the Fall of 1985. At that time, connections between statistical physics and the Jones polynomial were just beginning to appear, and it seemed to be a good idea to write a book of lecture notes entitled **Knots and Physics**.

The subject of knot polynomials was opening up, with the Jones polynomial as the first link polynomial able to distinguish knots from their mirror images. *We were looking at the tip of an iceberg!* The field has grown by leaps and bounds with remarkable contributions from mathematicians and physicists – a wonderful interdisciplinary interplay.

In writing this book I wanted to preserve the flavor of those old Bologna/Torino notes, and I wanted to provide a pathway into the more recent events. After a good deal of exploration, I decided, in 1989, to design a book divided into two parts. The first part would be combinatorial, elementary, devoted to the bracket polynomial as state model, partition function, vacuum-vacuum amplitude, Yang-Baxter model. The bracket also provides an entry point into the subject of quantum groups, and it is the beginning of a significant generalization of the Penrose spin-networks (see Part II, section 13^0.) Part II is an exposition of a set of related topics, and provides room for recent developments. In its first incarnation, Part II held material on the Potts model and on spin-networks.

Part I grew to include expositions of Yang-Baxter models for the Homfly and Kauffman polynomials – as discovered by Jones and Turaev, and a treatment of the Alexander polynomial based on work of Francois Jaeger, Hubert Saleur and the author. By using Yang-Baxter models, we obtain an induction-free introduction to the existence of the Jones polynomial and its generalizations. Later, Part I grew some more and picked up a chapter on the 3-manifold invariants of Reshetikhin and Turaev as reformulated by Raymond Lickorish. The Lickorish model is completely elementary, using nothing but the bracket, trickery with link diagrams, and the tangle diagrammatic interpretation of the Temperley-Lieb algebra. These 3-manifold invariants were foretold by Edward Witten in his landmark paper on quantum field theory and the Jones polynomial. Part I ends with an introduction to Witten's functional integral formalism, and shows how the knot polynomials arise in that context. An appendix computes the Yang-Baxter solutions for spin-preserving R-matrices in dimension two. From this place the Jones and Alexander polynomials appear as twins!

Part II begins with Bayman's theory of hitches – how to prove that your horse can not get away if you tie him with a well constructed clove hitch. Then follows a discussion of the experiments available with a rubber band. In sections 3^0 and 4^0 we discuss attempts to burrow beneath the usual Reidemeister moves for link diagrams. There are undiscovered realms beneath our feet. Then comes a discussion of the Penrose chromatic recursion (for colorations of planar three-valent graphs) and its generalizations. This provides an elementary entrance into spin networks, coloring and recoupling theory. Sections 7^0 and 8^0 on coloring and the Potts model are taken directly from the course in Torino. They show how the bracket model encompasses the dichromatic polynomial, the Jones polynomial and the Potts model. Section 9^0 is a notebook on spin, quantum mechanics, and special relativity. Sections 10^0 and 11^0 play with the quaternions, Cayley numbers and the Dirac string trick. Section 11^0 gives instructions for building your very own topological/mechanical quaternion demonstrator with nothing but scissors, paper and tape. Sections 12^0 and 13^0 discuss spin networks and q-deformed spin networks. The end of section 13^0 outlines work of the author and Sostenes Lins, constructing 3-manifold invariants of Turaev-Viro type via q-spin networks. Part II ends with three essays: Strings, DNA, Lorenz Attractor. These parting essays contain numerous suggestions for further play.

Much is left unsaid. I would particularly like to have had the space to discuss Louis Crane's approach to defining Witten's invariants using conformal field theory, and Lee Smolin's approach to quantum gravity – where the states of the theory are functionals on knots and links. This will have to wait for the next time!

It gives me great pleasure to thank the following people for good conversation and intellectual sustenance over the years of this project.

Corrado Agnes	Steve Bryson
Jack Armel	Scott Carter
Randy Baadhio	Naomi Caspe
Gary Berkowitz	John Conway
Joan Birman	Paolo Cotta-Ramusino
Joe Birman	Nick Cozzarelli
Herbert Brun	Louis Crane
William Browder	Anne Dale

Tom Etter

Massimo Ferri

David Finkelstein

Michael Fisher

James Flagg

George Francis

Peter Freund

Larry Glasser

Seth Goldberg

Fred Goodman

Arthur Greenspoon

Bernard Grossman

Victor Guggenheim

Ivan Handler

John Hart

Mark Hennings

Kim Hix

Tom Imbo

Francois Jaeger

Herbert Jehle

Vaughan Jones

Alice Kauffman

Akio Kawauchi

Milton Kerker

Doug Kipping

Robion Kirby

Peter Landweber

Ruth Lawrence

Vladimir Lepetic

Anatoly Libgober

Raymond Lickorish

Sostenes Lins

Roy Lisker

Mary Lupa

Maurizio Martellini

John Mathias

Ken Millett

John Milnor

Jose Montesinos

Hugh Morton

Kunio Murasugi

Jeanette Nelson

Frank Nijhoff

Pierre Noyes

Kathy O'Hara

Eddie Oshins

Gordon Pask

Annetta Pedretti

Mary-Minn Peet

Roger Penrose

Mario Rasetti

Nicolai Reshetikhin

Dennis Roseman

Hubert Saleur

Dan Sandin

Jon Simon

Isadore Singer

Milton Singer

Diane Slaviero

Lee Smolin

David Solzman

George Spencer-Brown

Sylvia Spengler

Joe Staley

Stan Tenen

Tom Tieffenbacher

Volodja Turaev

Sasha Turbiner

David Urman   Randall Weiss

Jean-Michel Vappereau   Steve Winker

Rasa Varanka   Edward Witten

Francisco Varela   John Wood

Oleg Viro   Nancy Wood

Heinz von Foerster   David Yetter

Miki Wadati

(As with all lists, this one is necessarily incomplete. Networks of communication and support span the globe.) I also thank the Institute des Hautes Études Scientifiques of Bures-Sur-Yvette, France for their hospitality during the academic year 1988-1989, when this book was begun.

Finally, I wish to thank Ms. Shirley Roper, Word Processing Supervisor in the Department of Mathematics, Statistics, and Computer Science at the University of Illinois at Chicago, for a great typesetting job and for endurance above and beyond the call of duty.

This project was partially supported by NSF Grant Number DMS-8822602 and the Program for Mathematics and Molecular Biology, University of California at Berkeley, Berkeley, California.

Chicago, Illinois, January 1991

Table of Contents

PART I

PART I. A SHORT COURSE OF KNOTS AND PHYSICS.

This first part of the book is a short course on knots and physics. It is a rapid penetration into key ideas and examples. The second half of the book consists in a series of excursions into special topics (such as the Potts Model, map coloring, spin-networks, topology of 3-manifolds, and special examples). These topics can be read by themselves, but they are informed by Part I.

Acts of abstraction, changes of mathematical mood, shifts of viewpoint are the rule rather than the exception in this short course. The course is a rapid guided ascent straight up the side of a cliff! Later, we shall use this perspective - both for the planning of more leisurely walks, and for the plotting of more complex ascents.

Here is the plan of the short course. First we discuss the diagrammatic model for knot theory. Then we construct the bracket polynomial [LK4], and obtain from it the original Jones polynomial. We then back-track to the diagrams and uncover abstract tensors, the Yang-Baxter equation and the concept of the quantum group. With state and tensor models for the bracket in hand, we then introduce other (generalized) link polynomials. Then comes a sketch of the Witten invariants, first via combinatorial models, then using functional integrals and gauge field theory. The ideas and techniques spiral outward toward field theory and three-dimensional topology. Statistical mechanics models occur in the middle, and continue to weave in and out throughout the play. Underneath all this structure of topology and physical ideas there sounds a theme of combinatorics - graphs, matroids, coloring problems, chromatic polynomials, combinatorial structures, enumerations. And finally, throughout there is a deep underlying movement in the relation between space and sign, geometry and symbol, the source common to mathematics, physics, language and all.

And so we begin. It is customary to begin physics texts either with mathematical background, or with the results of experiments. But this is a book of knots. Could we then begin with the physics of knots? Frictional properties? The design of the clove-hitch and the bow-line? Here is the palpable and practical physics of the knots.

1^0. Physical Knots.

The **clove-hitch** is used to secure a line to a post or tree-trunk. The illustration in Figure 1 shows how it is made.

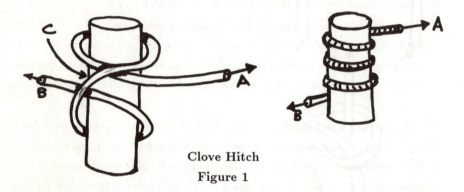

Clove Hitch
Figure 1

A little experimentation with the clove hitch shows that it holds very well indeed, with tension at A causing the line at C to grab B - and keep the assemblage from slipping.

Some physical mathematics can be put in back of these remarks - a project that we defer to Part II. At this stage, I wish to encourage experimentation. For example, just wind the rope around the post (as also in Figure 1). You will discover easily that the amount of tension needed at A to make the rope slide on the post increases exponentially with the number of windings. (The rope is held lightly at B with a fixed tension.) With the introduction of weaving - as in the clove hitch - interlocking tensions and frictions can produce an excellent bind.

In Figure 2 we find a sequence of drawings that show the tying of the **bowline**.

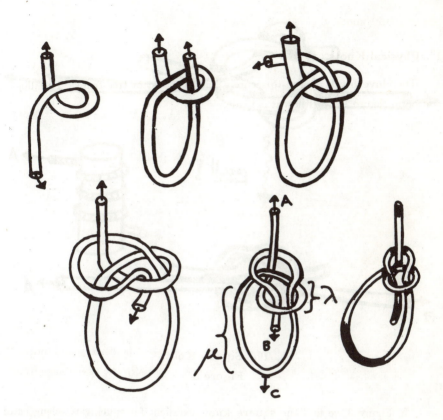

The Bowline

Figure 2

In the bowline we have an excellent knot for securing a line. The concentrated part of the knot allows the loop μ to be of arbitrary size. This loop is secure under tension at A. Pulling at A causes the loop λ to shrink and grab tightly the two lines passing through it. This secures μ. Relax tension at A and the knot is easily undone, making loop adjustment easy.

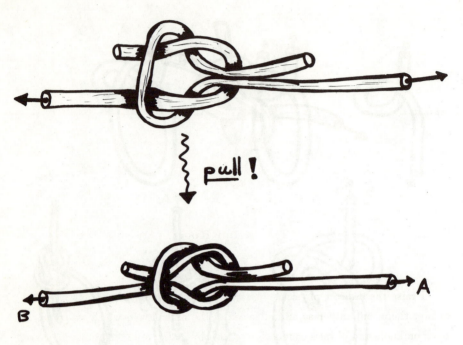

The Square Knot
Figure 3

In Figure 3 we find the **square knot**, excellent for splicing two lengths of rope. Tension at A and B creates an effect of mutual constriction - keeping the knot from slipping.

And finally in this rogue's gallery of knots and hitches we have the **granny** (Figure 4), a relative of the square knot, but quite treacherous. The granny will not hold. No mutual interlock here. A pull facilitates a flow of rope through the knot's pattern until it is undone.

It is important to come to some practical understanding of how these knots work. The facts that the square knot holds, and that the granny does not hold are best observed with actual rope models. It is a tremendous challenge to give a good mathematical analysis of these phenomena. Tension, friction and topology conspire to give the configuration a form - and within that form the events of slippage or interlock can occur.

The Granny Knot
Figure 4

I raise these direct physical questions about knotting, not because we shall answer them, but rather as an indication of the difficulty in which we stand. A book on knots and physics cannot ignore the physicality of knots of rope in space. Yet all of our successes will be more abstract, more linguistic, patterned and poised between the internal logic of patterns and their external realizations. And yet these excursions will strike deeply into topological and physical questions for knots and other realms as well.

There are other direct physical questions for knotting. We shall return to them in the course of the conversation.

2^0. Diagrams and Moves.

The transition from matters of practical knotwork to the associated topological problems is facilitated by deciding that a **knot** shall correspond to a **single closed loop of rope** (A **link** is composed of a number of such loops.). Thus the square knot is depicted as the closed loop shown in Figure 5. This figure also shows the loop-form for the granny knot.

Square Knot

Trefoil Knot

Granny Knot

Figure Eight Knot

Two Unknots

Figure 5

The advantage for topological exploration of the loop-form is obvious. One can make a knot, splice the ends, and then proceed to experiment with various deformations of it - without fear of losing the pattern as it slips off the end of the rope.

On the other hand, some of the original motivation for using the knot may be lost as we go to the closed loop form. The square knot is used to secure two lengths of rope. This is nowhere apparent in the loop form.

For topology, the mathematical advantages of the closed form are overwhelming. We have a uniform definition of a knot or link and correspondingly, a definition of **unknottedness**. A standard ring, as shown in Figure 5 is the canonical unknot. Any knot that can be continuously deformed to this ring (without tearing the rope) is said to be unknotted.

Note that we have not yet abstracted a mathematical definition of knot and link (although this is now relatively easy). Rather, I wish to emphasize the advantages of the closed loop form in doing experimental topological work. For example, one can form both the trefoil T, as shown in Figures 5 and 6, and its mirror image T^* (as shown in Figure 6). The trefoil T **cannot** be continuously deformed into its mirror image T^*. It is a remarkably subtle matter to prove this fact. One should try to actually create the deformation with a model - to appreciate this problem.

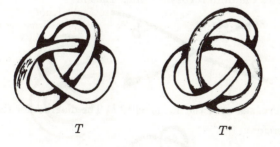

T T^*

Trefoil and its Mirror Image
Figure 6

A knot is said to be **chiral** if it is topologically distinct from its mirror image, and **achiral** if it can be deformed into its mirror image. Remarkably, the figure eight knot (see Figure 5) **can** be deformed into its mirror image (Exercise: Create this deformation.). Thus the figure eight knot is achiral, while the trefoil is chiral.

Yet, before jumping ahead to the chirality of the trefoil, we need to have a mathematical model in which facts such as chirality or knottedness can be proved or disproved.

One way to create such a model is through the use of **diagrams** and **moves**. A **diagram** is an abstract and schematized picture of a knot or link composed of curves in the plane that cross transversely in 4-fold vertices:

Each vertex is equipped with extra structure in the form of a deleted segment:

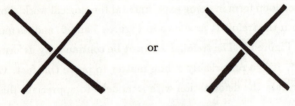

The deletion indicates the **undercrossing line**. Thus in

the line labelled a **crosses over** the line labelled b. The diagram for the trefoil T (of Figure 6) is therefore:

T

Because we have deleted small segments at each crossing, the knot or link diagram can be viewed as an interrelated collection of **arcs**. Thus for the trefoil T, I have labelled below the arcs a, b, c on the diagram:

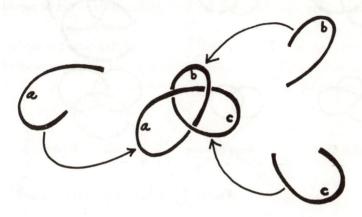

Each arc begins at an undercrossing, and continues, without break, until it reaches another undercrossing.

The diagram is a compact **weaving pattern** from which one can make an unambiguous rope model. It has the appearance of a projection or shadow of the knot in the plane. I shall, however, take **shadow** as a technical term that indicates the knot/link diagram without the breaks, hence without under or over-crossing information. Thus in Figure 7, we have shown the trefoil, the Hopf link, the Borromean rings and their shadows. The shadows are plane graphs with 4-valent vertices. Here I mean a graph in the sense of graph theory (technically a multi-graph). A graph G consists of two sets: a set of **vertices** $V(G)$ and a set of edges $E(G)$. To each edge is associated a set of one or two vertices (the **endpoints** of that edge). Graphs are commonly realized by letting the edges be curve segments, and the endpoints of these edges are the vertices.

Trefoil   Hopf Link   Borromean Rings

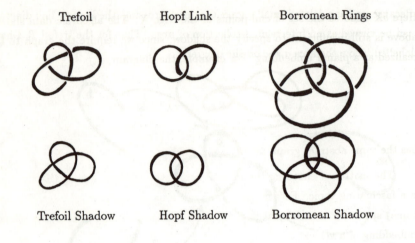

Trefoil Shadow   Hopf Shadow   Borromean Shadow

Figure 7

In the trefoil shadow, the graph G that it delineates can be described via the labelled diagram shown below:

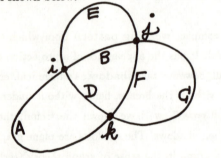

$$E(G) = \{A, B, C, D, E, F\}$$
$$V(G) = \{i, j, k\}$$

$$bA = \{i, k\} \qquad bE = \{i, j\}$$
$$bB = \{i, j\} \qquad bF = \{j, k\}$$
$$bC = \{j, k\}$$
$$bD = \{i, k\}.$$

Here bX denotes the set of end-points of the edge X. The algebraic data shown above is still insufficient to specify the shadow, since we require the graph to be realized as a planar embedding. For example, the diagram

has the same abstract graph as the trefoil shadow.

The matter of encoding the information in the shadow, and in the diagram is a fascinating topic. For the present I shall let the diagrams (presented in the plane) speak for themselves. It is, however, worth mentioning, that the planar embedding of a vertex

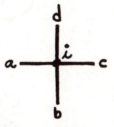

can be specified by giving the (counter-clockwise) order of the edges incident to that vertex. Thus $(i : a, b, c, d)$ can denote the vertex as drawn above, while $(i : a, d, c, b)$ indicates the vertex

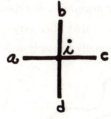

Note that both $(i : a, b, c, d)$ and $(i : a, d, c, b)$ have the same cross-lines ($a \to c$ and $b \to d$). One is obtained from the other by picking up the pattern and putting it back in the plane after turning it over. An augmentation of this code can indicate the over and undercrossing states. Thus we can write $(i : \bar{a}, b, \bar{c}, d)$ to indicate that the segment ac crosses over bd.

$$(i : \bar{a}, b, \bar{c}, d) \leftrightarrow \quad a \underset{b}{\overset{d}{\rule{0pt}{0pt}\!\!\!\!}} c$$

But this is enough said about graph-encodement.

Let the diagram take precedence. The diagram is to be regarded as a notational device at the same level as the letters and other symbols used in mathematical writing. Now I ask that you view these diagrams with the eyes of a topologist. Those topological eyes see no difference between

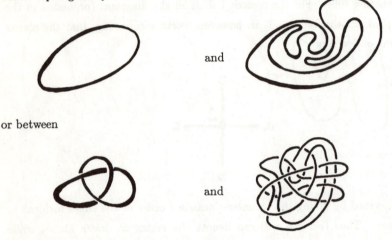

and

or between

and

You will find that for a smoothly drawn simple closed curve in the plane (no self-intersections) there is a direction so that lines perpendicular to this direction $(\vec{v})$ intersect the curve either transversally (linearly independent tangent vectors) or tangentially.

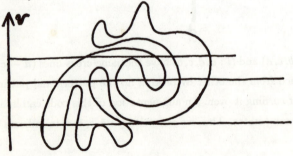

With respect to this special direction the moves

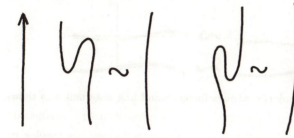

can be used repeatedly to simplify the curve until it has only one (local) maximum and one (local) minimum:

Later, this way of thinking will be very useful to us. We have just sketched a proof of the Jordan Curve Theorem [ST] for smooth (differentiable, non-zero tangent vector) curves in the plane. The Jordan Curve Theorem states that such a curve divides the plane into two disjoint regions, each homeomorphic to a disk. By moving the curve to a standard form, we accomplish these homeomorphisms.

Notationally the Jordan curve theorem is a fact about the plane upon which we write. It is the fundamental underlying fact that makes the diagrammatics of knots and links correspond to their mathematics. This is a remarkable situation – a fundamental theorem of mathematics is the underpinning of a notation for that same mathematics.

In any case, I shall refer to the basic topological deformations of a plane curve as **Move Zero**:

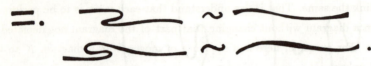

We then have the **Reidemeister Moves** for knot and link diagrams - as shown in Figure 8.

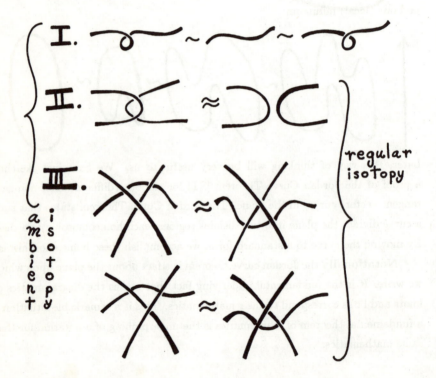

The Reidemeister Moves

Figure 8

The three moves on diagrams (I, II, III) change the graphical structure of the diagram while leaving the topological type of the embedding of the corresponding knot or link the same. That is, we understand that each move is to be performed locally on a diagram without changing that part of the diagram not depicted in the move. For example, here is a sequence of moves from a diagram K to the standard unknot U:

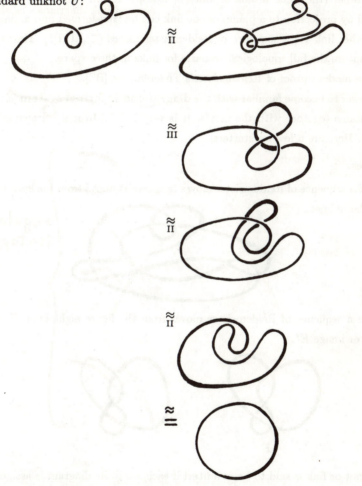

In this sequence I have only used the moves II and III (and the "zero move"). The equivalence relation generated by II and III is called **regular isotopy**. Because it is possible to cancel curls () of opposite type - as shown above - I single out the regular isotopy relation.

The equivalence relation on diagrams that is generated by all four Reidemeister moves is called **ambient isotopy**. In fact, Reidemeister [REI] showed that two knots or links in three-dimensional space can be deformed continuously one into the other (the usual notion of ambient isotopy) if and only if any diagram (obtained by projection to a plane) of one link can be transformed into a diagram for the other link via a sequence of Reidemeister moves ($\leftrightarrows$,I,II,III). Thus these moves capture the full topological scenario for links in three-space.

For a modern proof of Reidemeister's Theorem, see [BZ].

In order to become familiar with the **diagrammatic formal system** of these link diagrams together with the moves, it is very helpful to make drawings and exercises. Here are a few for starters:

Exercises.

1. Give a sequence of Reidemeister moves (ambient isotopy) from the knot below to the unknot.

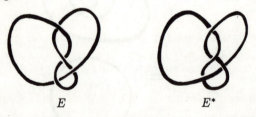

2. Give a sequence of Reidemeister moves from the figure eight knot E to its mirror image E^*.

$$E \qquad\qquad E^*$$

3. A knot or link is said to be **oriented** if each arc in its diagram is assigned a

direction so that at each crossing the orientations appear either as

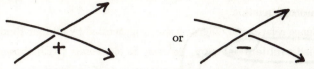

or

Note that each of these oriented crossings has been labelled with a sign of plus (+) or minus (−). Call this the **sign of the (oriented) crossing**.

Let $L = \{\alpha, \beta\}$ be a link of two components α and β. Define the **linking number** $\ell k(L) = \ell k(\alpha, \beta)$ by the formula

$$\ell k(\alpha, \beta) = \frac{1}{2} \sum_{p \in \alpha \sqcap \beta} \epsilon(p)$$

where $\alpha \sqcap \beta$ denotes the set of crossings of α with β (no self-crossings) and $\epsilon(p)$ denotes the sign of the crossing.

Thus

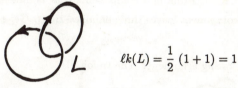

$$\ell k(L) = \frac{1}{2}\,(1+1) = 1$$

and

$$\ell k(L') = -1.$$

Prove: If L_1 and L_2 are oriented two-component link diagrams and if L_1 is ambient isotopic to L_2, then $\ell k(L_1) = \ell k(L_2)$. (Check the oriented picture for oriented moves.)

4. Let K be any oriented link diagram. Let the **writhe of K** (or **twist number of K**) be defined by the formula $w(K) = \sum_{p \in \mathcal{C}(K)} \epsilon(p)$ where $\mathcal{C}(K)$ denotes the set of crossings in the diagram K. Thus $w(\ \ ) = +3$.

Show that regularly isotopic links have the same writhe.

5. Check that the link W below has zero linking number - no matter how you orient its components.

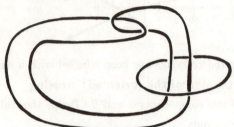

6. The Borromean rings (shown here and in Figure 7) have the property that they are linked, but the removal of any component leaves two unlinked rings. Create a link of 4 components that is linked, and so that the removal of any component leaves three unlinked rings. Generalize to n components.

7. What is wrong with the following argument? The trefoil T

has no Reidemeister moves except ones that make the diagram more complex such as

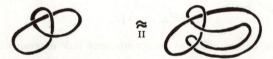

Thus the diagram T is in a minimal form, and therefore T is truly knotted (not equivalent to the unknot).

8. What is lacking in the following argument? The trefoil T has writhe $w(T) = +3$ (independent of how T is oriented), while its mirror image T^* has writhe $w(T^*) = -3$. Therefore T is not ambient isotopic to T^*.

9. A diagram is said to be **alternating** if the pattern of over and undercrossings

alternates as one traverses a component.

Take any knot or link shadow

Attempt to draw an alternating diagram that overlies this shadow.

Will this always work? When are alternating knot diagrams knotted? (Investigate empirically.) Does every knot have an alternating diagram? (Answer: No. The least example has eight crossings.)

Discussion.

Exercises 1. → 4. are integral to the rest of the work. I shall assume that the reader has done these exercises. I shall comment later on 5. and 6. The fallacy in 7. derives from the fact that it is possible to have a diagram that represents an unknot admitting no simplifying moves.

We shall discuss 8. and 9. in due course.

The Trefoil is Knotted.

I conclude this section with a description of how to prove that the trefoil is knotted. Here is a trefoil diagram with its arcs colored (labelled) in three distinct colors (*R*-red, *B*-blue, *P*-purple).

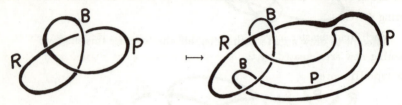

I claim that, with an appropriate notion of coloring, this property of being three-colored can be preserved under the Reidemeister moves. For example:

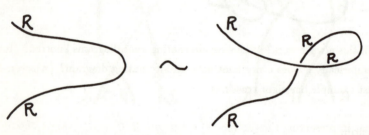

However, note that under a type I move we may be forced to retain only one color at a vertex:

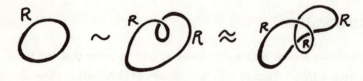

Thus I shall say that a knot diagram K is **three-colored** if each arc in K is assigned one of the three colors (R, B, P), all three colors occur on the diagram **and** each crossing carries either **three colors or one color**. (Two-colored crossings are not allowed.)

It follows at once from these coloring rules, that we can never transform a singly colored diagram to a three-colored diagram by local changes (obeying the **three or one rule**). For example,

The arc in question must be colored R since two R's already occur at each neighboring vertex.

Each Reidemeister move affords an opportunity for a local color change - by coloring new arcs created by the move, or deleting some colored arcs. Consider the type III move:

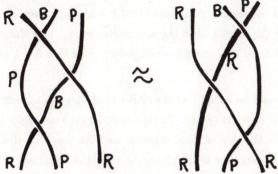

Note that in preserving the (non-local) inputs (R, B, P) and output (R, P, R) we were forced to introduce a single-color vertex in the resultant of the triangle move. Check (exercise) that all color inputs do re-configure under the type III move - and that three-coloration is preserved. Finally, you must worry about the following scenario:

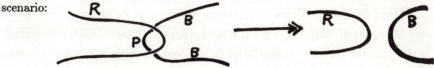

Upon performing a simplifying type II move, a colored arc is lost! Could we also

24

lose three-coloration as in the link below?

Well, knots do not allow this difficulty. In a knot the two arcs labelled R and B must eventually meet at a crossing:

The transversal arc at this crossing will be colored P in a three-colored diagram. Thus, while purple could be lost from the local arc, it will necessarily occur elsewhere in the diagram.

This completes the sketch of the proof that the trefoil is knotted. If T were ambient isotopic to the unknot, then the above observations plus the sequence of Reidemeister moves from T to unknot would yield a proof that $3 = 1$ and hence a contradiction. //

A coloration could be called a **state** of the knot diagram in analogy to the energetic states of a physical system. In this case the system admits topological deformations, and in the case of three-coloring, we have seen that there is a way to preserve the state structure as the system is deformed. Invariant properties of states then become topological invariants of the knot or link.

It is also possible to obtain topological invariants by considering all possible states (in some interpretation of that term-state) of a given diagram. Invariants emerge by summing (averaging) or integrating over the set of states for one diagram. We shall take up this viewpoint in the next section.

These two points of view - topological evolution of states versus integration over the space of states for a given system - appear to be quite complementary in studying the topology of knots and links. Keep watching this theme as we go along. The evolution of states is most closely related to the fundamental group of the knot and allied generalizations (section 13^{0}). The integration over states is the fundamental theme of this book.

3^0. States and the Bracket Polynomial.

Consider a crossing in an unoriented link diagram: Two associated diagrams can be obtained by **splicing the crossing**:

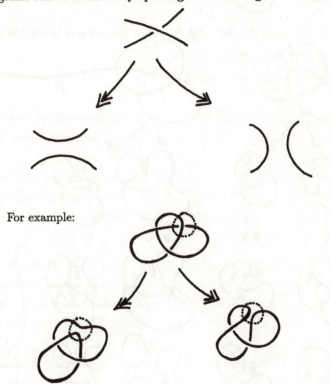

For example:

One can repeat this process, and obtain a whole family of diagrams whose one ancestor is the original link diagram. (Figure 9).

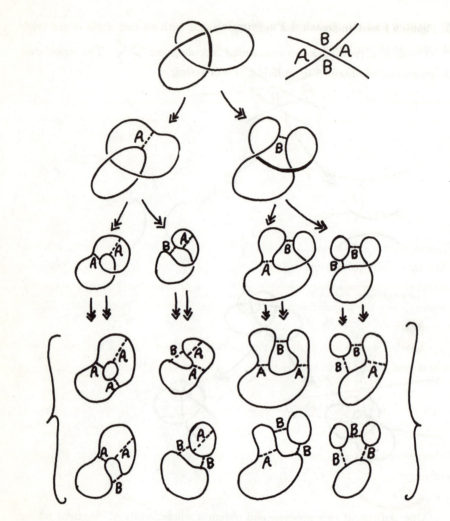

Figure 9

In Figure 9 I have indicated this splitting process with an indication of the type of split - as shown below

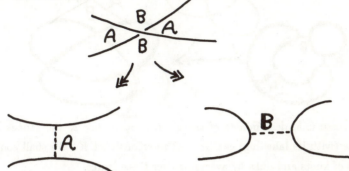

That is, a given split is said to be of **type A** or **type B** according to the convention that an *A*-**split joins the regions labelled *A* at the crossing.** The **regions labelled *A* are those** that appear on the **left** to an observer walking toward the crossing along the undercrossing segments. The *B*-regions appear on the right for this observer.

Another way to specify the *A*-regions is that **the *A*-regions are swept out when you turn the overcrossing line counter-clockwise:**

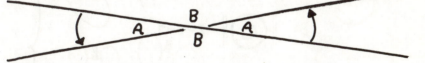

In any case, a split site labelled *A* or *B* can be **reconstructed** to form its ancestral crossing:

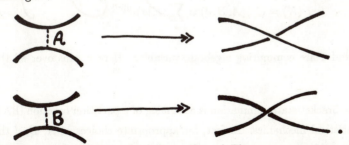

Therefore, by keeping track of these *A*'s and *B*'s we can reconstruct the ancestor link from any of its descendants. The most primitive descendants are collections

of Jordan curves in the plane. Here all crossings have been spliced. In Figure 9 we see eight such descendants. A sample reconstruction is:

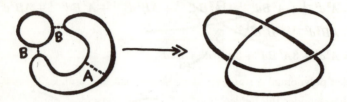

Call these final descendants of the given knot or link K the **states** of K. Each state (with its labelling) can be used to reconstruct K. We shall construct invariants of knots and links by averaging over these states.

The specific form of the averaging is as follows: Let σ be a state of K. Let $\langle K|\sigma \rangle$ **denote the product** (commutative labels) **of the labels attached to** σ. Thus

[Note that we deduce the labels from the structure of the state in relation to K.] Let $\|\sigma\|$ denote one less than the number of loops in σ. Thus

$$\left\|\;\bigcirc\;\right\| = 2 - 1 = 1.$$

Definition 3.1. We define the **bracket polynomial** by the formula

$$\langle K \rangle = \langle K \rangle(A, B, d) = \sum_{\sigma} \langle K|\sigma \rangle d^{\|\sigma\|}$$

where A, B, and d are commuting algebraic variables. Here σ runs over all the states of K.

Remark. The bracket state summation is an analog of a partition function [BA1] in discrete statistical mechanics. In fact, for appropriate choices of A, B, d the bracket can be used to express the partition function for the Potts model. See Part II sections 7^0 and 8^0 for this connection.

Example. From Figure 9 we see that the bracket polynomial for the trefoil diagram is given by the formula:

$$\langle K \rangle = A^3 d^{2-1} + A^2 B d^{1-1} + A^2 B d^{1-1} + AB^2 d^{2-1} + A^2 B d^{1-1} + AB^2 d^{2-1}$$
$$AB^2 d^{2-1} + B^3 d^{3-1}$$

$$\langle K \rangle = A^3 d^1 + 3A^2 B d^0 + 3AB^2 d^1 + B^3 d^2.$$

This bracket polynomial is **not** a topological invariant as it stands. We investigate how it behaves under the Reidemeister moves - and determine conditions on A, B and d for it to become an invariant.

Proposition 3.2.

$$\left\langle \times \right\rangle = A \left\langle \asymp \right\rangle + B \left\langle \supset \subset \right\rangle$$

Remark. The meaning of this statement rests in regarding each small diagram as part of a larger diagram, so that the three larger diagrams are identical except at the three local sites indicated by the small diagrams. Thus a special case of Proposition 3.2 is

$$\left\langle \bigodot \right\rangle = A \left\langle \bigcirc \!\!\! \bigcirc \right\rangle + B \left\langle \bigcirc\!\!\!\!\varphi \right\rangle$$

The labels A and B label A and B - splits, respectively.

Proof. Since a given crossing can be split in two ways, it follows that the states of a diagram K are in one-to-one correspondence with the union of the states of K' and K'' where K' and K'' are obtained from K by performing A and B splits at a given crossing in K. It then follows at once from the definition of $\langle K \rangle$ that $\langle K \rangle = A \langle K' \rangle + B \langle K'' \rangle$. This completes the proof of the proposition.   //

Remark. The above proof actually applies to a more general bracket of the form

$$\langle K \rangle = \sum_{\sigma} \langle K | \sigma \rangle \langle \sigma \rangle$$

where $\langle \sigma \rangle$ is any well-defined state evaluation. Here we have used $\langle \sigma \rangle = d^{\|\sigma\|}$ as above. We shall see momentarily that this form of state-evaluation is demanded by the topology of the plane.

Remark. Proposition 3.2 can be used to compute the bracket. For example,

$$\left\langle \vcenter{\hbox{⌾⌾}} \right\rangle = A\left\langle \vcenter{\hbox{◯◯}} \right\rangle + B\left\langle \vcenter{\hbox{◯◯}} \right\rangle$$

$$= A\left\{ A\left\langle \vcenter{\hbox{◯◯}} \right\rangle + B\left\langle \vcenter{\hbox{◯◯}} \right\rangle \right\} +$$

$$B\left\{ A\left\langle \vcenter{\hbox{◯◯}} \right\rangle + B\left\langle \vcenter{\hbox{◯◯}} \right\rangle \right\}$$

$$= A^2 d^{2-1} + ABd^{1-1} + BAd^{1-1} + B^2 d^{2-1}$$

$$= A^2 d^1 + 2ABd^0 + B^2 d^1 .$$

Proposition 3.3.

(a)
$$\left\langle \vcenter{\hbox{⊃⊂}} \right\rangle = AB\left\langle \vcenter{\hbox{⊃ ⊂}} \right\rangle + AB\left\langle \vcenter{\hbox{⊃⊂}} \right\rangle$$

$$+ (A^2 + B^2)\left\langle \vcenter{\hbox{≍}} \right\rangle$$

(b)
$$\left\langle \vcenter{\hbox{⌒}} \right\rangle = (Ad + B)\left\langle \vcenter{\hbox{∼}} \right\rangle$$

$$\left\langle \vcenter{\hbox{⌒}} \right\rangle = (A + Bd)\left\langle \vcenter{\hbox{∼}} \right\rangle$$

Proof. (a)

$$\left\langle \text{⟩} \right\rangle = A\left\langle \right\rangle + B\left\langle \right\rangle$$

$$= A\left\{ A\left\langle \right\rangle + B\left\langle \right\rangle \right\} +$$

$$B\left\{ A\left\langle \right\rangle + B\left\langle \right\rangle \right\}$$

$$= AB\left\langle \supset \subset \right\rangle + AB\left\langle \right\rangle$$

$$+ (A^2 + B^2)\left\langle \right\rangle.$$

Part (b) is left for the reader. Note that $\left\langle \bigcirc \right\rangle = d\left\langle \right\rangle$ and, in general, $\langle OK \rangle = d\langle K \rangle$ where OK denotes any addition of a disjoint circle to the diagram K. Thus

$$\left\langle \right\rangle = d\left\langle \right\rangle. \qquad //$$

Corollary 3.4. If $B = A^{-1}$ and $d = -A^2 - A^{-2}$, then

(a)
$$\left\langle \right\rangle = \left\langle \supset \subset \right\rangle$$

(b)
$$\left\langle \right\rangle = (-A^3)\left\langle \sim \right\rangle$$

$$\left\langle \right\rangle = (-A^{-3})\left\langle \sim \right\rangle.$$

Proof. The first part follows at once from 3.3. The second part follows from 3.3 and the calculation $Ad + B = A(-A^2 - A^{-2}) + A^{-1} = -A^3$. $\qquad //$

32

Remark. The formula 3.3 (a) shows that just on the basis of the assumptions

$$\left\langle \asymp \right\rangle = A\left\langle \asymp \right\rangle + B\left\langle \supset \subset \right\rangle$$

and

$$\left\langle \asymp \right\rangle = \left\langle \supset \subset \right\rangle,$$

we need that $AB = 1$ and

$$AB\left\langle \bigcirc \right\rangle = (-A^2 - B^2)\left\langle \asymp \right\rangle$$

whence

$$\left\langle \bigcirc \right\rangle = -\left(\frac{A}{B} + \frac{B}{A}\right)\left\langle \asymp \right\rangle$$

or

$$\left\langle \bigcirc \right\rangle = d\left\langle \asymp \right\rangle.$$

This shows that the rule we started with for evaluating all simple closed curves is necessary for type II invariance. Thus from the topological viewpoint, the bracket unfolds completely from the recursive relation

$$\left\langle \asymp \right\rangle = A\left\langle \asymp \right\rangle + B\left\langle \supset \subset \right\rangle.$$

We have shown that $\langle K \rangle$ can be adjusted to be invariant under II. In fact,

Proposition 3.5. Suppose $B = A^{-1}$ and $d = -A^2 - A^{-2}$ so that $\left\langle \asymp \right\rangle = \left\langle \supset \subset \right\rangle$. Then

$$\left\langle \asymp \right\rangle = \left\langle \asymp \right\rangle.$$

Proof.

$$\left\langle \asymp \right\rangle = A\left\langle \asymp \right\rangle + B\left\langle \asymp \right\rangle$$

$$= A\left\langle \asymp \right\rangle + B\left\langle \asymp \right\rangle$$

$$= A\left\langle \asymp \right\rangle + B\left\langle \asymp \right\rangle$$

$$= \left\langle \asymp \right\rangle. \qquad //$$

The bracket with $B = A^{-1}$, $d = -A^2 - A^{-2}$ is invariant under the moves II and III. If we desire an invariant of ambient isotopy (I, II and III), this is obtained by a normalization:

Definition 3.6. Let K be an oriented link diagram. Define the **writhe of K**, $w(K)$, by the equation $w(K) = \sum_p \epsilon(p)$ where p runs over all crossings in K, and $\epsilon(p)$ is the sign of the crossing:

$$\epsilon = +1 \qquad\qquad \epsilon = -1$$

(Compare with exercises 3 and 4). Note that $w(K)$ is an invariant of regular isotopy (II, III) and that

$$
\begin{cases}
w\left(\;\right) = 1 + w\left(\;\right) \\
w\left(\;\right) = 1 + w\left(\;\right)
\end{cases}
$$

$$
\begin{cases}
w\left(\;\right) = -1 + w\left(\;\right) \\
w\left(\;\right) = -1 + w\left(\;\right)
\end{cases}
$$

Thus, we can define a **normalized bracket**, $\mathcal{L}_K$, for oriented links K by the formula

$$\mathcal{L}_K = (-A^3)^{-w(K)} \langle K \rangle$$

and we have

Proposition 3.7. The normalized bracket polynomial $\mathcal{L}_K$ is an invariant of ambient isotopy.

Proof. Since $w(K)$ is a regular isotopy invariant, and $\langle K \rangle$ is also a regular isotopy invariant, it follows at once that $\mathcal{L}_K$ is a regular isotopy invariant. Thus we need only check that $\mathcal{L}_K$ is invariant under type I moves. This follows at once. For example

$$\mathcal{L}_{\overrightarrow{\text{⌁}}} = (-A^3)^{-w(\overrightarrow{\text{⌁}})} \langle \overrightarrow{\text{⌁}} \rangle$$

$$= (-A^3)^{-[1+w(\overrightarrow{\frown})]}(-A^3)\langle \frown \rangle$$

$$= (-A^3)^{-w(\overrightarrow{\frown})}\langle \frown \rangle$$

$$= \mathcal{L}_{\overrightarrow{\frown}} \qquad\qquad //$$

Finally, we have the particularly important behavior of $\langle K \rangle$ and $\mathcal{L}_K$ under mirror images:

[**Note:** Unless otherwise specified, we assume that $B = A^{-1}$ and $d = -A^2 - A^{-2}$ in the bracket.]

Proposition 3.8. Let K^* denote the mirror image of the (oriented) link K that is obtained by switching all the crossings of K. Then $\langle K^* \rangle(A) = \langle K \rangle(A^{-1})$ and

$$\mathcal{L}_{K^*}(A) = \mathcal{L}_K(A^{-1}).$$

Proof. Reversing all crossings exchanges the roles of A and A^{-1} in the definition of $\langle K \rangle$ and $\mathcal{L}_K$. $\qquad\qquad$ QED//

Remark. In section 5^0 (Theorem 5.2) we show that $\mathcal{L}_K$ is the original Jones polynomial after a change of variable.

Examples. It follows from 3.8 that if we calculate $\mathcal{L}_K$ and find that $\mathcal{L}_K(A) \neq \mathcal{L}_K(A^{-1})$, then K is not ambient isotopic to K^*. Thus $\mathcal{L}_K$ has the potential to detect chirality. In fact, this is the case with the trefoil knot as we shall see.

(i)
$$\left\langle \underset{L}{\text{⦿}} \right\rangle = A\left\langle \infty \right\rangle + A^{-1}\left\langle \text{⊘} \right\rangle$$
$$= A(-A^3) + A^{-1}(-A^{-3})$$
$$\langle L \rangle = -A^4 - A^{-4}.$$

(ii) $\left\langle \vcenter{\hbox{\includegraphics{trefoil}}}_{T} \right\rangle = A\left\langle \vcenter{\hbox{\includegraphics}} \right\rangle + A^{-1}\left\langle \vcenter{\hbox{\includegraphics}} \right\rangle$

$$= A(-A^4 - A^{-4}) + A^{-1}(-A^{-3})^2$$

$$\langle T \rangle = -A^5 - A^{-3} + A^{-7}$$

$w(T) = 3$ (independent of the choice of orientation

$\qquad$ since T is a knot)

$$\therefore \mathcal{L}_T = (-A^3)^{-3}\langle T \rangle$$

$$= -A^{-9}(-A^5 - A^{-3} + A^{-7})$$

$$\therefore \mathcal{L}_T = A^{-4} + A^{-12} - A^{-16}$$

$$\therefore \mathcal{L}_{T^*} = A^4 + A^{12} - A^{16}.$$

Since $\mathcal{L}_{T^*} \neq \mathcal{L}_T$, we conclude that the trefoil is not ambient isotopic to its mirror image. Incidentally, we have also shown that the trefoil is knotted and that the link L is linked.

(iii)   (The Figure Eight Knot)

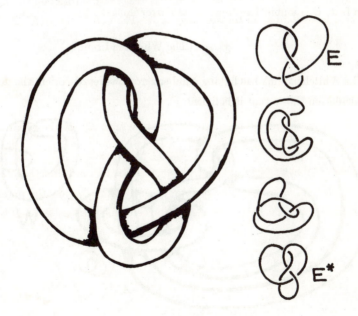

$$\langle E \rangle = A \left\langle \vcenter{\hbox{⬮}} \right\rangle + A^{-1} \left\langle \vcenter{\hbox{⬮}} \right\rangle$$

$$= A(-A^3) \left\langle \vcenter{\hbox{⬮}} \right\rangle + A^{-1} \left\langle \vcenter{\hbox{⬮}} \right\rangle$$

$$= -A^4[-A^4 - A^{-4}] + A^{-1}[-A^5 - A^{-3} + A^{-7}]$$

$$= A^8 + 1 - A^4 - A^{-4} + A^{-8}$$

$$\langle E \rangle = A^8 - A^4 + 1 - A^{-4} + A^{-8}.$$

Since $w(E) = 0$, we see that

$$\mathcal{L}_E = \langle E \rangle \text{ and } \mathcal{L}_{E^*} = \mathcal{L}_E.$$

In fact, E is ambient isotopic to its mirror image.

(iv) W (The Whitehead Link).

The Whitehead link has linking number zero however one orients it. A bracket calculation shows that it is linked:

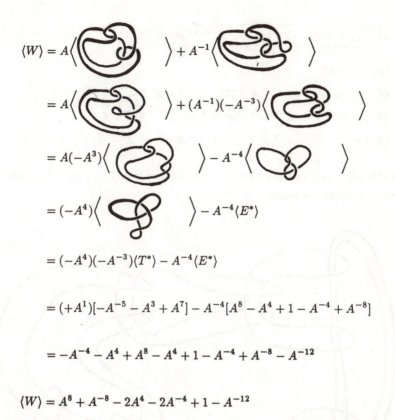

$$\langle W \rangle = A \left\langle \right\rangle + A^{-1} \left\langle \right\rangle$$

$$= A \left\langle \right\rangle + (A^{-1})(-A^{-3}) \left\langle \right\rangle$$

$$= A(-A^3) \left\langle \right\rangle - A^{-4} \left\langle \right\rangle$$

$$= (-A^4) \left\langle \right\rangle - A^{-4} \langle E^* \rangle$$

$$= (-A^4)(-A^{-3}) \langle T^* \rangle - A^{-4} \langle E^* \rangle$$

$$= (+A^1)[-A^{-5} - A^3 + A^7] - A^{-4}[A^8 - A^4 + 1 - A^{-4} + A^{-8}]$$

$$= -A^{-4} - A^4 + A^8 - A^4 + 1 - A^{-4} + A^{-8} - A^{-12}$$

$$\langle W \rangle = A^8 + A^{-8} - 2A^4 - 2A^{-4} + 1 - A^{-12}$$

(v)

$K_1 \qquad K_2 \qquad K_3 \qquad K_4$

Here K_n is a **torus link of type** $(2, n)$. (I will explain the terminology below.)

$$\left\langle \right\rangle = A \left\langle \right\rangle$$

$$+ A^{-1} \left\langle \right\rangle$$

$$\Rightarrow \langle K_n \rangle = A \langle K_{n-1} \rangle + A^{-1}(-A^{-3})^{n-1}.$$

Thus $\boxed{\langle K_n \rangle = A \langle K_{n-1} \rangle + (-1)^{n-1} A^{-3n+2}}$.

$\langle K_n \rangle = -A^3$

$\langle K_2 \rangle = A(-A^3) + (-1)^1 A^{-3 \cdot 2 + 2} = -A^4 - A^{-4}$

$\langle K_3 \rangle = A(-A^4 - A^{-4}) + (-1)^2 A^{-7} = -A^5 - A^{-3} + A^{-7}$

$\langle K_4 \rangle = A(-A^5 - A^{-3} + A^{-7}) + (-1)^3 A^{-10} = -A^6 - A^{-2} + A^{-6} - A^{-10}$.

Use this procedure to show that no torus knot of type $(2, n)$ $(n > 1)$ is ambient isotopic to its mirror image.

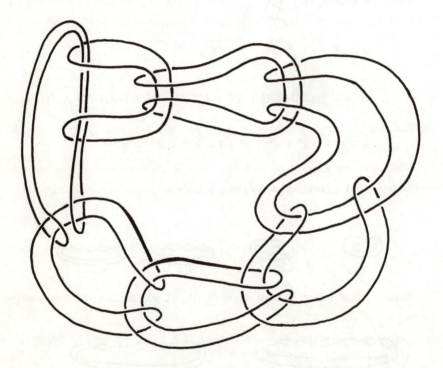

4⁰. Alternating Links and Checkerboard Surfaces.

This is a long example. To begin with, take any link shadow. (No over or undercrossings indicated.)

U

Shade the regions with two colors (black and white) so that the outer region is shaded white, and so that regions sharing an edge receive opposite colors.

You find that this shading is possible in any example that you try. Prove that it always works!

While you are working on that one, here is another puzzle. Thicken the shadow U so that it looks like a network of roadways with street-crossings:

Now, find a route that traverses every street in this map **once**, and so that it always turns at each cross-street.

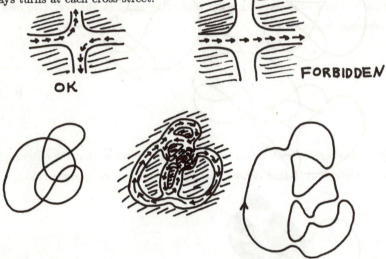

You see that the roadway problem can be rephrased as: **split each crossing of the diagram so that the resulting state has a single component.**

Now it is **easy** to prove that there exists such a splitting. **Just start splitting the diagram making sure that you maintain connectivity at each step.** For example

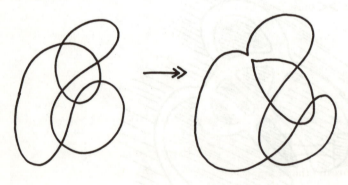

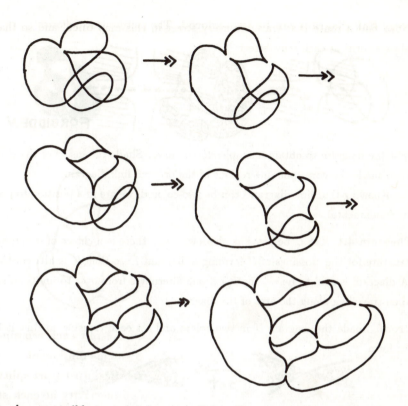

It is always possible to maintain connectivity. For suppose that >< is discon-nected. Then >< must have the form

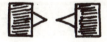

and hence ![connected bar] is connected. Here, I am implicitly using the **Jordan Curve Theorem** - that a simple closed curve divides the plane into two connected pieces.

Knowing that the **roadway problem** can be solved, we in fact know that

every link diagram can be two-colored. The picture is as follows:

Split the diagram to obtain a simple closed curve. Shade the inside of the curve. Then shade the corresponding regions of the original link diagram!

Knowing that link diagrams can be two-colored, we are in a position to prove the fundamental.

Theorem 4.1. Let U be any link shadow. Then there is a choice of over/under structure for the crossings of U forming a diagram K so that K is **alternating.** (A diagram is said to be alternating if one alternates from over to under to over when travelling along the arcs of the diagram.)

Proof. Shade the diagram U in two colors and set each crossing so that it has the form

that is - so that the A-regions at this crossing are shaded. The picture below should convince you that K (as set above) is alternating:

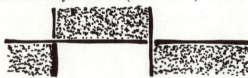

This completes the proof. //

Example.

Now we come to the center of this section. Consider the bracket polynomial, $\langle K \rangle$, for an alternating link diagram K. If we shade K as in the proof above, so that every pair of A-regions is shaded, then the state S obtained by splicing each shading

will contribute

$$A^{\#(\text{crossings})}(-A^2 - A^{-2})^{\ell(S)-1}$$

where $\ell(S)$ is the number of loops in this state. Let $V(K)$ denote the number of crossings in the diagram K. Thus the **highest power term contributed by S is**

$$(-1)^{\ell(S)}A^{V(K)+2\ell(S)-2}.$$

I claim that **this is the highest degree term in $\langle K \rangle$ and that it occurs with exactly this coefficient** $(-1)^{\ell(S)}$.

In order to see this assertion, take a good look at the state S:

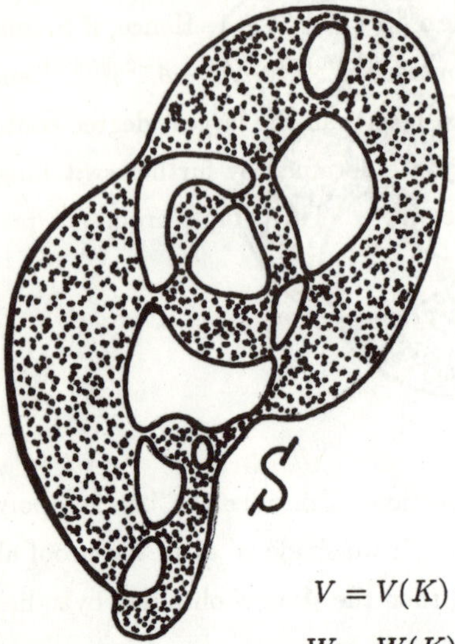

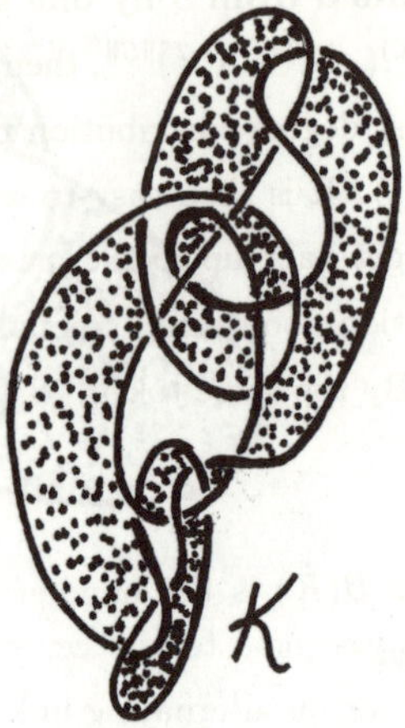

$$\left.\begin{aligned} V &= V(K) = 17 \\ W &= W(K) = 10 \\ B &= B(K) = \ \ 9 \\ R &= R(K) = 19 \end{aligned}\right\}$$

By construction we see that

$$\|S\| = \ell(S) - 1 = W - 1$$

where $W = W(K)$ is the number of white (unshaded) regions in the two-coloring of the diagram. Thus we are asserting that the maximum degree of the bracket is given by the formula

$$\max\deg\langle K\rangle = V(K) + 2W(K) - 2$$

where $V(K)$ denotes the number of crossings in K, and $W(K)$ denotes the number of white regions in the checkerboard shading of K.

To see the truth of this formula, consider any other state S' of K. Then S' can be obtained from S by switching some of the splices in $S[\asymp \overset{\text{switch}}{\longmapsto})(]$. **I am assuming that the diagram K is reduced.** This means that K is **not** in the form with a 2-strand bridge between pieces that contain crossings. It follows from this assumption, and the construction of S, that **if S' is obtained from S by one switch then** $\|S'\| = \|S\| - 1$. Hence, if S contributes $A^{V(K)}(-A^2 - A^{-2})^{\|S\|}$, then S' contributes $A^{V(K)-2}(-A^2 - A^{-2})^{\|S\|-1}$ and so the largest degree contribution from S' is **4 less** than the largest degree contribution from S. It is then easy to see that no state obtained by further switching S' can ever get back up to the maximal degree for S. This simple argument proves our assertion about the maximal degree.

By the same token, the minimal degree is given by the formula

$$\min \deg \langle K \rangle = -V(K) - 2B(K) + 2$$

where $B(K)$ is the number of black regions in the shading. Formally, we have proved.

Proposition 4.2. Let K be a reduced alternating link diagram, shaded in (white and black) checkerboard form with the unbounded region shaded white. Then the maximal and minimal degrees of $\langle K \rangle$ are given by the formulas

$$\max \deg \langle K \rangle = V + 2W - 2$$
$$\min \deg \langle K \rangle = -V - 2B + 2$$

where V is the number of crossings in the diagram K, W is the number of white regions, B is the number of black regions.

For example, we have seen that the right-handed trefoil knot T has bracket $\langle T \rangle = -A^5 - A^{-3} + A^{-7}$ and in the shading we have

$$V = 3$$
$$W = 2$$
$$B = 3$$

so that

$$V + 2W - 2 = \quad 3 + 4 - 2 = \quad 5$$
$$-V - 2B + 2 = -3 - 6 + 2 = -7$$

and these are indeed the maximal and minimal degrees of $\langle K \rangle$.

Definition 4.3. The **span** of an unoriented knot diagram K is the difference between the highest and lowest degrees of its bracket. Thus

$$\text{span}(K) = \max \deg \langle K \rangle - \min \deg \langle K \rangle.$$

Since $\mathcal{L}_K = (-A^3)^{-w(K)} \langle K \rangle$ is an ambient isotopy invariant, **we know that** $\text{span}(K)$ **is an ambient isotopy invariant.**

Now let K be a reduced alternating diagram. We know that

$$\max \deg \langle K \rangle = \quad V + 2W - 2$$
$$\min \deg \langle K \rangle = -V - 2B + 2.$$

Hence $\text{span}(K) = 2V + 2(W + B) - 4$.

However, $W + B = R$, the total number of regions in the diagram - and it is easy to see that $R = V + 2$. Hence

$$\text{span}(K) = 2V + 2(V + 2) - 4$$
$$\text{span}(K) = 4V.$$

Theorem 4.4. ([LK4], [MUR1]). Let K be a reduced alternating diagram. Then the number of crossings $V(K)$ in K is an ambient isotopy invariant of K.

This is an extraordinary application of the bracket (hence of the Jones polynomial). The topological invariance of the number of crossings was conjectured since the tabulations of Tait, Kirkman and Little in the late 1800's. The bracket is remarkably adapted to the proof.

Note that in the process we have (not surprisingly) shown that the reduced alternating diagrams with crossings represent non-trivial knots and links. See [LICK1] for a generalization of Theorem 4.4 to so-called adequate links.

Exercise 4.5. Using the methods of this section, obtain the best result that you can about the chirality of alternating knots and links. (Hint: If K^* is the mirror image of K, then $\mathcal{L}_{K^*}(A) = \mathcal{L}_K(A^{-1})$.)

Note that $\mathcal{L}_K(A) = (-A^3)^{-w(K)} \langle K \rangle$. Hence

$$\max \deg \mathcal{L}_K = -3w(K) + \max \deg \langle K \rangle$$
$$\min \deg \mathcal{L}_K = -3w(K) + \min \deg \langle K \rangle.$$

If $\mathcal{L}_K(A) = \mathcal{L}_K(A^{-1})$, then $\max \deg \mathcal{L}_K = -\min \deg \mathcal{L}_K$, whence

$$6w(K) = +\max \deg \langle K \rangle + \min \deg \langle K \rangle.$$

Thus if K is reduced and alternating, we have, by Proposition 4.2, that

$$6w(K) = 2(W - B)$$

whence $3w(K) = W - B$.

For example, if K is the trefoil  then $B = 3$, $W = 2$, $w(K) = 3$. Since $W - B = -1 \neq 9$ we conclude that K is not equivalent to its mirror image. We have shown that **if K is reduced and alternating, then K achiral implies that $3w(K) = W - B$.** One can do better than this result, but it is a good easy start. (Thistlethwaite [TH4] and Murasugi [MUR2] have shown that for K reduced and alternating, $w(K)$ is an ambient isotopy invariant.) Note also that in the event that $w(K) = 0$ and K is achiral, we have shown that $B = W$. See [LICK1] and [TH3] for generalizations of this exercise to adequate links.

Example 4.6. Define the graph $\Gamma(K)$ of a link diagram K as follows: Shade the diagram as a white/black checkerboard with the outer region shaded white. Choose a vertex for each black region and connect two vertices by an edge of $\Gamma(K)$ whenever the corresponding regions of K meet at a crossing.

The corresponding graph $\Gamma^*(K)$ that is constructed from the white regions is the planar dual of $\Gamma(K)$. That is, $\Gamma^*(K)$ is obtained from Γ by assigning a vertex of Γ^* to each region of Γ, and an edge of Γ^* whenever two regions of Γ have a common edge:

In the case of the figure eight knot E we see that $\Gamma(E)$ and $\Gamma^*(E)$ are isomorphic graphs:

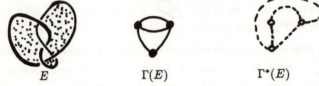

E $\qquad\qquad\qquad$ $\Gamma(E)$ $\qquad\qquad\qquad$ $\Gamma^*(E)$

The fact that these graphs are isomorphic is related to the fact that the knot E is ambient isotopic to its mirror image. In fact, if $\Gamma(K)$ and $\Gamma^*(K)$ are isomorphic graphs on the two-sphere S^2 (that is on the plane with a point at infinity added), then the associated **alternating** knots K and K^* (its mirror image) are ambient isotopic. (Exercise.) One possible converse to this statement is the **Mirror Conjecture**: If K is an alternating knot, and K is ambient isotopic to its mirror image, then there exists an alternating diagram **K** for K such that $\Gamma(\mathbf{K})$ and $\Gamma^*(\mathbf{K})$ are isomorphic graphs on the two-sphere S^2.

5^0. The Jones Polynomial and its Generalizations.

The original 1-variable Jones polynomial was discovered [JO2] via a new representation of the Artin braid group. We shall see this in section 6^0.

It is easy to show how the bracket polynomial gives rise to the Jones polynomial via the following definition.

Definition 5.1. The 1-variable Jones polynomial, $V_K(t)$, is a Laurent polynomial in the variable t (i.e. finitely many positive and negative powers of t) assigned to an oriented link K. The polynomial satisfies the properties:

(i) If K is ambient isotopic to K', then $V_K(t) = V_{K'}(t)$.

(ii) $V_{\bigcirc} = 1$

(iii) $t^{-1}V_{\overcrossing} - tV_{\undercrossing} = \left(\sqrt{t} - \frac{1}{\sqrt{t}}\right)V_{\smoothing}$

In the last formula, the three small diagrams stand for three larger link diagrams that differ only as indicated by the smaller diagrams.

For example, $V_{\mathbf{8}} = V_{\mathbf{8}} = V_{\bigcirc} = 1$ by (i) and (ii). Therefore,

$$t^{-1} - t = \left(\sqrt{t} - \frac{1}{\sqrt{t}}\right)V_{\bigodot} \quad \text{by (iii).}$$

Let

$$\delta = \frac{t^{-1} - t}{\sqrt{t} - 1/\sqrt{t}} = \frac{(t^{-1} - t)(\sqrt{t} + 1/\sqrt{t})}{t - t^{-1}}$$

$$\therefore \delta = -(\sqrt{t} + 1/\sqrt{t}).$$

This definition gives sufficient information to compute the Jones polynomial, recursively on link diagrams. (We will discuss this shortly.) The definition is not obviously well-defined, nor is it obvious that such an invariant exists. Jones proved (i), (ii), (iii) as theorems about his invariant. However,

Theorem 5.2. Let $\mathcal{L}_K(A) = (-A^3)^{-w(K)}\langle K \rangle$ as in section 4^0. Then

$$\mathcal{L}_K(t^{-1/4}) = V_K(t).$$

Thus the normalized bracket yields the 1-variable Jones polynomial.

50

Remark. In this context, I take this theorem to mean that if we let $V_K(t) = \mathcal{L}_K(t^{-1/4})$, then $V_K(t)$ satisfies the properties (i), (ii) and (iii) of Definition 5.1. Thus this theorem proves the existence (and well-definedness) of the 1-variable Jones polynomial. For the reader already familiar with $V_K(t)$ as satisfying (i), (ii), and (iii) the theorem draws the connection of this polynomial with the bracket.

Proof of 5.2. Keeping in mind that $B = A^{-1}$, we have the formulas for the bracket

$$\left\langle \times \right\rangle = A\left\langle \asymp \right\rangle + B\left\langle \supset \subset \right\rangle$$

$$\left\langle \times \right\rangle = B\left\langle \asymp \right\rangle + A\left\langle \supset \subset \right\rangle$$

Hence

$$B^{-1}\left\langle \times \right\rangle - A^{-1}\left\langle \times \right\rangle = \left(\frac{A}{B} - \frac{B}{A}\right)\left\langle \asymp \right\rangle.$$

Thus

$$A\left\langle \times \right\rangle - A^{-1}\left\langle \times \right\rangle = (A^2 - A^{-2})\left\langle \rightrightarrows \right\rangle.$$

Let $w = w(\rightrightarrows)$ so that $w(\nearrow\!\!\!\!\searrow) = w+1$ and $w(\times) = w-1$. Let $\alpha = -A^3$. Then

$$A\left\langle \times \right\rangle \alpha^{-w} - A^{-1}\left\langle \times \right\rangle \alpha^{-w} = (A^2 - A^{-2})\left\langle \rightrightarrows \right\rangle \alpha^{-w}.$$

Hence

$$A\alpha\left\langle \times \right\rangle \alpha^{-(w+1)} - A^{-1}\alpha^{-1}\left\langle \times \right\rangle \alpha^{-(w-1)} = (A^2 - A^{-2})\left\langle \rightrightarrows \right\rangle \alpha^{-w}$$

$$A\alpha\mathcal{L}_{\times} - A^{-1}\alpha^{-1}\mathcal{L}_{\times} = (A^2 - A^{-2})\mathcal{L}_{\rightrightarrows}$$

$$-A^4\mathcal{L}_{\times} + A^{-4}\mathcal{L}_{\times} = (A^2 - A^{-2})\mathcal{L}_{\rightrightarrows}.$$

Letting $A = t^{-1/4}$, we conclude that

$$t^{-1}\mathcal{L}_{\times} - t\mathcal{L}_{\times} = \left(\sqrt{t} - \frac{1}{\sqrt{t}}\right)\mathcal{L}_{\rightrightarrows}.$$

This proves property (iii) of Proposition 5.1. Properties (i) and (ii) follow directly from the corresponding facts about $\mathcal{L}_K$. This completes the proof. //

One effect of this approach to the Jones polynomial is that we get an immediate and simple proof of the **reversing property**.

Proposition 5.3. Let K and K' be two oriented links, so that K' is obtained by reversing the orientation of a component $K_1 \subset K$. Let $\lambda = \ell k(K_1, K - K_1)$ denote the total linking number of K_1 with the remaining components of K. (That is, λ is the sum of the linking number of K_1 with the remaining components of K.) Then

$$V_{K'}(t) = t^{-3\lambda} V_K(t).$$

Proof. It is easy to see that the writhes of K and K' are related by the formula: $w(K') = w(K) - 4\lambda$ where $\lambda = \ell k(K_1, K - K_1)$ as in the statement of the proposition. Thus

$$\mathcal{L}_{K'}(A) = (-A^3)^{-w(K')} \langle K' \rangle$$
$$= (-A^3)^{-w(K')} \langle K \rangle$$
$$= (-A^3)^{-w(K)+4\lambda} \langle K \rangle$$
$$\therefore \mathcal{L}_{K'}(A) = (-A^3)^{4\lambda} \mathcal{L}_K(A).$$

Thus, by 5.2, $V_{K'}(t) = \mathcal{L}_{K'}(t^{-1/4}) = t^{-3\lambda} \mathcal{L}_K(t^{-1/4}) = t^{-3\lambda} V_K(t)$. This completes the proof. //

The reversing result for the Jones polynomial is a bit surprising (see [MO1] and [LM2]) if the polynomial is viewed from the vantage of the relation

$$t^{-1} V_{\overset{\nwarrow}{\diagup\!\!\!\!\diagdown}} \quad - t V_{\overset{\nwarrow}{\diagup\!\!\!\!\diagdown}} \quad = \left(\sqrt{t} - \frac{1}{\sqrt{t}} \right) V_{\rightrightarrows} \quad .$$

This relation is structurally similar to the defining relations for the **Alexander-Conway polynomial** $\nabla_K(z) \in \mathbf{Z}[z]$ (polynomials in z with integer coefficients):

$$
\begin{cases}
\text{(i)} \quad \nabla_K(z) = \nabla_{K'}(z) \text{ if the oriented links } K \\
\qquad \text{and } K' \text{ are ambient isotopic} \\
\\
\text{(ii)} \quad \nabla_{\bigcirc} = 1 \\
\\
\text{(iii)} \quad \nabla_{\overset{\nwarrow}{\diagup\!\!\!\!\diagdown}} - \nabla_{\overset{\nwarrow}{\diagup\!\!\!\!\diagdown}} = z \nabla_{\rightrightarrows}
\end{cases}
$$

These properties are John H. Conway's [CON] reformulation (and generalization) of the original Alexander polynomial [A]. We discuss the Alexander polynomial at some length and from various points of view in sections 12^0 and 13^0. It was Conway who perceived that the three properties ((i), (ii), (iii)) above characterize this polynomial. Moreover, $\nabla_K(z)$ is very sensitive to changes in orientation. It does not simply multiply by a power of a linking number.

Upon juxtaposing the exchange identities for the Jones and Conway-Alexander polynomials, one is led to ask for a common generalization. This generalization exists, and is called the Homfly polynomial after its many discoverers ([F], [PT]):

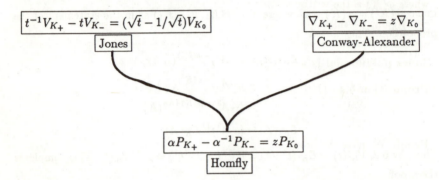

In the diagram above, I have indicated the basic exchange relation for the **oriented 2-variable polynomial** $P_K(\alpha, z)$. For $\alpha = t^{-1}$, $z = \sqrt{t} - 1/\sqrt{t}$, it specializes to the Jones polynomial. For $\alpha = 1$, it specializes to the Conway-Alexander polynomial. The original proofs of the existence of the Homfly polynomial were by induction on knot diagrams. We shall deduce its existence via generalizations of the bracket state model. Another approach to the Homfly polynomial is given by Jones in [JO3], that generalizes his original methods for $V_K(t)$.

Along with this oriented 2-variable polynomial there is also a **semi-oriented 2-variable polynomial** $F_K(\alpha, z)$ [LK8] (the Kauffman polynomial) that generalizes the bracket and original Jones polynomial. This polynomial is a normalization of a polynomial, L_K, defined for unoriented links and satisfying the properties:

(i) If K is **regularly isotopic** to K', then $L_K(\alpha, z) = L_{K'}(\alpha, z)$.

(ii) $L_{\bigcirc} = 1$

(iii) $L_{\times} + L_{\times} = z(L_{\asymp} + L_{\supset\subset})$

(iv) $L_{\partial\!\!\!\frown} = \alpha L$

$L_{\frown\!\!\!\sigma} = \alpha^{-1} L$

See Figure 8 and the discussion in sections 2^0 and 3^0 for more information about regular isotopy. Just as with the bracket, the L-polynomial is multiplicative under a type I Reidemeister move. The F-polynomial is defined by the formula

$$F_K(\alpha, z) = \alpha^{-w(K)} L_K(\alpha, z)$$

where $w(K)$ is the writhe of the oriented link K.

Again, we take up the existence of this invariant in section 14^0. Given its existence, we can see easily that **the bracket is a special case of L and the Jones polynomial is a special case of F.** More precisely,

Proposition 5.4. $\langle K\rangle(A) = L_K(-A^3, A + A^{-1})$ and

$$V_K(t) = F_K(-t^{-3/4}, t^{-1/4} + t^{1/4}).$$

Proof. We have the bracket identities

$$\left\langle \times \right\rangle = A\left\langle \asymp \right\rangle + A^{-1}\left\langle \supset\subset \right\rangle$$

$$\left\langle \times \right\rangle = A^{-1}\left\langle \asymp \right\rangle + A\left\langle \supset\subset \right\rangle.$$

Hence $\left\langle \times \right\rangle + \left\langle \times \right\rangle = (A + A^{-1})\left[\left\langle \asymp \right\rangle + \left\langle \supset\subset \right\rangle\right].$

Therefore $\langle K\rangle(A) = L_K(-A^3, A + A^{-1})$. Since $V_K(t) = \mathcal{L}_K(t^{-1/4})$ and

$$\mathcal{L}_K(A) = (-A^3)^{-w(K)}\langle K\rangle(A)$$

we have that

$$V_K(t) = (-t^{-3/4})^{-w(K)}\langle K\rangle(t^{-1/4}).$$

Therefore $V_K(t) = (-t^{-3/4})^{-w(K)} L_K(-t^{-3/4}, t^{-1/4} + t^{1/4})$. Hence

$$V_K(t) = F_K(-t^{-3/4}, t^{-1/4} + t^{1/4}).$$

This completes the proof. $//$

Thus we can form the chart

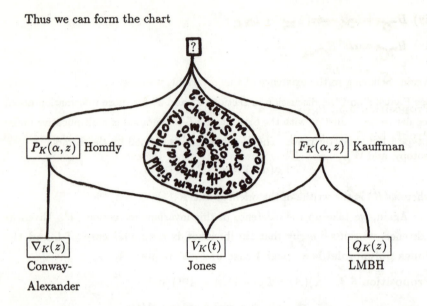

$P_K(\alpha, z)$ Homfly

$F_K(\alpha, z)$ Kauffman

$\nabla_K(z)$
Conway-
Alexander

$V_K(t)$
Jones

$Q_K(z)$
LMBH

where the question mark denotes an unknown unification of the oriented and semi-oriented 2-variable polynomials. Here the invariant $Q_K(z)$ is the unoriented polynomial invariant of ambient isotopy discovered by Lickorish, Millett, Brandt and Ho ([BR], [HO]). It is, in this context, obtained by setting $\alpha = 1$ in the L-polynomial.

Both of these 2-variable polynomials - P_K and F_K are good at distinguishing mirror images. F_K seems a bit better at this game. (See [LK8].)

In this description of knot polynomial generalizations of the Jones polynomial I have avoided specific calculations and examples. However, it is worth mentioning that the oriented invariant $P_K(\alpha, z)$ can also be regarded as the normalization of a regular isotopy invariant. In this way we define the **regular isotopy homfly polynomial**, $H_K(\alpha, z)$, by the properties

(i) If the oriented links K and K' are regularly isotopic, then
$H_K(\alpha, z) = H_{K'}(\alpha, z)$.

(ii) $H_{\text{\o}} = 1$

(iii) $H_{\nearrow\!\!\!\nwarrow} - H_{\nwarrow\!\!\!\nearrow} = z H_{\rightrightarrows}$

(iv) $H_{\mathbf{\nearrow}} = \alpha H_{\mathbf{\nearrow}}$

$\quad H_{\mathbf{\nwarrow}} = \alpha^{-1} H_{\mathbf{\nearrow}}.$

Again, believing in the existence of this invariant, we have

Proposition 5.5. $P_K(\alpha, z) = \alpha^{-w(K)} H_K(\alpha, z).$

Proof. Let $W_K(\alpha, z) = \alpha^{-w(K)} H_K(\alpha, z)$. Then W_K is an invariant of ambient isotopy, and $W_{\mathbf{\bigcirc}} = 1$. Thus it remains to check the exchange identity:

$$H_{\mathbf{\nearrow}} \quad - H_{\mathbf{\searrow}} \quad = z H_{\mathbf{\rightrightarrows}}$$

$$\Rightarrow \alpha^{+1} \alpha^{-(w+1)} H_{\mathbf{\nearrow}} \quad - \alpha^{-1} \alpha^{-(w-1)} H_{\mathbf{\searrow}} \quad = z\alpha^{-w} H_{\mathbf{\rightrightarrows}}$$

where $w = w(\mathbf{\rightrightarrows})$. Hence

$$\alpha W_{\mathbf{\nearrow}} \quad - \alpha^{-1} W_{\mathbf{\searrow}} \quad = z W_{\mathbf{\rightrightarrows}} \quad .$$

This completes the proof. $/\!/$

There is a conceptual advantage in working with the regular isotopy Homfly polynomial. For one thing, we see that the second variable originates in measuring the writhing of the diagrams involved in a given calculation. One can think of this as a measure of actual topological twist (rather than diagrammatic curl) if we **replace the link components by embedded bands**. Thus a trefoil knot can be replaced by a knotted band

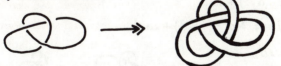

Then the rule $H_{\mathbf{\nearrow}} = \alpha H_{\mathbf{\nearrow}}$ becomes the rule:

$$H_{} = H_{} = \alpha H_{} \quad .$$

That is, the polynomial H may be interpreted as **an ambient isotopy invariant of embedded, oriented band-links.** (No Möbius bands here. Compare [HE1] and [HE2].) The exchange identity remains essentially the same:

With this interpretation the variables z and α acquire separate meanings. The z-variable measures the splicing and shifting operations. The α-variable measures twisting in the bands. This point of view can then be reformulated to work with **framed links** - links with an associated normal vector field.

To underline this discussion, let's compute the specific examples of Hopf link and trefoil knot:

(a) $H_{\text{⊘○}} = \alpha^{-1} H_{\text{○}}$; $H_{\text{⊙}} = \alpha^{-1} H_{\text{○}} = \alpha^{-1}$.

(b) $H_{\text{8}} - H_{\text{8}} = z H_{\text{8}}$

$\alpha - \alpha^{-1} = z H_{\text{○○}}$

Let $\delta = (\alpha - \alpha^{-1})/z$.

The same reasoning shows that

$H_{\text{⊙}K} = \delta H_K$ for any K.

(c) $H_{\text{⊙}} - H_{\text{⊙}} = z H_{\text{⊙}}$,

$H_L - \delta = z\alpha$

$H_L = z^{-1}(\alpha - \alpha^{-1}) + z\alpha$

$P_L = \alpha^{-2} H_L = z^{-1}(\alpha^{-1} - \alpha^{-3}) + z\alpha^{-1}$.

(d) $H_T - H_{T'} = z H_L$

$H_T - \alpha = z[z^{-1}(\alpha - \alpha^{-1}) + z\alpha]$

$H_T = (2\alpha - \alpha^{-1}) + z^2 \alpha$

$P_T = \alpha^{-3} H_T = 2\alpha^{-2} - \alpha^{-4} + z^2 \alpha^{-2}$.

Since $P_{K^*}(\alpha, z) = P_K(\alpha^{-1}, -z)$ when K^* is the **mirror image** of K, we see the chirality of T reflected in this calculation of P_T.

In these calculations we have used a tree-like decomposition into simpler knots and links. Thus in the case of the trefoil, the full calculation can be hung on the tree in Figure 10.

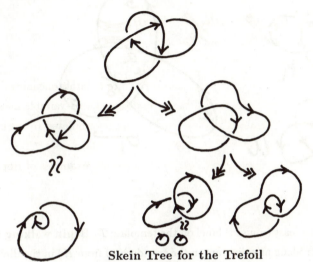

Skein Tree for the Trefoil

Figure 10

Each branching of this **skein-tree** takes the form of splicing or switching a given crossing:

We can make this tree-generation process completely automatic by the following algorithm ([JA1], [LK13]):

Skein-Template Algorithm.

(i) Let K be a given oriented link and let U be the universe underlying K. (U is the diagram for K without any indicated over or undercrossings). **Label each**

edge of U with a distinct positive integer. Call this labelled U the **template T for the algorithm.**

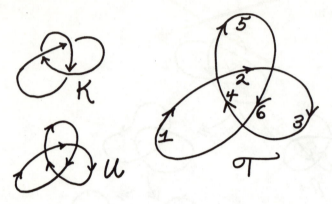

(ii) Choose the least (unused) label on the template T. **Begin walking** along the diagram K (or along any diagram in the tree derived from K) in the direction of the orientation and **from the center of this least labelled edge.**

Continue walking across an overpass, and **decorate** it (see below).

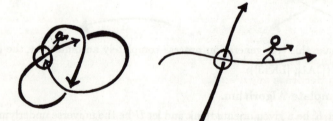

Stop walking at an underpass. **If you have stopped walking at an**

underpass, **generate two branches of the tree** by switch and splice.

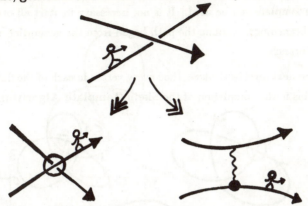

Decorate the switched and spliced crossings as indicated above and also below:

These decorations are designed to indicate that the state of the site has been decided. Note that the dot ⬤➤ for a spliced crossing is placed along the arc of passage for the walker.

In the switched diagram a walker would have **passed over** as indicated on the decorated crossing (using the given template). In the spliced diagram the walker can **continue walking** along the segment with the dot.

Also decorate any overpass that you walk along in the process of the algorithm.

(iii) If you have completed (ii) for a given diagram, go to the next level in the tree and apply the algorithm (ii) again – using the same template. [Since switched and spliced descendants of K have the same edges (with sometimes doubled vertices), the same template can be used.] It is not necessary to start all over again at any of these branchings. Assume the path began from the generation node for your particular branch.

If there is no next level in the tree, then every vertex in each of the final nodes is decorated. This is the completion of the **Skein-Template Algorithm**.

Example.

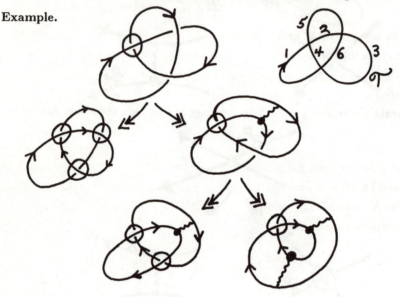

Regard the **output** of the Skein-Template Algorithm as the set

$$\text{STA}(K, \mathcal{T}) = \left\{ D \;\middle|\; \begin{array}{l} D \text{ is a decorated final node (diagram)} \\ \text{in the tree generated by the Skein-} \\ \text{Template Algorithm with template } \mathcal{T}. \end{array} \right\}$$

Thus we have

Note that (at least in this example) every element $D \in \mathrm{STA}(K, \mathcal{T})$ is an unknot or unlink. This is always the case - it is a slight generalization of the fact that you will always draw an unknot if you follow the rule: **first crossing is an overcrossing**, while traversing a given one-component universe.

start

The tree generated by the skein-template algorithm is sufficient to calculate $H_K(\alpha, z)$ via the exchange identity and the basic facts about H_K.

In fact, with our decorations of the elements $D \in \mathrm{STA}(K, \mathcal{T})$ we can write a formula:

$$H_K(\alpha, z) = \sum_{D \in \mathrm{STA}(K, \mathcal{T})} \langle K|D \rangle \delta^{\|D\|}$$

where $\delta = (\alpha - \alpha^{-1})/z$, $\|D\|$ denotes one less than the number of link components in D, and $\langle K|D \rangle$ is **a product of vertex weights** depending upon the crossing types in K and the corresponding decorations in D. The local rules for these vertex weights are as follows:

$$\left\langle \, \vcenter{\hbox{}} \; \middle| \; \vcenter{\hbox{}} \, \right\rangle = z$$

$$\left\langle \, \vcenter{\hbox{}} \; \middle| \; \vcenter{\hbox{}} \, \right\rangle = -z$$

$$\left\langle \, \vcenter{\hbox{}} \; \middle| \; \vcenter{\hbox{}} \, \right\rangle = 0$$

$$\left\langle \, \vcenter{\hbox{}} \; \middle| \; \vcenter{\hbox{}} \, \right\rangle = 0$$

Positive crossings only allow first passage on the lower leg, while negative crossings only allow first passage on the upper leg. These rules correspond to our generation process. The D's generated by the algorithm will yield no zeros.

A different and simpler rule applies to the encircled vertices:

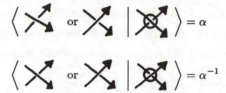

The product of these weights will give the writhe of D. Thus, we have the formula

$$\langle K|D\rangle = [(-1)^{t_-(D)} z^{t(D)} \alpha^{w(D)}]$$

where $t_-(D)$ denotes the number of splices of negative crossings (of K) in D and $t(D)$ denotes the number of spliced crossings in D.

Here it is assumed that D is obtained from $\mathrm{STA}(K,T)$ so that there are no zero-weighted crossings.

I leave it as an exercise for the reader to check that the formula

$$H_K = \sum_{D \in \mathrm{STA}(K,T)} \langle K|D\rangle \delta^{\|D\|}$$

is indeed correct (on the basis of the axioms for H_K). This formula shows that the tree-generation process also generates a state-model for H_K that is very similar in its form to the bracket. The real difference between this **skein model for H_K** and the state model for the bracket is that it is not at all obvious how to use the skein model as a logical or conceptual foundation for H_K. It is really a **computational expression**. The bracket model, on the other hand, gives us a direct entry into the inner logic of the Jones polynomial. I do not yet know how to build a model of this kind for $H_K(\alpha, z)$ as a whole.

Remark. The first skein model was given by Francois Jaeger [JA1], using a matrix inversion technique. In [LK13] I showed how to interpret and generalize this model as a direct consequence of skein calculation.

Example. Returning to the trefoil, we calculate $H_T(\alpha, z)$ using the skein model.

$$\mathrm{STA}(T,T) = \left\{ \begin{array}{ccc} & & \end{array} \right\}$$

$$H_T = \alpha + z\delta + \alpha z^2 = (2\alpha - \alpha^{-1}) + z^2\alpha.$$

Example.

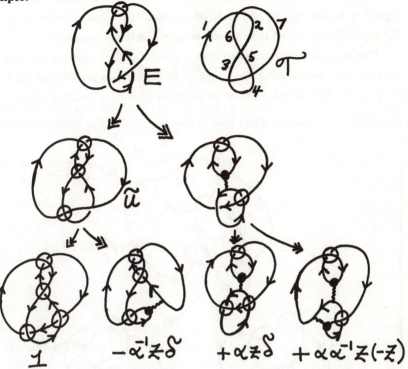

$$\Rightarrow H_E = 1 - \alpha^{-1} z\delta + \alpha z\delta - z^2$$
$$= 1 - \alpha^{-1}(\alpha - \alpha^{-1}) + \alpha(\alpha - \alpha^{-1}) - z^2$$
$$H_E = (\alpha^{-2} + \alpha^2 - 1) - z^2$$

Note that for this representation of the figure eight knot E, we have $w(E) = 0$. Thus $H_E = P_E$. $H_E(\alpha, z) = H_E(\alpha^{-1}, -z)$ reflects the fact that $E \sim E^*$. That is, E is an achiral knot. Of course the fact that $P_K(\alpha, z) = P_K(\alpha^{-1}, -z)$ does not necessarily imply that K is achiral. The knot 9_{42} (next example) is the first instance of an anomaly of this sort.

This example also indicates how the Skein Template Algorithm is a typically dumb algorithm. At the first switch, we obtained the unknot $\widetilde{U}$ with $H_{\widetilde{U}} = \alpha^{-2}$. The algorithm just went about its business switching and splicing $\widetilde{U}$ - this is reflected in the calculation $1 - \alpha^{-1}z\delta = \alpha^{-2}$. A more intelligent version of the algorithm would look for unknots and unlinks. This can be done in practice (Compare [EL].), but no theoretically complete method is available.

The example illustrates how, after a splice, the path continuation may lead through a previously decorated crossing. Such crossings are left undisturbed in the process - just as in the process of return in drawing a standard unlink.

Example. $K = 9_{42}$, the 42-nd knot of nine crossings in the Reidemeister tables:

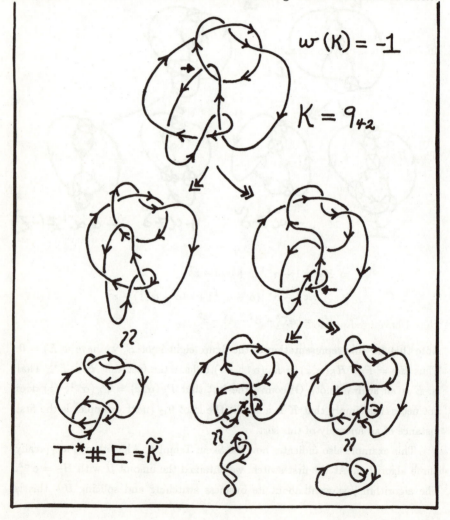

Here I have created (the beginning of) a skein tree for 9_{42} by making choices and regular isotopies by hand. In this case the hand-choice reduces the calculation by an enormous amount. We see that

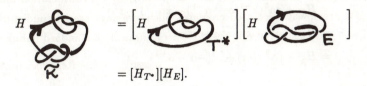

$$= [H_{T^*}][H_E].$$

In general, it is easy to see that

$$H_{\boxed{A}=\boxed{B}} = (H_{\boxed{A}})(H_{\boxed{B}}).$$
$$\underset{A \# B}{} \qquad \underset{A}{} \qquad \underset{B}{}$$

Thus $H_K = H_{\widetilde{K}} + z\left[\alpha^2 H_{\bigcirc\!\!\!\bigcirc} - z\alpha^{-1}\right]$, using the three diagrams in the tree.

$$\bigcirc\!\!\!\bigcirc = L^* \text{ so that}$$
$$H_{L^*} = z^{-1}(\alpha - \alpha^{-1}) + z\alpha^{-1}$$
$$H_{T^*} = (2\alpha^{-1} - \alpha) + z^2\alpha^{-1}$$
$$H_E = (\alpha^{-2} + \alpha^2 - 1) - z^2.$$

Here I have used previous calculations in this section.

Putting this information together, we find

$$\begin{aligned}
H_{\widetilde{K}} &= H_{T^*}H_E \\
&= ((2\alpha^{-1} - \alpha) + z^2\alpha^{-1})(\alpha^{-2} + \alpha^2 - 1 - z^2) \\
&= 2\alpha^{-3} + 2\alpha - 2\alpha^{-1} - 2\alpha^{-1}z^2 \\
&\quad - \alpha^{-1} - \alpha^3 + \alpha + \alpha z^2 \\
&\quad + z^2\alpha^{-3} + z^2\alpha - z^2\alpha^{-1} - z^4\alpha^{-1}
\end{aligned}$$

$$H_{\widetilde{K}} = (2\alpha^{-3} - 3\alpha^{-1} + 3\alpha - \alpha^3) + (\alpha^{-3} - 3\alpha^{-1} + 2\alpha)z^2 - \alpha^{-1}z^4$$

$$\begin{aligned}
\therefore H_K &= H_{\widetilde{K}} + z\alpha^2 H_{L^*} - z^2\alpha^{-1} \\
&= H_{\widetilde{K}} + z\alpha^2(z^{-1}(\alpha - \alpha^{-1}) - z\alpha^{-1}) - z^2\alpha^{-1} \\
&= H_{\widetilde{K}} + \alpha^3 - \alpha - z^2\alpha - z^2\alpha^{-1} \\
&= H_{\widetilde{K}} + (\alpha^3 - \alpha) + z^2(-\alpha - \alpha^{-1})
\end{aligned}$$

$$H_K = (2\alpha^{-3} - 3\alpha^{-1} + 2\alpha) + (\alpha^{-3} - 4\alpha^{-1} + \alpha)z^2 - \alpha^{-1}z^4$$

$$\boxed{P_K = \alpha H_K = (2\alpha^{-2} - 3 + 2\alpha^2) + (\alpha^{-2} - 4 + \alpha^2)z^2 - z^4}.$$

This calculation reveals that $P_K(\alpha, z) = P_K(\alpha^{-1}, -z)$ when K is the knot 9_{42}. Thus P_K does not detect any chirality in 9_{42}. Nevertheless, 9_{42} **is** chiral! (See [LK3], p. 207.)

This computation is a good example of how ingenuity can sometimes overcome the cumbersome combinatorial complexity inherent in the skein template algorithm.

Summary. This section has been an introduction to the original Jones polynomial $V_K(t)$ and its 2-variable generalizations. We have shown how the recursive form of calculation for these polynomials leads to formal state models, herein called skein models, and we have discussed sample computations for the oriented 2-variable generalization $P_K(\alpha, z)$, including a direct computation of this polynomial for the knot 9_{42}.

Exercise. (i) Show that

is inequivalent to its mirror image.

(ii) Compute H_K for K as shown below:

(iii) Investigate H_K when K has the form

Exercise. Let K be an oriented link diagram and U its corresponding universe. Let $\mathcal{T}$ be a template for U. Let $\widetilde{D}$ be any diagram obtained from U by splicing some subset of the crossings of U. Decorate $\widetilde{D}$ by using the template: That is, traverse $\widetilde{D}$ by always choosing the least possible base-point from $\mathcal{T}$, and decorate each crossing as you first encounter it:

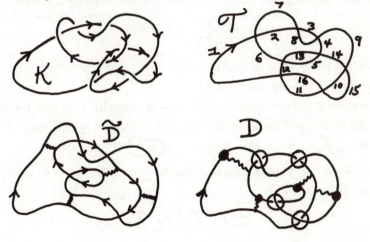

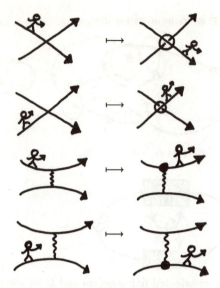

Let D denote the diagram $\widetilde{D}$ after decoration by this procedure. Let $\langle K|D\rangle$ be defined as in the Skein-Template-Algorithm. (It is now possible that $\langle K|D\rangle = 0$.)

Show. $\boxed{\langle K|D\rangle \neq 0 \Leftrightarrow D \in \mathrm{STA}(K,\mathcal{T}).}$

This exercise shows that we can characterize the skein-model for H_K via

$$H_K = \sum_{D \in \mathcal{D}(K,\mathcal{T})} \langle K|D\rangle \delta^{\|D\|}$$

where $\mathcal{D}(K,\mathcal{T})$ is the set of all diagrams obtained as in the exercise (splice a subset of crossings of U and decorate using $\mathcal{T}$).

The result of this exercise shifts the viewpoint about H_K away from the skein-tree to the structure of the states themselves. For example, in computing H_L for L the link, ⟨⟨⟩⟩ , we choose a template ⟨⟨⟩⟩ $\mathcal{T}$, and then consider the set $\widetilde{D}$ of all diagrams obtained from L by splicing or projecting crossings. These are:

$$\left\{ \text{⟨⟨⟩⟩} , \text{⟨⟨⟩⟩} , \text{⟨⟨⟩⟩} , \text{⟨⟨⟩⟩} \right\} = \widetilde{D}.$$

Each element of $\widetilde{\mathcal{D}}$ contributes, after decoration, a term in H_K. This procedure goes as follows:

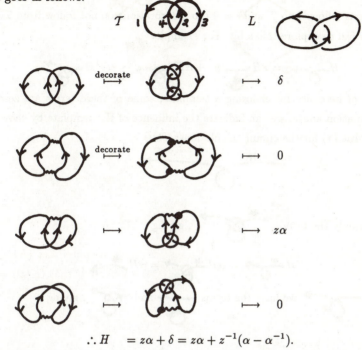

$$\therefore H \quad = z\alpha + \delta = z\alpha + z^{-1}(\alpha - \alpha^{-1}).$$

Example/Exercise. Finally, it is worth remarking that we can summarize our version of the skein model for H_K by an expansion formula similar to that for the bracket. Thus we write:

$$H \bigtimes \;\; = \;\; zH \longmapsto \;\; + \alpha H \bigotimes \;\; + \alpha^{-1} H \bigotimes$$

$$H \bigtimes \;\; = \;\; -zH \longmapsto \;\; + \alpha H \bigotimes \;\; + \alpha^{-1} H \bigotimes$$

to summarize the vertex weights, and

$$H_{\bigcirc K} = \delta H_K, \qquad \delta = (\alpha - \alpha^{-1})/z$$

to summarize the loop behavior. Here it is understood that the resulting states must be compared to a given template $\mathcal{T}$ for admissibility. Thus, if σ is a state (i.e. a decorated diagram without crossings that is obtained from the expansion) then $H\sigma = \epsilon\delta^{\|\sigma\|}$ ($\|\sigma\|$ = number of loops in σ minus one) where $\epsilon = 1$ if σ's decorations follow from $\mathcal{T}$ and $\epsilon = 0$ if σ's decorations do not follow from $\mathcal{T}$.

It is fun to explore this a bit. For example:

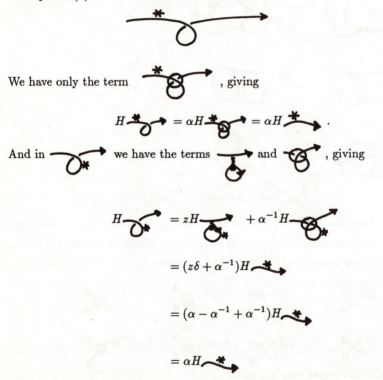

Because of necessity for choosing a template, some of these terms are zero. In the calculation above, we can indicate the influence of the template by showing a **base-point** $(*)$ for the circuit calculations: Thus in

This calculation is the core of the well-definedness of the skein model.

In fact, we can sketch this proof of well-definedness. It is a little easier (due to the direct nature of the model), but of essentially the same nature as the original induction proofs for the Homfly polynomial [LM1]. The main point is to see that H_K is independent of the choice of template. We can first let the $*$ (above) stand for a template label that is **least** among all template labels. Then going from

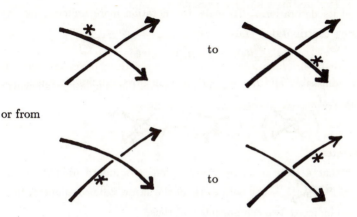

to

or from

to

can denote an exchange of template labels. Let's concentrate on the case

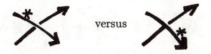

versus

We want to show that these two choices of template will give identical calculations in the state summations. Now we have (assuming that $\searrow$ and $\nearrow$ are on the same component)

$$H\;{*}\!\!\!\diagdown\!\!\!\nearrow = \alpha H\;{*}\!\!\!\bigotimes\!\!\!\nearrow$$

and

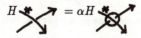

Each right hand term is a sum over states with the given local configurations.

72

Thus it will suffice to show that

$$\alpha H \;\rlap{\raisebox{0.2em}{\large$\nearrow$}}{\rlap{\large$\searrow$}{\large$\times$}} \;=\; zH \;\rlap{\raisebox{-0.2em}{$\bullet$}}{\rlap{\large$\nearrow$}{\large$\succ$}} \;+\; \alpha^{-1} H \;\rlap{\raisebox{0.2em}{\large$\nearrow$}}{\rlap{\large$\searrow$}{\large$\times$}}$$

holds for states with respect to these templates. Each state is weighted by the contributions of its decorations. To make the notation more uniform, let $\langle\sigma\rangle$ denote the state's contribution. Thus

$$\langle\sigma\rangle = z^{t(\sigma)}(-1)^{t-(\sigma)}\alpha^{w(\sigma)}\delta^{\|\sigma\|}$$

if the state σ is admissible, and $H_K = \sum_\sigma \langle\sigma\rangle$, summing over admissible states. We then must show that

$$\left\langle \rlap{\large$\times$}{} \right\rangle = \left\langle \rlap{\large$\succ$}{} \right\rangle + \left\langle \rlap{\large$\times$}{} \right\rangle$$

for individual states.

Remember that a state σ is, by construction, a standard unknot or unlink. Thus, if $\rlap{\large$\times$}{}$ has both edges in the **same** state component, then $\rlap{\large$\succ$}{}$ indicates two components and thus,

$$\left\langle \rlap{\large$\succ$}{} \right\rangle = z\delta \left\langle \rlap{\large$\times$}{} \right\rangle$$

$$\left\langle \rlap{\large$\times$}{} \right\rangle = \alpha^{-1} \left\langle \rlap{\large$\times$}{} \right\rangle, \quad \left(z\delta + \alpha^{-1} = \alpha \right)$$

$$\therefore \left\langle \rlap{\large$\succ$}{} \right\rangle + \left\langle \rlap{\large$\times$}{} \right\rangle = \alpha \left\langle \rlap{\large$\times$}{} \right\rangle$$

$$= \alpha \left\langle \rlap{\large$\times$}{} \right\rangle = \left\langle \rlap{\large$\times$}{} \right\rangle.$$

[$\rlap{\large$\times$}{}$ denotes a neutral node that is simply carried along through all the calculations.]

If $\rlap{\large$\times$}{}$ denotes a crossing of two different components, then the same form of calculation ensues. Thus, we have indicated the proof that H_K is independent of the template labelling.

Given this well-definedness, the proof of regular isotopy invariance is easy. Just position the least label correctly:

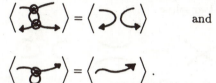

Remark. I have here adopted the notation ⊗ to indicate a neutral node - i.e. a node whose vertex contribution has been already cataloged. Thus

$$\left\langle \mathord{\overset{\displaystyle\oplus}{\rightleftarrows}} \right\rangle = \left\langle \mathord{\supset\subset} \right\rangle \qquad \text{and}$$

$$\left\langle \mathord{\overset{\oplus}{\rightarrow}} \right\rangle = \left\langle \mathord{\rightsquigarrow} \right\rangle .$$

Speculation. In the skein models we may imagine particles moving on the diagram in trajectories whose initial points are dictated by the chosen template. Each state σ is a possible set of trajectories - with the specific vertex weights (hence state evaluation) determined by these starting points. Nevertheless, the entire summation H_K is independent of the choice of starting points. This independence is central for the topological meaning of the model. What does this mean physically?

In a setting involving the concept of local vertex weights determined by the global structure of trajectories (answering the question "Who arrived first?" where first refers to template order) the independence of choices avoids a multiplicity of "times". Imagine a spacetime universe with a single self-interacting trajectory moving forward and backward in time. A particle set moving on this track must be assigned a mathematical "meta-time" - but any physical calculation related to the trajectory (e.g. a vacuum-vacuum expectation) must be independent of meta-time parameters for that particle.

Later, we shall see natural interpretations for the knot polynomials as just such vacuum-vacuum expectations. In that context the skein calculation is akin to using meta-time coordinates in such a way that the dependence cancels out in the averaging.

6^0. An Oriented State Model for $V_K(t)$.

We have already seen the relationship (Theorem 5.2) between the Jones polynomial $V_K(t)$ and the bracket polynomial:

$$V_K(t) = (-t^{3/4})^{w(K)} \langle K \rangle (t^{-1/4}).$$

In principle this gives an oriented state model for $V_K(t)$. Nevertheless, it is interesting to design such a model directly on the oriented diagrams. In unoriented diagrams we created two splits:

In oriented diagrams these splits become

The second split acquires orientations outside the category of link diagrams.

It is useful to **think of these new vertices as abstract Feynman diagrams with a local arrow of time that is coincident with the direction of the diagrammatic arrows.** Then ⇄ represents an interaction (a pass-by) and and ⤚ ⤙ represent creations and annihilations. **In order to have topological invariance** we need to be able to cancel certain combinations of creation and annihilation. Thus

$$\left[\uparrow\right] \qquad \text{(diagram)} \quad \approx \quad \text{(diagram)}$$

This leads to the formalism

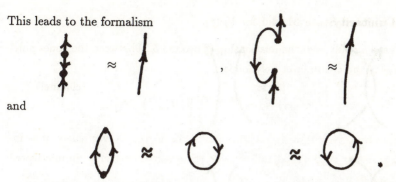

and

In other words, the topological diagrams know nothing (at this level of approximation) about field effects such as

radiation

and the probability of each of the infinity of loops must be the same.

This background for diagrammatic interactions is necessary just to begin the topology. Then we want the interactions and to satisfy **channel and cross-**

channel unitarity:

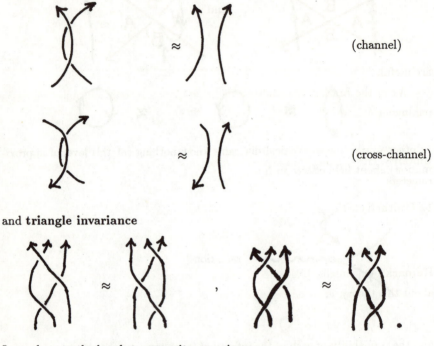

$\approx$ (channel)

$\approx$ (cross-channel)

and **triangle invariance**

$\approx$, $\approx$

In analogy to the bracket, we posit expansions:

$$V \nearrow \!\!\! = AV \rightrightarrows \; + BV \succ\!\!\cdot\!\!\prec$$

$$V \searrow\!\!\!\nwarrow = A'V \rightrightarrows \; + B'V \succ\!\!\cdot\!\!\prec$$

Each crossing interaction is regarded as a superposition of "pass-by" and "annihilate-create". We can begin the model with arbitrary weights, A, B for the positive crossing, and A' and B' for the negative crossing. The mnemonics

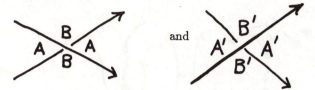 and

are useful.

As in the bracket, we shall assume that an extra loop multiplies V_K by a parameter δ:

$$V \bigcirc {}_K = \delta V_K.$$

Unitarity, cross-channel unitarity and triangle invariance will determine these parameters.

1. Unitarity.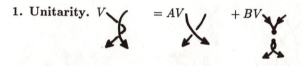

Remark. One might think that, given the tenor of the initial topological remarks about this theory, it would be required that

However, this would trivialize the whole enterprise. Thus **the "field-effects" occur at the level of the interaction of creation or annihilation with the crossing interaction.** As we continue, it will happen that

but, fortunately,

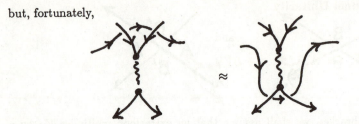

This is in keeping with the virtual character of the radiative part of this inter-action. (All of these remarks are themselves virtual - to be taken as a metaphor/mathematico-physical fantasy on the diagrammatic string.)

To return to the calculation:

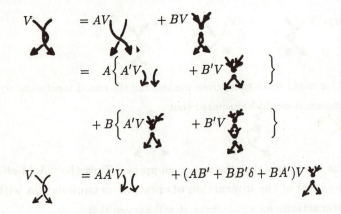

Therefore, unitarity requires

$$\boxed{\begin{array}{c} AA' = 1 \\ \hline \delta = -\left(\dfrac{A}{B} + \dfrac{A'}{B'}\right) \end{array}}$$

2. Cross-Channel Unitarity.

Therefore, cross-channel unitarity requires

$$
\boxed{
\begin{aligned}
BB' &= 1 \\
\delta &= -\left(\frac{B'}{A'} + \frac{B}{A}\right)
\end{aligned}
} \quad .
$$

Thus we now have (1. and 2.)

$$
\boxed{
\begin{aligned}
A' &= 1/A, \qquad B' = 1/B \\
\delta &= -\left(\frac{A}{B} + \frac{B}{A}\right)
\end{aligned}
} \quad .
$$

3. Annihilation and Crossing.

$$V \;=\; A'V \qquad + B'V$$

$$=\; A'\left\{ A'V \qquad + B'V \right\}$$

$$+\, B'\left\{ A'V \qquad + B'V \right\}$$

$$=\; ((A')^2 + (B')^2 + \delta_{A'B'})V \qquad + A'B'V$$

$$\therefore\; V \;=\; A'B'V \;=\; \frac{1}{AB}V$$

Similarly,

$$\boxed{\,V \;=\; ABV\,}$$

and

$$\boxed{\,V \;=\; ABV\,}$$

$$\boxed{\,V \;=\; \frac{1}{AB}V\,}$$

$$\therefore\; V \;=\; V$$

This is the annihilation-creation version of the triangle move, and it implies **triangle invariance:**

$$V \;=\; AV \qquad + BV \;=\; AV \qquad + BV$$

$$A\nabla \text{ } + B\nabla \text{ } = V \text{ } .$$

4. Oriented Triangles.

There are two types of oriented triangle moves. Those like

where the central triangle does **not** have a cyclic orientation, and those like

where the central triangle does have a cyclic orientation. Call the latter a **cyclic** triangle move.

Fact. In the presence of non-cyclic triangle moves and the two orientations of type II moves, cyclic triangle moves can be generated from non-cyclic triangle moves.

Proof. (Sketch) Observe the pictures below:

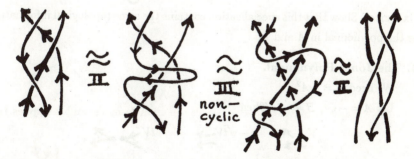

Exercise. Show that triangle moves involving crossings of different signs can be accomplished by triangle moves all-same-sign by using type II moves.

With the help of these remarks and the exercise we see that we have now done enough work to ensure that the model:

$$V\,\diagdown\!\!\!\diagup = AV\rightrightarrows + BV\,\asymp$$

$$V\,\diagup\!\!\!\diagdown = \frac{1}{A}V\rightrightarrows + \frac{1}{B}V\,\asymp$$

is a regular isotopy invariant for oriented link diagrams.

5. Type I Invariance.

$$V\,\sigma\!\!\!\rightarrow = AV\,\bigcirc\!\!\!\rightarrow + BV\,\sigma\!\!\!\rightarrow$$

$$= (A\delta + B)V\longrightarrow$$

$$A\delta + B = A\left(-\frac{A}{B} - \frac{B}{A}\right) + B = -A^2/B.$$

Thus

$$V\,\sigma\!\!\!\rightarrow = V\,{\frown} \quad\Leftrightarrow\quad B = -A^2.$$

It is easy to see that this condition implies that $V\,\sigma\!\!\!\rightarrow = V\,{\frown}$ as well.

Exercise. Show that this specialization contains the same topological information as the polynomial in A and B.

6. The Jones Polynomial.

We need $B = -A^2$.

Let $A = -\sqrt{t}$, $B = -t$. Then

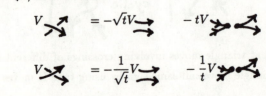

$$V\,\diagdown\!\!\!\diagup = -\sqrt{t}\,V\rightrightarrows \quad - tV\,\asymp$$

$$V\,\diagup\!\!\!\diagdown = -\frac{1}{\sqrt{t}}V\rightrightarrows \quad - \frac{1}{t}V\,\asymp$$

All the work of this section assures us that this expansion yields an invariant of ambient isotopy for oriented links. Multiplying by t^{-1} and by t, and taking the difference, we find

$$t^{-1}V_{\diagup\!\!\!\!\diagdown} - tV_{\diagup\!\!\!\!\diagdown} = \left(-\frac{\sqrt{t}}{t} + \frac{t}{\sqrt{t}}\right)V_{\rightrightarrows}$$

$$= \left(\sqrt{t} - \frac{1}{\sqrt{t}}\right)V_{\rightrightarrows}$$

Therefore $V_K(t)$ is the 1-variable Jones polynomial.

The interest of this model is that it does not depend upon writhe-normalization, and it shows how the parameters t, $\sqrt{t}$ are intrinsic to the knot-theoretic structure of this invariant. Finally, all the issues about oriented invariants (channel and cross-channel unitarity, triangle moves, orientation conventions) will reappear later in our work with Yang-Baxter models (beginning in section 8^0.)

It is rather intriguing to compare the art of calculating within this oriented model with the corresponding patterns of the bracket. A key lemma is

Lemma 6.1.

$$V_{\diagdown} = t\sqrt{t}\, V_{\rightarrow}$$

$$V'_{\diagdown} = \frac{1}{t\sqrt{t}}\, V_{\rightarrow}$$

$$V_{\diagup} = t\sqrt{t}\, V_{\rightarrow}$$

$$V_{\diagdown} = \frac{1}{t\sqrt{t}}\, V_{\rightarrow}$$

Proof.

$$V_{\diagdown} = -\sqrt{t}V_{\rightarrow} - tV_{\rightarrow}\bigcirc$$

$$= \left(-\sqrt{t} - t\left(-\sqrt{t} - \frac{1}{\sqrt{t}}\right)\right)V_{\rightarrow}$$

$$= t\sqrt{t}\, V_{\rightarrow}$$

The rest proceeds in the same way.   //

Thus the model is an ambient isotopy invariant, but regular isotopy patterns of calculation live inside it!

Remark. It is sometimes convenient to use another parameterization of the Jones polynomial. Here we write

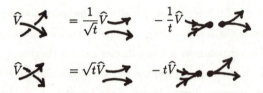

Note that the loop variable $\widehat{\delta} = (\sqrt{t} + \frac{1}{\sqrt{t}})$ for this model. The expansion is given by the formulas:

In the corresponding state expansion we have

$$\widehat{V}_K = \sum_{\overline{\sigma}} \langle K | \overline{\sigma} \rangle \left(\sqrt{t} + \frac{1}{\sqrt{t}} \right)^{\|\overline{\sigma}\|}$$

where $\langle K | \overline{\sigma} \rangle$ is the product of the vertex weights $(\sqrt{t}, 1/\sqrt{t}, -t, -1/t)$ and $\|\overline{\sigma}\|$ is the number of loops in the oriented state. For each state, the **sign** is $(-1)^\pi$ where π **is the parity of the number of creation-annihilation splices in the state.** This version of the Jones polynomial may lead some insight into the vexing problem of cancellations in the state summation.

Unknot Problem. For a knot K, does $V_K = 1$ $(\widehat{V}_K = 1)$ imply that K is ambient isotopic to the unknot?

7^0. Braids and the Jones Polynomial.

In this section I shall demonstrate that the normalized bracket $\mathcal{L}_K(A) = (-A^3)^{-w(K)}\langle K \rangle$ is a version of the original Jones polynomial $V_K(t)$ by way of the theory of braids. The Jones polynomial has been subjected to extraordinary generalizations since it was first introduced in 1984 [JO2]. These generalizations will emerge in the course of discussion. Here we stay with the story of the original Jones polynomial and its relation with the bracket.

Jones constructed the invariant $V_K(t)$ by a route involving braid groups and von Neumann algebras. Although there is much more to say about von Neumann algebras, it is sufficient here to consider a sequence of algebras A_n $(n = 2, 3, \ldots)$ with multiplicative generators $e_1, e_2, \ldots, e_{n-1}$ and relations:

1) $e_i^2 = e_i$

2) $e_i e_{i\pm 1} e_i = \tau e_i$

3) $e_i e_j = e_j e_i \quad |i - j| > 2$

(τ is a scalar, commuting with all the other elements.) For our purposes we can let A_n be the free additive algebra on these generators viewed as a module over the ring $C[\tau, \tau^{-1}]$ (C denotes the complex numbers.). The scalar τ is often taken to be a complex number, but for our purposes is another algebraic variable commuting with the e_i's. This algebra arose in the theory of classification of von Neumann algebras [JO7], and it can itself be construed as a von Neumann algebra.

In this von Neumann algebra context it is natural to study a certain tower of algebras associated with an inclusion of algebras $N \subset M$. With $M_0 = N$, $M_1 = M$ one forms $M_2 = \langle M_1, e_1 \rangle$ where $e_1 : M_1 \to M_0$ is projection to M_0 and $\langle M_1, e_1 \rangle$ denotes an algebra generated by M_1 with e_1 adjoined. Thus we have the pattern

$$M_0 \subset M_1 \subset M_2 = \langle M_1, e_1 \rangle, \qquad e_1 : M_1 \to M_0, \quad e_1^2 = e_1.$$

This pattern can be iterated to form a tower

$$M_0 \subset M_1 \subset M_2 \subset M_3 \subset \ldots \subset M_n \subset M_{n+1} \subset \ldots$$

with $e_i : M_i \to M_{i-1}$, $e_i^2 = e_i$ and $M_{i+1} = \langle M_i, e_i \rangle$. Jones constructs such a tower of algebras with the property that $e_i e_{i\pm 1} e_i = \tau e_i$ and $e_i e_j = e_j e_i$ for $|i - j| > 1$.

Here he found that $\tau^{-1} = [M_1 : M_0]$ a generalized notion of index for these algebras. Furthermore, he defined a **trace** $\mathrm{tr} : M_n \to \mathbf{C}$ (complex numbers) such that it satisfied the

> **Markov Property:** $\mathrm{tr}(we_i) = \tau\, \mathrm{tr}(w)$
> for w in the algebra generated by $M_0, e_1, \ldots, e_{i-1}$

A function from an algebra $\mathcal{A}$ to the complex numbers is said to be a **trace** (tr) if it satisfies the identity

$$\mathrm{tr}(ab) = \mathrm{tr}(ba)$$

for $a, b, \in \mathcal{A}$. Thus ordinary matrix traces are examples of trace functions.

The **tower construction** is very useful for studying the index $[M_1 : M_0]$ for special types of von Neumann algebras. While I shall not discuss von Neumann algebras in this short course, we shall construct a model of such a tower that is directly connected with the bracket polynomial. Thus the combinatorial structure of the tower construction will become apparent from the discussion that follows.

Now to return to the story of the Jones polynomial: Jones was struck by the analogy between the relations for the algebra A_n and the generating relations for the n-**strand Artin braid group** B_n. View Figure 11 for a comparison of these sets of relations

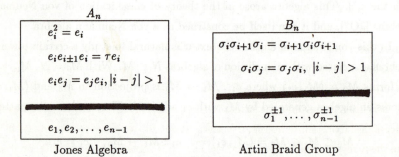

$$A_n$$

$$e_i^2 = e_i$$
$$e_i e_{i\pm 1} e_i = \tau e_i$$
$$e_i e_j = e_j e_i, |i - j| > 1$$

$$e_1, e_2, \ldots, e_{n-1}$$

Jones Algebra

$$B_n$$

$$\sigma_i \sigma_{i+1} \sigma_i = \sigma_{i+1} \sigma_i \sigma_{i+1}$$
$$\sigma_i \sigma_j = \sigma_j \sigma_i, |i - j| > 1$$

$$\sigma_1^{\pm 1}, \ldots, \sigma_{n-1}^{\pm 1}$$

Artin Braid Group

Figure 11

Jones constructed a representation $\rho_n : B_n \to A_n$ of the Artin Braid group to the algebra A_n. The representation has the form

$$\rho_n(\sigma_i) = ae_i + b$$

with a and b chosen appropriately. Since A_n has a trace $\text{tr} : A_n \to \mathbf{C}[t, t^{-1}]$ one can obtain a mapping $\text{tr} \circ \rho : B_n \to \mathbf{C}[t, t^{-1}]$. Upon appropriate normalization this mapping is the Jones polynomial $V_K(t)$. It is an ambient isotopy invariant for oriented links. While the polynomial $V_K(t)$ was originally defined only for braids, it follows from the theorems of Markov (see [B2]) and Alexander [ALEX1] that (due to the Markov property of the Jones trace) it is well-defined for arbitrary knots and links.

These results of Markov and Alexander are worth remarking upon here. First of all there is **Alexander's Theorem: Each link in three-dimensional space is ambient isotopic to a link in the form of a closed braid.**

Braids. A **braid** is formed by taking n points in a plane and attaching strands to these points so that parallel planes intersect the strands in n points. It is usually assumed that the braid begins and terminates in the same arrangement of points so that it has the diagrammatic form

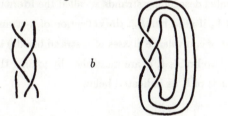

Here I have illustrated a 3-strand braid $b \in B_3$, and its closure $\bar{b}$. The closure $\bar{b}$ of a braid b is obtained by connecting the initial points to the end-points by a collection of parallel strands.

It is interesting and appropriate to think of the braid as a diagram of a physical process of particles interacting or moving about in the plane. In the diagram, we take the arrow of time as moving up the page. Each plane (spatial plane) intersects the page perpendicularly in a horizontal line. Thus successive slices give a picture of the motions of the particles whose world-lines sweep out the braid.

Of course, from the topological point of view one wants to regard the braid as a purely spatial weave of descending strands that are fixed at the top and the bottom of the braid. Two braids in B_n are said to be **equivalent** (and we write $b = b'$ for this equivalence) if there is an ambient isotopy from b to b' that keeps the end-points fixed and does not move any strands outside the space between the top and bottom planes of the braids. (It is assumed that b and b' have identical input and output points.)

For example, we see the following equivalence

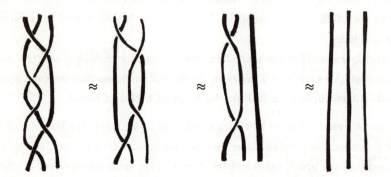

The braid consisting in n parallel descending strands is called the **identity braid** in B_n and is denoted by 1 or 1_n if need be. B_n, the collection of n-strand braids, up to equivalence, (i.e. the set of equivalence classes of n-strand braids) is a group - the **Artin Braid Group**. Two braids b, b' are multiplied by joining the output strands of b to the input strands of b' as indicated below:

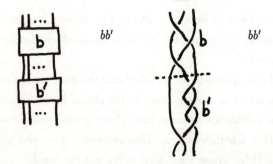

Every braid can be written as a product of the **generators** $\sigma_1, \sigma_2, \ldots, \sigma_{n-1}$ and their **inverses** $\sigma_1^{-1}, \sigma_2^{-1}, \ldots, \sigma_{n-1}^{-1}$. These elementary braids σ_i and σ_i^{-1} are obtained by interchanging only the i-th and $(i+1)$-th points in the row of inputs.

Thus

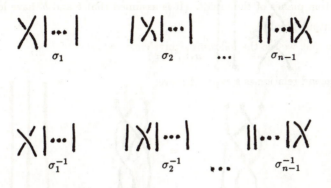

These generators provide a convenient way to catalog various weaving patterns. For example

$$= \sigma_1^{-1}\sigma_2\sigma_1^{-1}\sigma_2\sigma_1^{-1}\sigma_2$$

A 360° twist in the strands has the appearance

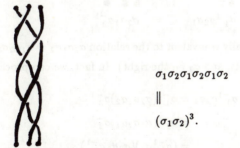

$$\sigma_1\sigma_2\sigma_1\sigma_2\sigma_1\sigma_2$$
$$\|$$
$$(\sigma_1\sigma_2)^3.$$

The braid group B_n is completely described by these generators and relations. The relations are as follows:

$$\begin{cases} \sigma_i\sigma_i^{-1} = 1, & i = 1,\ldots,n-1 \\ \sigma_i\sigma_{i+1}\sigma_i = \sigma_{i+1}\sigma_i\sigma_{i+1}, & i = 1,\ldots,n-2 \\ \sigma_i\sigma_j = \sigma_j\sigma_i, & |i-j| > 1. \end{cases}$$

90

Note that the first relation is a version of the type II move,

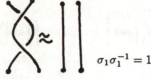

$$\sigma_1 \sigma_1^{-1} = 1$$

while the second relation is a type III move:

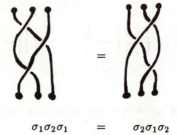

$$\sigma_1 \sigma_2 \sigma_1 \qquad = \qquad \sigma_2 \sigma_1 \sigma_2$$

Note that since $\sigma_1 \sigma_2 \sigma_1 = \sigma_2 \sigma_1 \sigma_2$ is stated in the **group** B_3, we also know that $(\sigma_1 \sigma_2 \sigma_1)^{-1} = (\sigma_2 \sigma_1 \sigma_2)^{-1}$, whence $\sigma_1^{-1} \sigma_2^{-1} \sigma_1^{-1} = \sigma_2^{-1} \sigma_1^{-1} \sigma_2^{-1}$.

There are, however, a few other cases of the type III move. For example:

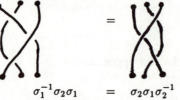

$$\sigma_1^{-1} \sigma_2 \sigma_1 \qquad = \qquad \sigma_2 \sigma_1 \sigma_2^{-1}$$

However, this is algebraically equivalent to the relation $\sigma_2 \sigma_1 \sigma_2 = \sigma_1 \sigma_2 \sigma_1$ (multiply both sides by σ_1 on the left, and σ_2 on the right). In fact, we can proceed directly as follows:

$$\begin{aligned}
\sigma_1^{-1} \sigma_2 \sigma_1 &= \sigma_1^{-1}(\sigma_2 \sigma_1 \sigma_2)\sigma_2^{-1} \\
&= \sigma_1^{-1}(\sigma_1 \sigma_2 \sigma_1)\sigma_2^{-1} \\
&= (\sigma_1^{-1}\sigma_1)(\sigma_2 \sigma_1 \sigma_2^{-1}) \\
&= \sigma_2 \sigma_1 \sigma_2^{-1}.
\end{aligned}$$

I emphasize this form of the equivalence because it shows that the type III move with a mixture of positive and negative crossings can be accomplished via a combination of type II moves and type III moves where all the crossings have the same sign.

Note that there is a homomorphism π of the braid group B_n onto the permutation group S_n on the set $\{1,2,\dots,n\}$. The map $\pi : B_n \to S_n$ is defined by taking the permutation of top to bottom rows of points afforded by the braid. Thus

$$\pi \begin{bmatrix} 1 & 2 & 3 \\ & \times\times & \\ 1 & 2 & 3 \end{bmatrix} = \begin{pmatrix} 1 & 2 & 3 \\ 3 & 1 & 2 \end{pmatrix}$$

where the notation on the right indicates a permutation $\rho : \{1,2,3\} \to \{1,2,3\}$ with $\rho(1) = 3$, $\rho(2) = 1$, $\rho(3) = 2$. If $\rho = \pi(b)$ for a braid b, then $\rho(i) = j$ where j is the lower endpoint of the braid strand that begins at point i.

Letting $\tau_k : \{1,2,\dots,n\} \to \{1,2,\dots,n\}$ denote the transposition of k and $k+1$: $\tau_k(i) = i$ if $i \ne k$, $k+1$, $\tau_k(k) = k+1$, $\tau_k(k+1) = k$. We have that $\pi(\sigma_i) = \tau_i$, $i = 1,\dots,n-1$. In terms of these transpositions, S_n has the presentation

$$S_n = (\tau_1,\dots,\tau_{n-1} \mid \tau_i^2 = 1, \ \tau_i\tau_{i+1}\tau_i = \tau_{i+1}\tau_i\tau_{i+1}).$$

The permutation group is the quotient of the braid group B_n, obtained by setting the squares of all the generators equal to the identity.

Alexander's Theorem.

As we mentioned a few paragraphs ago, Alexander proved [ALEX1] that any knot or link could be put in the form of a closed braid (via ambient isotopy). Alexander proved this result by regarding a closed braid as a looping of the knot around an axis:

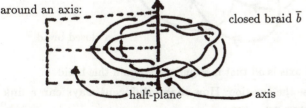

closed braid $\overline{b}$

half-plane   axis

Thinking of three-dimensional space as a union of half-planes, each sharing the axis, we require that $\overline{b}$ intersect each half-plane in the same number of points – (the number of braid-strands). As you move along the knot or link, you are circulating the axis in either a clockwise or counterclockwise orientation. Alexander's method was to choose a **proposed braid axis**. Then follow along the knot or link,

throwing the strand over the axis whenever it began to circulate incorrectly. Eventually, you have the link in braid form.

Figure 12 illustrates this process for a particular choice of axis. Note that it is clear that this process will not always produce the most efficient braid representation for a given knot or link. In the example of Figure 12 we would have fared considerably better if we had taken the axis at a different location - as shown below.

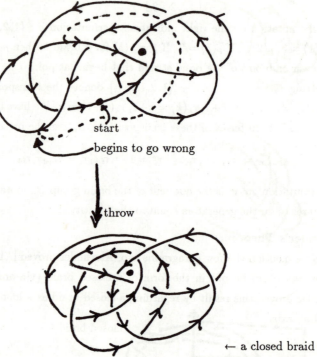

start

begins to go wrong

throw

← a closed braid

One throw over the new axis is all that is required to obtain this braid.

These examples raise the question: **How many different ways can a link be represented as a closed braid?**

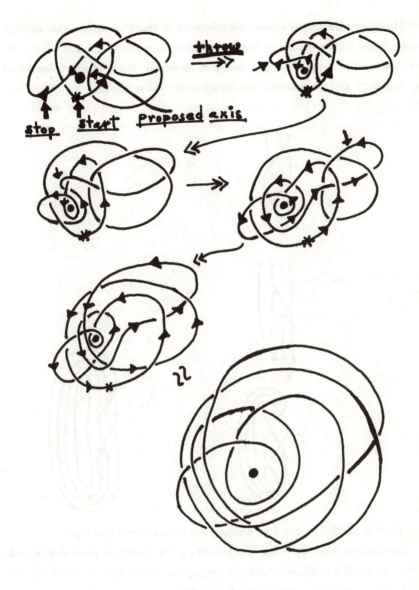

Alexander's Theorem

Figure 12

There are some simple ways to modify braids so that their closures are ambient isotopic links. First there is the **Markov move**: Suppose β is a braid word in B_n (hence a word in $\sigma_1, \sigma_2, \ldots, \sigma_{n-1}$ and their inverses). Then the three braids β, $\beta\sigma_n$ and $\beta\sigma_n^{-1}$ all have ambient isotopic closures. For example,

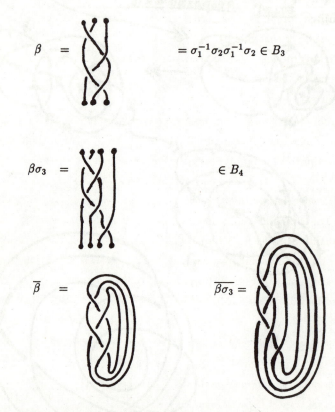

$$\beta = \qquad = \sigma_1^{-1}\sigma_2\sigma_1^{-1}\sigma_2 \in B_3$$

$$\beta\sigma_3 = \qquad \in B_4$$

$$\overline{\beta} = \qquad \overline{\beta\sigma_3} =$$

Thus $\overline{\beta\sigma_n^{\pm 1}}$ is obtained from β by a type I Reidemeister move.

A somewhat more diabolical way to make a braid with the same closure is to choose any braid g in B_n and take the **conjugate braid** $g\beta g^{-1}$. When we close $g\beta g^{-1}$ to form $\overline{g\beta g^{-1}}$ the braid g and its inverse g^{-1} can cancel each other out by interacting through the closure strands. The fundamental theorem that relates the theory of knots and the theory of braids is the

Markov Theorem 7.1. Let $\beta_n \in B_n$ and $\beta'_m \in B_m$ be two braids in the braid groups B_n and B_m respectively. Then the links (closures of the braids β, β') $L = \overline{\beta_n}$ and $L' = \overline{\beta'_m}$ are ambient isotopic if and only if β'_m can be obtained from β_n by a series of

1) **equivalences** in a given braid group.

2) **conjugation** in a given braid group. (That is, replace a braid by some conjugate of that braid.)

3) **Markov moves:** (A Markov move replaces $\beta \in B_n$ by $\beta\sigma_n^{\pm 1} \in B_{n+1}$ or the inverse of this operation - replacing $\beta\sigma_n^{\pm 1} \in B_{n+1}$ by $\beta \in B_n$ if β has no occurrence of σ_n.)

For a proof of the Markov theorem the reader may wish to consult [B2].

The reader may enjoy pondering the question: How can Alexander's technique for converting links to braids be done in an algorithm that a computer can perform? (See [V].)

With the Markov theorem, we are in possession of the information needed **to use the presentations of the braid groups B_n to extract topological information about knots and links.** In particular, it is now possible to explain how the Jones polynomial works in relation to braids. For suppose that we are given a commutative ring R (polynomials or Laurent polynomials for example), and functions $J_n : B_n \to R$ from the n-strand braid group to the ring R, defined for each $n = 2, 3, 4, \ldots$ Then the Markov theorem assures us that the family of functions $\{J_n\}$ can be used to construct link invariants if the following conditions are satisfied:

1. If b and b' are equivalent braid words, then $J_n(b) = J_n(b')$. (This is just another way of saying that J_n is well-defined on B_n.)

2. If $g, b \in B_n$ then $J_n(b) = J_n(gbg^{-1})$.

3. If $b \in B_n$, then there is a constant $\alpha \in R$, independent of n, such that
$$J_{n+1}(b\sigma_n) = \alpha^{+1} J_n(b)$$
$$J_{n+1}(b\sigma_n^{-1}) = \alpha^{-1} J_n(b).$$

We see that for the closed braid $\overline{b'} = \overline{b\sigma_n}$ the result of the Markov move $b \mapsto b'$ is to perform a type I move on $\overline{b}$. Furthermore, $\overline{b\sigma_n}$ corresponds to a type I move of positive type, while $\overline{b\sigma_n^{-1}}$ corresponds to a type I move of negative type. It is for this reason that I have chosen the conventions for α and α^{-1} as above. Note also that, orienting a braid downwards, as in

has positive crossings corresponding to σ_i's with positive exponents.

With these remarks in mind, let's define the **writhe of a braid**, $w(b)$, to be its **exponent sum**. That is, we let $w(b) = \sum_{\ell=1}^{k} a_\ell$ in any braid word

$$\sigma_{i_1}^{a_1} \sigma_{i_2}^{a_2} \ldots \sigma_{i_k}^{a_k}$$

representing b. From our previous discussion of the writhe, it is clear that $w(b) = w(\overline{b})$ where $\overline{b}$ is the oriented link obtained by closing the braid b (with downward-oriented strands). Here $w(\overline{b})$ is the writhe of the oriented link $\overline{b}$.

Definition 7.2. Let $\{J_n : B_n \to R\}$ be given with properties 1., 2., 3. as listed above. Call $\{J_n\}$ a **Markov trace** on $\{B_n\}$. For any link L, let $L \sim \overline{b}$, $b \in B_n$ via Alexander's theorem. Define $J(L) \in R$ via the formula

$$J(L) = \alpha^{-w(b)} J_n(b).$$

Call $J(L)$ the **link invariant for the Markov trace** $\{J_n\}$.

Proposition 7.3. Let J be the link invariant corresponding to the Markov trace $\{J_n\}$. Then J is an invariant of ambient isotopy for oriented links. That is, if $L \sim L'$ ($\sim$ denotes ambient isotopy) then $J(L) = J(L')$.

Proof. Suppose, by Alexander's theorem, that $L \sim \overline{b}$ and $L' \sim \overline{b'}$ where $b \in B_n$ and $b' \in B_n$ are specific braids. Since L and L' are ambient isotopic, it follows that $\overline{b}$ and $\overline{b'}$ are also ambient isotopic. Hence $\overline{b'}$ can be obtained from $\overline{b}$ by a sequence of Markov moves of the type 1., 2., 3. Each such move leaves the function $\alpha^{-w(b)} J_n(b)$ ($b \in B_n$) invariant since the exponent sum is invariant under conjugation, braid moves, and it is used here to cancel the effect of the type 3. Markov move. This completes the proof. //

The Bracket for Braids.

Having discussed generalities about braids, we can now look directly at the bracket polynomial on closed braids. In the process, the structure of the Jones polynomial and its associated representations of the braid groups will naturally emerge.

In order to begin this discussion, let's define $\langle \ \rangle : B_n \to \mathbf{Z}[A, A^{-1}]$ via $\langle b \rangle = \langle \bar{b} \rangle$, the evaluation of the bracket on the closed braid b. In terms of the Markov trace formalism, I am letting $J_n : B_n \to \mathbf{Z}[A, A^{-1}] = R$ via $J_n(b) = \langle \bar{b} \rangle$. In fact, given what we know about the bracket from section 3^0, it is obvious that $\{J_n\}$ is a Markov trace, with $\alpha = -A^3$.

Now consider the **states of a braid**. That is, consider the states determined by the recursion formula for the bracket:

$$\left\langle \; \times \; \right\rangle = A \left\langle \;)(\; \right\rangle + A^{-1} \left\langle \; \asymp \; \right\rangle .$$

In terms of braids this becomes

$$\left\langle \; \| \cdots | \times | \cdots \| \; \right\rangle = A \left\langle \; \| \cdots \| \| | \cdots \| \; \right\rangle + A^{-1} \left\langle \; \| \cdots | \asymp | \cdots \| \; \right\rangle$$

$$\langle \sigma_i \rangle = A \langle 1_n \rangle + A^{-1} \langle U_i \rangle$$

where 1_n denotes the identity element in B_n (henceforth denoted by 1), and U_i is a **new** element written in braid input-output form, but with a cup-$\cup$ cap-$\cap$ combination $\genfrac{}{}{0pt}{}{\cup}{\cap}$ at the i-th and $(i+1)$-th strands:

$$U_1 \qquad\qquad , \qquad U_2 \qquad\qquad , \qquad U_3 \qquad\qquad \text{(for 4-strands)}$$

Since a state for $\bar{b}$, is obtained by choosing splice direction for each crossing of b, we see that **each state of $\bar{b}$ can be written as the closure of an (input-output) product of the elements U_i.** (See section 3^0 for a discussion of bracket states.)

For example, let $L = \bar{b}$ be the link

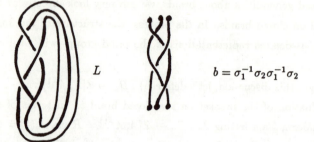

$$b = \sigma_1^{-1}\sigma_2\sigma_1^{-1}\sigma_2$$

Then the state s of L shown below corresponds to the product $U_1^2 U_2$.

$$\overline{U_1^2 U_2} = s$$

In fact it is clear that we can use the following formalism: Write

$$\sigma_i \equiv A + A^{-1}U_i, \qquad \sigma_i^{-1} \equiv A^{-1} + AU_i$$

Given a braid word b, write $b \equiv \mathcal{U}(b)$ where $\mathcal{U}(b)$ is a sum of products of the U_i's, obtained by performing the above substitutions for each σ_i. Each product of U_i's, when closed gives a collection of loops. Thus if U is such a product, then $\langle U \rangle = \langle \overline{U} \rangle = \delta^{\|U\|}$ where $\|U\| = \#(\text{of loops in}) \ \overline{U} - 1$ and $\delta = -A^2 - A^{-2}$. Finally if $\mathcal{U}(b)$ is given by

$$\mathcal{U}(b) = \sum_s \langle b|s \rangle U_s$$

where s indexes all the terms in the product, and $\langle b|s \rangle$ is the product of A's and A^{-1}'s multiplying each U-product U_s, then

$$\langle b \rangle = \langle \mathcal{U}(b) \rangle = \sum_s \langle b|s \rangle \ \langle U_s \rangle$$

$$\langle b \rangle = \sum_s \langle b|s \rangle \delta^{\|s\|}.$$

This is the braid-analog of the state expansion for the bracket.

Example. $b = \sigma_1^2$.

Then

$$\mathcal{U}(b) = (A + A^{-1}U_1)(A + A^{-1}U_1)$$
$$\mathcal{U}(b) = A^2 + 2U_1 + A^{-2}U_1^2$$
$$\langle b \rangle = \langle \mathcal{U}(b) \rangle = A^2\langle 1_2 \rangle + 2\langle U_1 \rangle + A^{-2}\langle U_1^2 \rangle$$

$$\therefore \langle L \rangle = A^2(-A^2 - A^{-2}) + 2 + A^{-2}(-A^2 - A^{-2})$$
$$= -A^4 - 1 + 2 - 1 - A^{-4}$$
$$\langle L \rangle = -A^4 - A^{-4}.$$

This is in accord with our previous calculation of the bracket for the simple link of two components.

The upshot of these observations is that in calculating the bracket for braids in B_n it is useful to have the **free additive algebra** $\mathcal{A}_n$ with generators $U_1, U_2, \ldots, U_{n-1}$ and multiplicative relations coming from the interpretation of the U_i's as cup-cap combinations. This algebra $\mathcal{A}_n$ will be regarded as a module over the ring $\mathbf{Z}[A, A^{-1}]$ with $\delta = -A^2 - A^{-2} \in \mathbf{Z}[A, A^{-1}]$ the designated loop value. I shall call $\mathcal{A}_n$ the **Temperley-Lieb Algebra** (see [BA1], [LK4]).

What are the multiplicative relations in $\mathcal{A}_n$? Consider the pictures in Figure 13. They illustrate the relations:

$$[\mathcal{A}] \quad \left\{ \begin{array}{rcl} U_i U_{i\pm1} U_i & = & U_i \\ U_i^2 & = & \delta U_i \\ U_i U_j & = & U_j U_i, \quad \text{if } |i - j| > 1 \end{array} \right\}.$$

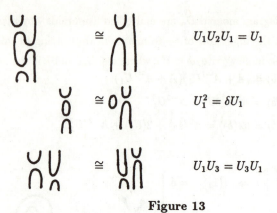

Figure 13

In fact, these are precisely the relations for The Temperley-Lieb algebra. Note that the Temperley-Lieb algebra and the Jones algebra are closely related. In fact, if we define $e_i = \delta^{-1} U_i$, then $e_i^2 = e_i$ and $e_i e_{i\pm 1} e_i = \tau e_i$ where $\tau = \delta^{-2}$. Thus, by considering the state expansion of the bracket polynomial for braids, we recover the formal structure of the original Jones polynomial.

It is convenient to view $\mathcal{A}_n$ in a more fundamental way: Let D_n denote the collection of all (topological) equivalence classes of diagrams obtained by connecting pairs of points in two parallel rows of n points. The arcs connecting these points must satisfy the following conditions:

1) All arcs are drawn in the space between the two rows of points.

2) No two arcs cross one another.

3) Two elements $a, b \in D_n$ are said to be **equivalent** if they are topologically equivalent via a planar isotopy through elements of D_n. (That is, if there is a continuous family of embeddings of arcs - giving elements $\mathcal{C}_t \in D_n$ ($0 \le t \le 1$) with $\mathcal{C}_0 = a$, $\mathcal{C}_1 = b$ and $\mathcal{C}_t$ the identity map on the subset of endpoints for each t, $0 \le t \le 1$).

Call D_n the **diagram monoid on $2n$ points**.

Example. For $n = 3$, D_3 has the following elements:

Elements of the diagram monoid D_n are multiplied like braids - by attaching the output row of a to the input row of b - forming ab. Multiplying in this way, closed loops may appear in ab. Write $ab = \delta^k c$ where $c \in D_n$, and k is the number of closed loops in the product.

For example, in D_3

$$rs = \quad \\quad = \delta \quad \\quad = \delta U_2.$$

Proposition 7.4. The elements $1, U_1, U_2, \dots, U_{n-1}$ generate D_n. If an element $x \in D_n$ is equivalent to two products, P and Q, of the elements $\{U_i\}$, then Q can be obtained from P by a series of applications of the relations $[\mathcal{A}]$.

See [LK8] for the proof of this proposition. The point of this proposition is that it lays bare the underlying combinatorial structure of the Temperley-Lieb algebra. And, for computational purposes, the multiplication table for D_n can be obtained easily with a computer program.

We can now define a mapping

$$\rho : B_n \rightarrow \mathcal{A}_n$$

by the formulas:

$$\rho(\sigma_i) = A + A^{-1} U_i$$
$$\rho(\sigma_i^{-1}) = A^{-1} + A U_i.$$

We have seen that for a braid b, $\langle \overline{b} \rangle = \sum_s \langle B | s \rangle \langle \overline{U}_s \rangle$ where $\rho(b) = \sum_s \langle b | s \rangle U_s$ is the explicit form of $\rho(b)$ obtained by defining $\rho(xy) = \rho(x)\rho(y)$ on products. (s runs through all the different products in this expansion.) Here $\langle \overline{U}_s \rangle$ counts one less than the number of loops in $\overline{U}_s$.

Define $\mathrm{tr} : \mathcal{A}_n \rightarrow \mathbf{Z}[A, A^{-1}]$ by $\mathrm{tr}(U) = \langle \overline{U} \rangle$ for $U \in D_n$. Extend tr linearly to $\mathcal{A}_n$. This mapping - by loop counts - is a realization of Jones' trace on the von Neumann algebra $\mathcal{A}_n$. We then have the formula: $\langle b \rangle = \mathrm{tr}(\rho(b))$.

This formalism explains directly how the bracket is related to the construction of the Jones polynomial via a trace on a representation of the braid group to the Temperley-Lieb algebra.

We need to check certain things, and some comments are in order. First of all, the trace on the von Neumann algebra A_n was not originally defined diagrammatically. It was, defined in [JO7] via normal forms for elements of the Jones algebra A_n. Remarkably, this version of the trace matches the diagrammatic loop count. In the next section, we'll see how this trace can be construed as a modified matrix trace in a representation of the Temperley-Lieb algebra.

Proposition 7.5. $\rho : B_n \rightarrow A_n$, as defined above, is a representation of the Artin Braid group.

Proof. It is necessary to verify that $\rho(\sigma_i)\rho(\sigma_i^{-1}) = 1$, $\rho(\sigma_i\sigma_{i+1}\sigma_i) = \rho(\sigma_{i+1}\sigma_i\sigma_{i+1})$ and that $\rho(\sigma_i\sigma_j) = \rho(\sigma_j\sigma_i)$ when $|i - j| > 1$. We shall do these in the order - first, third, second.

First.

$$
\begin{aligned}
\rho(\sigma_i)\rho(\sigma_i^{-1}) &= (A + A^{-1}U_i)(A^{-1} + AU_i) \\
&= 1 + (A^{-2} + A^2)U_i + U_i^2 \\
&= 1 + (A^{-2} + A^2)U_i + \delta U_i \\
&= 1 + (A^{-2} + A^2)U_i + (-A^{-2} - A^2)U_i \\
&= 1
\end{aligned}
$$

Third. Given that $|i - j| > 1$:

$$
\begin{aligned}
\rho(\sigma_i\sigma_j) &= \rho(\sigma_i)\rho(\sigma_j) \\
&= (A + A^{-1}U_i)(A + A^{-1}U_j) \\
&= (A + A^{-1}U_j)(A + A^{-1}U_i), \ [U_iU_j = U_jU_i \text{ if } |i - j| > 1] \\
&= \rho(\sigma_j\sigma_i).
\end{aligned}
$$

Second.

$$\rho(\sigma_i\sigma_{i+1}\sigma_i) = (A + A^{-1}U_i)(A + A^{-1}U_{i+1})(A + A^{-1}U_i)$$

$$= (A^2 + U_{i+1} + U_i + A^{-2}U_iU_{i+1})(A + A^{-1}U_i)$$

$$= A^3 + AU_{i+1} + AU_i + A^{-1}U_iU_{i+1} + A^{-1}U_i^2 + AU_i + A^{-1}U_{i+1}U_i$$

$$+ A^{-3}U_iU_{i+1}U_i$$

$$= A^3 + AU_{i+1} + (A^{-1}\delta + 2A)U_i + A^{-1}(U_iU_{i+1} + U_{i+1}U_i) + A^{-3}U_i$$

$$= A^3 + AU_{i+1} + (A^{-1}(-A^2 - A^{-2}) + 2A + A^{-3})U_i$$

$$+ A^{-1}(U_iU_{i+1} + U_{i+1}U_i)$$

$$= A^3 + A(U_{i+1} + U_i) + A^{-1}(U_iU_{i+1} + U_{i+1}U_i).$$

Since this expression is symmetric in i and $i + 1$, we conclude that

$$\rho(\sigma_i\sigma_{i+1}\sigma_i) = \rho(\sigma_{i+1}\sigma_i\sigma_{i+1}).$$

This completes the proof that $\rho : B_n \to \mathcal{A}_n$ is a representation of the Artin Braid Group. //

8^0. Abstract Tensors and the Yang-Baxter Equation.

In this section and throughout the rest of the book I begin a notational convention that I dub **abstract (diagrammatic) tensors**. It is really a diagrammatic version of matrix algebra where the matrices have many indices. Since tensors are traditionally such objects - endowed with specific transformation properties - the subject can be called abstract tensors.

The diagrammatic aspect is useful because it enables us to hide indices. For example, a matrix $M = (M_j^i)$ with entries M_j^i for i and j in an index set $\mathcal{I}$ will be a box with an upper strand for the upper index, and a lower strand for the lower index:

$$M = (m_j^i) \leftrightarrow \boxed{M}.$$

In general, a tensor-like object has some upper and lower indices. These become lines or strands emanating from the corresponding diagram:

$$T_{\ell m}^{ijk} \quad \leftrightarrow \quad \boxed{T}$$

Usually, I will follow standard typographical conventions with these diagrams. That is, the upper indices are in correspondence with upper strands and ordered from left to right, the lower indices correspond to lower strands and are also ordered from left to right. Note that because the boundary of the diagram may be taken to be a Jordan curve in the plane, the upper left-right order proceeds **clockwise** on the body of the tensor, while the lower left-right order proceeds **counter-clockwise**.

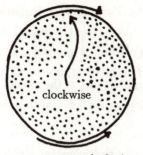

If we wish to discriminate indices in some way that is free of a given convention of direction, then the corresponding lines can be labelled. The simplest labelling scheme is to **put an arrow on the line**. Thus we may write

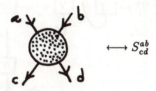

$$\longleftrightarrow S^{ab}_{cd}$$

to indicate "inputs" $\{a, b\}$ and "outputs" $\{c, d\}$. Here we see one of the properties of the diagrammatic notation: Diagrammatic notation acts as a mathematical metaphor. The picture

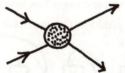

reminds us of a process (input/output, scattering). But we are free to take this reminder or leave it according to the demands of context and interpretation.

Normally, I take the **strand**

to denote a **Kronecker delta**:

$$\longleftrightarrow \delta_b^a = \begin{cases} 1 & \text{if } a = b \\ 0 & \text{if } a \neq b \end{cases}$$

There are contexts in which some bending of the strand, as in $a \bullet \frown \bullet b \leftrightarrow M_{ab}$ will not be a Kronecker delta, but we shall deal with these later.

Generalized Multiplication.

Recall the definition of matrix multiplication: $(MN)_j^i = \sum_k M_k^i N_j^k$. This is often written as $M_k^i M_j^k$ where it is assumed that one sums over all occurrences of a pair of repeated upper and lower indices (Einstein summation convention). The corresponding convention in the diagrammatics is that **a line that connects two index strands connotes the summation over all occurrences of an index value for that line.**

$$\sum_k M_k^i N_j^k \quad \leftrightarrow M_k^i N_j^k \leftrightarrow$$

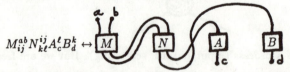

$$\text{trace}(M) \quad = \text{tr}(M) = \sum_i M_i^i \leftrightarrow M_i^i \leftrightarrow \boxed{M}$$

Thus matrix multiplication corresponds to plugging one box into another, and the standard matrix trace corresponds to plugging a box into itself.

We can now write out various products diagrammatically:

$$M_{ij}^{ab} N_{k\ell}^{ij} A_c^\ell B_d^k \leftrightarrow$$

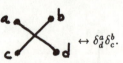

In this last example a crossover occurs - due to index reordering. Here I adopt the convention that

$$a \bullet \quad \bullet b \atop c \bullet \quad \bullet d \leftrightarrow \delta_d^a \delta_c^b.$$

The crossed lines are independent Kronecker deltas.

Exercise. Let the index set $\mathcal{I} = \{1, 2\}$ and define

$$\epsilon_{ij} = \begin{cases} 1 & i < j \\ -1 & i > j \\ 0 & \text{otherwise} \end{cases}$$

and $\epsilon^{ij} = \epsilon_{ij}$. Let $\prod$ be the diagram for ϵ_{ij} and $\coprod$ the diagram for ϵ^{ij}. Show that

$$\left[\text{Note } \frac{\coprod}{\prod} = \coprod \; \prod \right]$$

You will find that this is equivalent to saying

$$\boxed{\epsilon^{ab}\epsilon_{cd} = \delta^a_c \delta^b_d - \delta^a_d \delta^b_c}.$$

This identity will be useful later on.

To return to the crossed lines, they are not **quite** innocuous. They do represent a permutation. Thus, if $\mathcal{I} = \{1, 2\}$ as in the exercise, and

$$P^{ab}_{cd} = \delta^a_d \delta^b_c = \overset{a \quad b}{\underset{c \quad d}{\times}},$$

then P is a matrix representing a transposition. Similarly, for three lines, we have

$$\left\{ \; \times| \; , \; |\times \; , \; \times\!\!\times \; , \; \times\!\!\times \; , \times\!\!\times , ||| \; \right\}$$

represen*ing S_3, the symmetric group on three letters.

Knot Diagrams as Abstract Tensor Diagrams.

By now it must have become clear that we have intended all along to interpret a diagram of a knot or link as an abstract tensor diagram. How can this be done? Actually there is more than one way to do this, but the simplest is to use an oriented diagram and to associate two matrices to the two types of crossing. Thus,

$$\times \quad \leftrightarrow \quad \overset{a \quad b}{\underset{c \quad d}{\bullet}} \; = R^{ab}_{cd}$$

$$\times \quad \leftrightarrow \quad \overset{a \quad b}{\underset{c \quad d}{\circ}} \; = \overline{R}^{ab}_{cd}$$

With this convention, any **oriented link** diagram K is mapped to a specific contracted (no free lines) abstract tensor $T(K)$.

Example. 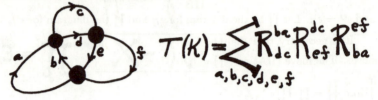 $K \longmapsto$ $T(K)$. Thus, if we label the lines of $T(K)$ with indices, then $T(K)$ corresponds to a formal product - or rather a **sum** of products, since repeated indices connote summations:

$$T(K) = \sum_{a,b,c,d,e,f} R_{dc}^{ba} R_{ef}^{dc} R_{ba}^{ef}$$

If there is a commutative ring $\mathcal{R}$ and an index set $\mathcal{I}$ such that $a, b, c, \ldots \in \mathcal{I}$ and $R_{cd}^{ab} \in \mathcal{R}$ then we can write

$$T(K) = \sum_{a,b,\ldots,f \in \mathcal{I}} R_{dc}^{ba} R_{ef}^{dc} R_{ba}^{ef},$$

and one can regard a choice of labels from $\mathcal{I}$ for the edges of $T(R)$ as a **state** of K. That is, a **state** σ **of** K is a mapping $\sigma : E(K) \to \mathcal{I}$ where E denotes the edge set of K and $\mathcal{I}$ is the given index set.

In this regard, we are seeing $T(K)$ as identical to the oriented graph that underlies the link diagram K with **labelled nodes** (black or white) corresponding to the crossing type in K. Such a diagram then translates directly into a formal product by associating R_{cd}^{ab} with positive crossings and $\overline{R}_{cd}^{ab}$ with negative crossings.

Using the concept of a state σ we can rewrite

$$T(K) = \sum_{\sigma} \langle K | \sigma \rangle$$

where σ runs over all states of K, and $\langle K | \sigma \rangle$ denotes the product of the **vertex weights** R_{cd}^{ab} (or $\overline{R}_{cd}^{ab}$) assigned to the crossings of K by the given state.

We wish to see, under what circumstances $T(K)$ will be invariant under the Reidemeister moves. In particular, the relevant question is invariance under moves of type II and type III (regular isotopy). We shall see that the model, using abstract tensors, is good for constructing representations of the Artin braid group. It requires modification to acquire regular isotopy invariance.

But let's take things one step at a time. There are **two** versions of the type II move:

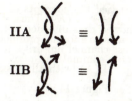

It is necessary that the model, as a whole, be invariant under both of these moves. The move (IIA) corresponds to a very simple matrix condition on R and $\overline{R}$:

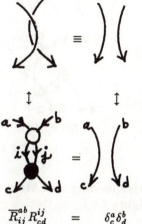

$$\overline{R}_{ij}^{ab} R_{cd}^{ij} = \delta_c^a \delta_d^b$$

} channel unitarity

That is, (IIA) will be satisfied if R and $\overline{R}$ are inverse matrices. Call this condition on R **channel unitarity**. The direct requirements imposed by move (IIB) could be termed **cross-channel unitarity**:

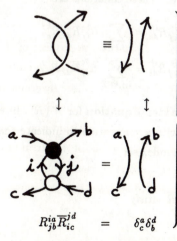

$$R_{jb}^{ia} \overline{R}_{ic}^{jd} = \delta_c^a \delta_b^d$$

} cross-channel unitarity

Recall that we already discussed the concepts of channel and cross-channel unitarity in section 6^0. We showed that in order to have type III invariance, in the presence of channel and cross-channel unitarity, it is sufficient to demand invariance under the move of type III(A) [with all crossings positive (as shown below) or all crossings negative]. See Figure 14.

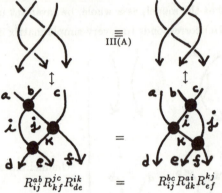

$$R_{ij}^{ab} R_{kf}^{jc} R_{de}^{ik} \qquad = \qquad R_{ij}^{bc} R_{dk}^{ai} R_{ef}^{kj}$$

The Yang-Baxter Equation Corresponding to a Move of type III(A)
Figure 14

As shown in Figure 14, there is a matrix condition that will guarantee invariance of $T(K)$ under the moves $III(A, +)$ and $III(A, -)$ where the $+$ or $-$ signs refer to the crossing types in these moves. These equations are:

$$III(A, +): \sum_{i,j,k \in \mathcal{I}} R_{ij}^{ab} R_{kf}^{jc} R_{de}^{ik} = \sum_{i,j,k \in \mathcal{I}} R_{ij}^{bc} R_{dk}^{ai} R_{ef}^{kj}$$

$$III(A, -): \sum_{i,j,k \in \mathcal{I}} \overline{R}_{ij}^{ab} \overline{R}_{kf}^{jc} \overline{R}_{de}^{ik} = \sum_{i,j,k \in \mathcal{I}} \overline{R}_{ij}^{bc} \overline{R}_{dk}^{ai} \overline{R}_{ef}^{kj}$$

and will be referred to as the **Yang-Baxter Equation** for R $(\overline{R})$. In the above forms, I have written these equations with a summation to indicate that the repeated indices are taken from the given index set $\mathcal{I}$.

Thus we have proved the

Theorem 8.1. If the matrices R and $\overline{R}$ satisfy

1) channel unitarity

2) cross-channel unitarity and

3) Yang-Baxter Equation

then $T(K)$ is a regular isotopy invariant for oriented diagrams K.

Remark. The Yang-Baxter Equation as it arises in mathematical physics involves extra parameters - sometimes called **rapidity** or **momentum**. In fact, if we interpret the vertex

as a particle interaction with incoming spins (or charges) a and b and outgoing spins (charges) c and d, then R_{cd}^{ab} can be taken to represent the scattering amplitude for this interaction. That is, it can be regarded as the probability amplitude for this particular combination of spins in and out.

Under these circumstances it is natural to also consider particle momenta and other factors.

For now it is convenient to consider only the spins. The conservation of spin (or charge) suggests the rule that $a+b = c+d$ whenever $R_{cd}^{ab} \neq 0$. This turns out to be a good starting place for the construction of solutions to the YBE (Yang-Baxter Equation).

Example. Let the index set $\mathcal{I} = \{1, 2, 3, \ldots, n\}$ and let $R_{cd}^{ab} = A\delta_c^a \delta_d^b + A^{-1}\delta^{ab}\delta_{cd}$, where $n = -A^2 - A^{-2}$. That is, A is chosen to satisfy the equation

$$nA^2 + A^4 + 1 = 0.$$

Here δ_b^a and δ^{ab} are Kronecker deltas:

$$\delta_b^a = \begin{cases} 1 & \text{if } a = b \\ 0 & \text{if } a \neq b, \end{cases}$$

$$\delta^{ab} = \begin{cases} 1 & \text{if } a = b \\ 0 & \text{if } a \neq b. \end{cases}$$

112

Here we are transcribing this solution from the bracket model (section 3^0) for the Jones polynomial!

Recall that the bracket is defined by the equations:

$$\left\langle \; \times \; \right\rangle = A \left\langle \;\right)\left(\; \right\rangle + A^{-1} \left\langle \; \asymp \; \right\rangle$$

$$\left\langle \; O \; \right\rangle = -A^2 - A^{-2}.$$

See section 3^0 for the other conventions for dealing with the bracket. This suggests a matrix model where

$$R_{cd}^{ab} = \times = A \;\Big)\Big(\; + A^{-1} \;\asymp$$

$$= A \delta_c^a \delta_d^b + A^{-1} \delta^{ab} \delta_{cd}.$$

In such a model the loop value, $\langle O \rangle$, is the trace of a Kronecker delta: $\langle O \rangle = \delta_a^a = n$ if $\mathcal{I} = \{1, 2, \dots, n\}$. Thus we require that A satisfies the equation

$$n = -A^2 - A^{-2}.$$

This completes the motivation picking the particular form of the R-matrix. Certain consistency checks are needed. For example, if without orientation we assigned

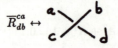 $R_{cd}^{ab} \leftrightarrow \qquad \uparrow$time's arrow

then (viewing the interaction with time's arrow turned by $90°$) we must have

$\overline{R}_{db}^{ca} \leftrightarrow$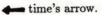

$\longleftarrow$ time's arrow.

Thus we need

$$\overline{R}_{db}^{ca} = R_{cd}^{ab}$$

whence

$$\overline{R}^{ab}_{cd} = R^{bd}_{ac}$$

$$= A\delta^b_a \delta^d_c + A^{-1}\delta^{bd}\delta_{ac}$$

$$= A^{-1}\delta^a_c \delta^b_d + A\delta^{ab}\delta_{cd}$$

$$= A^{-1} \smile\!\!\!\big)\big(\!\!\!\smile \ {}^b_d + A \ \overset{a \ b}{\underset{c \ d}{\smile}}$$

$$\updownarrow$$

Essentially the same checks that proved the regular isotopy invariance of $\langle K \rangle$ now go over to prove that $\overline{R}$ is the inverse of R, and that R and $\overline{R}$ satisfy the YBE. This is a case where R and $\overline{R}$ satisfy both channel and cross-channel unitarity.

As far as link **diagrams** are concerned, we could define an R-matrix via

$$R^{ab}_{cd} = A\delta^a_c \delta^b_d + B\delta^{ab}\delta_{cd}$$

with

If the index set is $\mathcal{I} = \{1, 2, \ldots, n\}$, then the resulting "tensor contraction" $T(K)$ will satisfy:

$$T(OK) = nT(K)$$

$$T(O) = n.$$

Thus $T(K) = \langle K \rangle$ where $\langle K \rangle$ is a generalized bracket with an integer value n for the loop.

A priori, specializing $B = A^{-1}$ and $n = -A^2 - A^{-2}$ does not guarantee that R will satisfy the YBE. But the fact that $\langle K \rangle$ is, under these conditions, invariant under the third Reidemeister move, provides strong motivation for a direct check on the R-matrix.

Checking the R-matrix.

Let

$$R_{cd}^{ab} = A \,)(\; + B \,\overset{\smile}{\underset{\frown}{}}\;$$

$$\boxed{\;\times = A \,)(\; + B \,\underset{}{\smile}\;} \quad \text{(notation)}$$

where it is understood that the arcs denote Kronecker deltas with indices from an index set $\mathcal{I} = \{1, 2, \ldots, n\}$. In this case, we can perform a direct expansion of the triangle configuration into eight terms:

$$\left[\times\!\!\times \right] = A^3 \left[\, \right] + A^2 B \left[\quad + \quad + \quad \right]$$

$$+ \, AB^2 \left[\quad + \quad + \quad \right]$$

$$+ \, B^3 \left[\quad \right]$$

and

$$\left[\times\!\!\times \right] = A^3 \left[\, \right] + A^2 B \left[\quad + \quad + \quad \right]$$

$$+ \, AB^2 \left[\quad + \quad + \quad \right]$$

$$+ \, B^3 \left[\quad \right]$$

Letting n denote the loop value, we then have an equation for the difference:

$$\left[\;\vcenter{\hbox{\includegraphics}}\;\right] - \left[\;\vcenter{\hbox{\includegraphics}}\;\right] = A^2 B \left[\;\vcenter{\hbox{\includegraphics}} - \vcenter{\hbox{\includegraphics}}\;\right]$$

$$+ AB^2 \left[\;\vcenter{\hbox{\includegraphics}} - \vcenter{\hbox{\includegraphics}}\;\right]$$

$$+ B^3 \left[\;\vcenter{\hbox{\includegraphics}} - \vcenter{\hbox{\includegraphics}}\;\right]$$

$$= (A^2 B + n AB^2 + B^3) \left[\;\vcenter{\hbox{\includegraphics}} - \vcenter{\hbox{\includegraphics}}\;\right]$$

Thus in order for R to satisfy the YBE it is sufficient for

$$B(A^2 + nAB + B^2) = 0.$$

Thus $B = 0$ gives a trivial solution to YBE (a multiple of the identity matrix). Otherwise, we require that $A^2 + nAB + B^2 = 0$ and so (assuming $A \neq 0 \neq B$) we have $n = -(\frac{A}{B} + \frac{B}{A})$ for the loop value.

Given this choice of loop value it is interesting to go back and examine the behavior of the bracket under a type II move:

$$\left\langle\;\vcenter{\hbox{\includegraphics}}\;\right\rangle = AB \left\langle\;\vcenter{\hbox{\includegraphics}}\;\right\rangle + (A^2 + B^2 + ABn) \left\langle\;\vcenter{\hbox{\includegraphics}}\;\right\rangle$$

$$= AB \left\langle\;\vcenter{\hbox{\includegraphics}}\;\right\rangle.$$

Thus, even if $B \neq A^{-1}$, this model for the bracket will be invariant under the type II move up to a multiplication by a power of AB. I leave it as an exercise for the reader to see that normalization to an ambient isotopy invariant yields the usual version of the Jones polynomial.

In any case, we have verified directly that

$$R_{cd}^{ab} = A\delta_c^a \delta_d^b + A^{-1}\delta^{ab}\delta_{cd}$$

$$n = -A^2 - A^{-2}$$

is a solution to the Yang-Baxter Equation.

In the course of this derivation I have used the notation

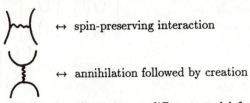

$$R = A \quad + B$$

to keep track of the composition of interactions. It is nice to think of these as **abstract Feynman diagrams**, where, with a vertical arrow of time [↑], we have

↔ spin-preserving interaction

↔ annihilation followed by creation

The next section follows this theme into a different model for the bracket and other solutions to the Yang-Baxter Equation.

9^0. Formal Feynman Diagrams, Bracket as a Vacuum-Vacuum Expectation and the Quantum Group $SL(2)q$.

The solutions to the Yang-Baxter Equation that we constructed in the last section are directly related to the bracket, but each solution gives the bracket at special values corresponding to solutions of $n + A^2 + A^{-2} = 0$ for a given positive integer n. We end up with Yang-Baxter state models for infinitely many specializations of the bracket.

In fact there is a way to construct a solution to the Yang-Baxter Equation and a corresponding state model for the bracket - giving the whole bracket polynomial in one model. In order to do this it is conceptually very pleasant to go back to the picture of creations, annihilations and interactions - taking it a bit more seriously.

If we take this picture seriously, then a **fragment** such as a maximum or a minimum, need no longer be regarded as a Kronecker delta. The fragment

$$\uparrow \qquad \overset{a \quad b}{\bigcup} \leftrightarrow M^{ab},$$

with the time's arrow as indicated, connotes a **creation** of spins a and b from the vacuum.

In any case, we now allow matrices M_{ab} and M^{ab} corresponding to **caps** and **cups** respectively. From the viewpoint of time's arrow running up the page, cups are creations and caps are annihilations. The matrix values M_{ab} and M^{ab} represent (abstract) amplitudes for these processes to take place.

Along with cups and caps, we have the R-matrices:

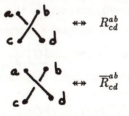

corresponding to our knot-theoretic interactions. Note that the crossings corresponding to R and $\overline{R}$ are now differentiated relative to time's arrow. Thus for R the over-crossing line goes from right to left as we go up the page.

·

A given link-diagram (**unoriented**) may be represented with respect to time's arrow so that it is naturally decomposed (via time as the height function) into cups, caps and interactions. There may also be a few residual Kronecker deltas (curves with no critical points vis-a-vis this height function):

$$\uparrow \qquad \qquad \qquad \qquad \delta_b^a.$$

Example.

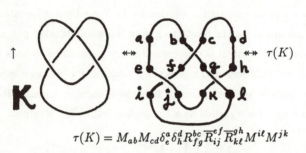

$$\tau(K) = M_{ab}M_{cd}\delta_e^a\delta_h^d R_{fg}^{bc}\overline{R}_{ij}^{ef}\overline{R}_{k\ell}^{gh}M^{i\ell}M^{jk}$$

In this example we have translated a trefoil diagram K into its corresponding expression $\tau(K)$ in the language of annihilation, creation and interaction. If the tensors are numerically valued (or valued in a commutative ring), then $\tau(K)$ represents a **vacuum-vacuum expectation** for the processes indicated by the diagram and this arrow of time.

The expectation is in accord with the principles of quantum mechanics. In quantum mechanics (see [FE]) **the probability amplitude for the concatenation of processes is obtained by summing the products of the amplitudes of the intermediate configurations in the process over all possible internal configurations.** Thus if we have a process $\underline{\quad\boxed{P}\quad}$ with "input" a and "output" b, and another process $\underline{\quad\boxed{Q}\quad}$, then the amplitude for $\underline{\quad\boxed{P}_i\boxed{Q}\quad}$ (given input a and output b) is the sum

$$\sum_i P_{ai}Q_{ib} = (PQ)_{ab}$$

and corresponds exactly to matrix multiplication. For a vacuum-vacuum expectation there are no inputs or outputs. The expression $\tau(K)$ (using Einstein summation convention) represents the sum over all internal configurations (spins on the lines) of the products of amplitudes for creation, annihilation and interaction.

Thus $\tau(K)$ can be considered as the basic form of vacuum-vacuum expectation in a highly simplified **quantum field theory of link diagrams.** We would like this to be a **topological quantum field theory** in the sense that $\tau(\mathbf{K})$ **should be an invariant of regular isotopy of K.** Let's look at what is required for this to happen.

Think of the arrow of time as a specified vertical bottom-to-top direction on the page. Regular isotopies of the link diagram are generated by the Reidemeister moves of type II and type III. A type II move can occur at various angles with respect to the time-arrow. Two extremes are illustrated below:

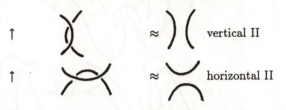

Similarly a type III move can occur at various angles, and we can **twist a crossing** keeping its endpoints fixed:

It is interesting to note that **vertical II plus the twist generates horizontal II.** See Figure 15. This generation utilizes the basic **topological move** of canceling pairs of critical points (maxima and minima) as shown below:

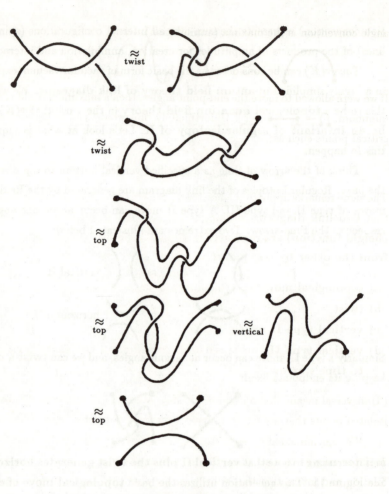

twist + vertical + top ⇒ horizontal

Figure 15

Note that in the deformation of Figure 15 we have kept the angles of the lines at their endpoints fixed. This allows a proper count of maxima and minima generated by the moves, and it means that these deformations can fit smoothly into larger diagrams of which the patterns depicted are a part. A simplest instance of this angle convention is the maxima (annihilation):

If we were allowed to move the end-point angles, then a maximum could turn into a minimum: $\cap \to \bullet\!\!-\!\!\bullet \to \cup$. Unfortunately, this would lead to non-differentiable critical points such as:

The same remarks apply to the horizontal version of the type III move, and we find that **two link diagrams arranged transversal to a given time direction (height function) are regularly isotopic if and only if one can be obtained from the other by a sequence of moves of the types:**

a) **topological move** (canceling maxima and minima)

b) **twist**

c) **vertical type II move**

d) **vertical type III move with all crossings of the same type relative to time's arrow.**

(Transversal means that a given level (of constant time) intersects the diagram in isolated points that are either transversal intersections, maxima or minima.)

We can immediately translate the conditions of these **relativized Reidemeister moves** to a set of conditions on the abstract tensors for creation, annihilation and interaction that will guarantee that $\tau(K)$ is an invariant of regular isotopy. Let's take the moves one at a time:

a) **topological move**

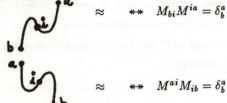

$$\approx \qquad \leftrightarrow \qquad M_{bi}M^{ia} = \delta_b^a$$

$$\approx \qquad \leftrightarrow \qquad M^{ai}M_{ib} = \delta_b^a$$

The matrices M_{ab} and M^{ab} are inverses of each other.

122

b) the twist

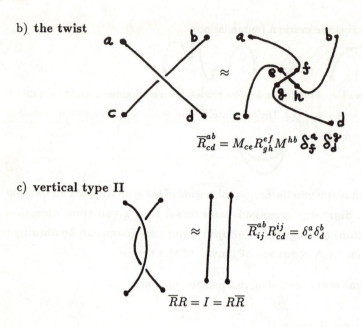

$$\overline{R}_{cd}^{ab} = M_{ce} R_{gh}^{ef} M^{hb} \delta_{f}^{a} \delta_{d}^{g}$$

c) vertical type II

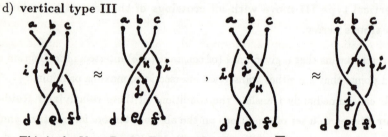

$$\overline{R}_{ij}^{ab} R_{cd}^{ij} = \delta_{c}^{a} \delta_{d}^{b}$$

$$\overline{R}R = I = R\overline{R}$$

d) vertical type III

This is the Yang-Baxter Equation for R and for $\overline{R}$.

Thus we see (compare with 8.1):

Theorem 9.1'. If the interaction matrix R and its inverse $\overline{R}$ satisfy the Yang-Baxter relation plus the interrelation with M_{ab} and M^{ab} specified by the twist and if M_{ab} and M^{ab} are inverse matrices then $\tau(K)$ will be an invariant of regular isotopy.

Braids.

The vacuum-vacuum-expectation viewpoint reflects very beautifully on the case of braids. For consider $\tau(K)$ when K is a closed braid:

$$K = \overline{B}.$$

If $K = \overline{B}$ where B is a braid, then each closure strand has one maximum and one minimum

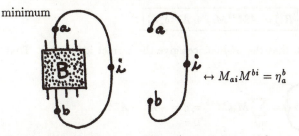

$$\leftrightarrow M_{ai}M^{bi} = \eta_a^b$$

The braid itself consists entirely of interactions, with no creations or annihilations. As we sweep time up from the vacuum state the minima from the closure strands give pair creations. The left hand member of each pair participates in the braid while its right hand twin goes up the trivial braid of the closure. At the top everybody pairs up and all cancel. Each braid-strand contributes a matrix of the form

$$\eta_a^b = M_{ai}M^{bi}$$

as illustrated above. If $\rho(B)$ denotes the interaction tensor (composed of R-matrices) coming from the braid, then we find that

$$\tau(K) = \text{Trace}(\eta^{\otimes n}\rho(B))$$

where there are n braid strands, and $\rho(B)$ is regarded as living in a tensor product of matrices as in section 7^0. Note that this version of $\tau(K)$ has the same form as the one we discussed vis-a-vis the Markov Theorem (Markov trace).

Obtaining the Bracket and its R-matrix.

If we wish $\tau(K)$ to satisfy the bracket equation

$$\tau\left(\times\right) = A\tau\left(\asymp\right) + A^{-1}\tau\left(\right)\left(\right)$$

then a wise choice for the R-matrix is

$$\times = R_{cd}^{ab} = A\;\underset{c\;d}{\overset{a\;b}{\cup}} + A^{-1}\;\underset{c\;d}{\overset{a\;b}{\big)\big(}}.$$

Hence

$$\boxed{R_{cd}^{ab} = AM^{ab}M_{cd} + A^{-1}\delta_c^a\delta_d^b}.$$

Suppose, for the moment, that the M-matrices give the correct loop value. That is, suppose that

$$a\;\bigcirc\;b = \sum_{a,b} M_{ab}M^{ab} = d = -A^2 - A^{-2}.$$

Letting $U = M^{ab}M_{cd} = \underset{\cap}{\cup}$, we have $U^2 = dU$ and $R = AU + A^{-1}I$. The proof that R (with the given loop value) satisfies the YBE is then identical to the proof given for the corresponding braiding relation in Proposition 7.5.

Thus, all we need, to have a model for the bracket **and** a solution to the Yang-Baxter Equation is a matrix pair (M_{ab}, M^{ab}) of inverse matrices whose loop value (d above) is $-A^2 - A^{-2}$. Let us assume that $M_{ab} = M^{ab}$ so that we are looking for a matrix M with $M^2 = I$.

$$\rotatebox{0}{$\sim$} \;\approx\; \rotatebox{0}{$/$} \quad \leftrightarrow\; M^2 = I$$

Then **the loop value, d, is the sum of the squares of the entries of M**:

$$d = \sum_{a,b}(M_{ab})^2.$$

If we let $M = \begin{bmatrix} 0 & \sqrt{-1} \\ -\sqrt{-1} & 0 \end{bmatrix}$, then $M^2 = I$ and $d = (-1) + (-1) = -2$. This is certainly a special case of the bracket, with $A = 1$ or $A = -1$. Furthermore, it is very easy to deform this M to obtain

$$M = \begin{bmatrix} 0 & \sqrt{-1}A \\ -\sqrt{-1}A^{-1} & 0 \end{bmatrix}$$

with $M^2 = I$ and $d = -A^2 - A^{-2}$, as desired.

This choice of creation/annihilation matrix M gives us a tensor model for $\langle K \rangle$, and a solution to the Yang-Baxter Equation.

In direct matrix language, we find

$$
M = \begin{array}{c c}
 & \begin{array}{cc} 1 & \quad\quad 2 \end{array}
\end{array}
$$

	1	2
$M = 1$	0	$\sqrt{-1}\,A$
2	$-\sqrt{-1}\,A^{-1}$	0

$U = M \otimes M =$

	11	12	21	22
11	0	0	0	0
12	0	$-A^2$	1	0
21	0	1	$-A^{-2}$	0
22	0	0	0	0

and $R = AM \otimes M + A^{-1}I \otimes I$.

$R =$

A^{-1}	0	0	0
0	$-A^3 + A^{-1}$	A	0
0	A	0	0
0	0	0	A^{-1}

There is much to say about this solution to the Yang-Baxter Equation. In order to begin this discussion, we first look at the relationship of the group $SL(2)$ and the cases $A = \pm 1$.

Spin-Networks, Binors and SL(2).

The special case of the bracket with $M = \begin{bmatrix} 0 & \sqrt{-1} \\ -\sqrt{-1} & 0 \end{bmatrix}$ is of particular significance. Let's write this matrix as

$$\begin{bmatrix} 0 & \sqrt{-1} \\ -\sqrt{-1} & 0 \end{bmatrix} = \sqrt{-1}\begin{bmatrix} 0 & 1 \\ -1 & 0 \end{bmatrix}$$
$$= \sqrt{-1}\epsilon$$

where ϵ_{ab} denotes the **alternating symbol**:

$$\epsilon_{ab} = \begin{cases} 1 & \text{if } a < b \\ -1 & \text{if } a > b \\ 0 & \text{if } a = b \end{cases}$$

$$a, b \in \{1, 2\} = \mathcal{I}.$$

Let us also use the following diagrammatic for this matrix:

$$\epsilon_{ab} = \prod_{a\ b}$$

$$\epsilon^{ab} = \coprod^{a\ b}$$

The alternating symbol is fundamental to a number of contexts.

Lemma 9.2. Let $P = \begin{bmatrix} a & b \\ c & d \end{bmatrix}$ be a matrix of commuting (associative) scalars. Then

$$P\epsilon P^T = \text{DET}(P)\epsilon$$

where P^T denotes the transpose of P.

Proof.

$$\begin{bmatrix} a & b \\ c & d \end{bmatrix} \begin{bmatrix} 0 & 1 \\ -1 & 0 \end{bmatrix} \begin{bmatrix} a & c \\ b & d \end{bmatrix} = \begin{bmatrix} a & b \\ c & d \end{bmatrix} \begin{bmatrix} b & d \\ -a & -c \end{bmatrix}$$

$$= \begin{bmatrix} ab - ba & ad - bc \\ cb - da & cd - dc \end{bmatrix}$$

$$= \begin{bmatrix} 0 & ad - bc \\ -ad + bc & 0 \end{bmatrix}$$

$$= (ad - bc) \begin{bmatrix} 0 & 1 \\ -1 & 0 \end{bmatrix}$$

//

Thus we have the definition.

Definition 9.3. Let $\mathcal{R}$ be a commutative, associative ring with unit. Then $SL(2, \mathcal{R})$ is defined to be the set of 2×2 matrices P with entries in $\mathcal{R}$ such that

$$P\epsilon P^T = \epsilon.$$

In the following, I shall write $SL(2)$ for $SL(2,\mathcal{R})$. $SL(2)$ is identified as the set of matrices P leaving the ϵ-symbol invariant.

Note also how this invariance appears diagrammatically: Let $P = (P^a_b) = (P_{ab})$

$$(P\epsilon P^T) = P^a_i \epsilon^{ij} P^b_j , \quad P^a_b$$

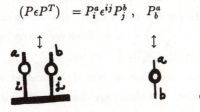

Thus the invariances $P\epsilon P^T = \epsilon$ and $P^T \epsilon P = \epsilon$ correspond to the diagrammatic equations

In this sense, **the link diagrams at $A = \pm 1$ correspond to $SL(2)$-invariant tensors.**

Also, the identity $\times$ = $\pm \cup \pm$)(for the R-matrix becomes **a specific $SL(2)$-invariant tensor identity:**

$$\# =)(- \times$$

$$\epsilon^{ab}\epsilon_{cd} = \delta^a_c \delta^b_d - \delta^a_d \delta^b_c$$

$$\Leftrightarrow -(\sqrt{-1}\,\text{II})(\sqrt{-1}\,\text{II}) =)(- \times$$

$$\Leftrightarrow - \asymp =)(- \times$$

$$\Leftrightarrow \times = \asymp +)(\qquad [A = +1].$$

In this calculus of $SL(2)$-invariant tensor diagrams there is no distinction between an over-crossing and an under-crossing. The calculus is a topologically invariant calculation for curves immersed in the plane. All Jordan curves receive the same value of -2. This form of a calculus for $SL(2)$-invariant tensors is the **binor calculus** of Roger Penrose [PEN1]. His binors are precisely the case for $A = -1$ so that $\times + \asymp +)(= 0$. Here each crossing receives a minus sign and we have the invariance $\underset{\wedge}{\times} = \cap$ as well.

The Penrose binors are a special case of the bracket, and hence a special case of the Jones polynomial. The binors are the underpinning for the Penrose theory of spin-networks.

For now, it is worth briefly re-tracing steps that led Penrose to discover the binors. Penrose began by considering **spinors** ψ^A. Now a spinor is actually a 2-vector over the complex numbers **C**. Thus $A \in \{1, 2\}$. $SL(2, \mathbf{C})$ acts (via matrix multiplication) on these spinors, and one wants an inner product $\psi\psi^* \in \mathbf{R}$ (real numbers) that is $SL(2, \mathbf{C})$ invariant. Let $\psi_A^* = \epsilon_{AB}\psi^B$ so that

$$\psi\psi^* = \psi^A \psi_A^* = \psi^A \epsilon_{AB} \psi^B.$$

If $(U_B^A) = U \in SL(2, \mathbf{C})$ then $\psi\psi^*$ is invariant under the action of U:

$$
\begin{aligned}
(U\psi)(U\psi)^* &= (U_I^A \psi^I)\epsilon_{AB}(U_J^B \psi^J) \\
&= (U_I^A U_J^B \epsilon_{AB})\psi^I \psi^J \\
&= \epsilon_{IJ}\psi^I \psi^J \\
&= \psi\psi^*.
\end{aligned}
$$

A calculus of diagrams that represents this inner product is suggested by the "natural" method of lowering an index:

$$\psi^A \leftrightarrow \;\square\!|\; , \; \psi_A^* \leftrightarrow \;\bullet\!\cap\!\square$$

Of course,

$$\psi\psi^* \leftrightarrow \;\square\!\cap\!\square$$

and one is led to take

$$\bigcap \leftrightarrow \epsilon_{ab}$$

$$\bigcup \leftrightarrow \epsilon^{ab}.$$

However, the resulting diagrams are **not** topologically invariant:

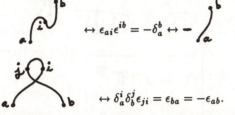

and

$$\leftrightarrow \delta_a^i \delta_b^j \epsilon_{ji} = \epsilon_{ba} = -\epsilon_{ab}.$$

Penrose solved this difficulty by associating a minus sign to each **minimum** and each **crossing**. This changes the loop from $+2$ to -2 and gives the **binor identity**

$$\times + \underset{\frown}{\smile} + \big)\big(= 0.$$

That the resulting calculus is indeed topologically invariant, we know well from our work with the bracket. Note that

$$\bigcup \leftrightarrow \sqrt{-1} \;\sqcup\!\sqcup$$

$$\bigcap \leftrightarrow \sqrt{-1} \;\top\!\top$$

works just as well as the Penrose convention - distributing the sign over maxima and minima.

It is amusing to do our usual topological verifications in the binor context:

$$\asymp \; = - \; \mathcal{C} \; - \; \mathcal{K}$$

$$= - \; \underset{\frown}{\smile} \; - \; \times$$

$$= - \; \underset{\smile}{\frown} \; + \; \underset{\frown}{\smile} \; + \; \big)\big($$

$$= \; \big)\big(\; .$$

130

There is much more to the theory of the binors and the spin networks. (See [PEN1], [PEN3], and sections 12^0 and 13^0 of Part II.)

The R-matrix.

We can now view the bracket calculation as stemming from the particular deformation $\tilde{\epsilon} = \begin{bmatrix} 0 & A \\ -A^{-1} & 0 \end{bmatrix}$ of the spinor epsilon (alternating symbol). We use the shorthand

$$M_{ab} = \sqrt{-1}\, \tilde{\epsilon}_{ab} = \sqrt{-1} \;\; \overset{\displaystyle\prod}{\underset{a\ \ b}{}} \;\; A^{\overset{\displaystyle\prod}{}_{ab}}$$

Thus

$$\boxed{\begin{array}{l} \cap \;\; \longleftrightarrow \;\; \sqrt{-1}\,\prod A^{\overset{\prod}{}} \\[2ex] \cup \;\; \longleftrightarrow \;\; \sqrt{-1}\,\underset{}{\text{⨆}}\, A^{\text{⨆}} \end{array}}$$

are the creation and annihilation matrices for the bracket. This symbolism, plus the Fierz identity

$$\#\!\# \; = \;)(\; - \times$$

make it easy for us to re-express the R-matrix:

$$R = A \; \overset{\cup}{\cap} \;\; + A^{-1} \,)\,($$

$$= A(\sqrt{-1})^2 \; \#\!\# \;\; A^{\amalg+\amalg} + A^{-1}\,)\,($$

$$= - \#\!\# \; A^{\amalg+\amalg+1} + A^{-1}\,)\,($$

$$= [\, \times -)(\,] A^{\amalg+\amalg+1} + [\,)(\,] A^{-1}$$

$$= (A^{-1} - A^{\amalg+\amalg+1})[\,)(\,] + A^{\amalg+\amalg+1}[\,\times\,].$$

From the viewpoint of the particle interactions, the Fierz identity lets us replace the annihilation ∿ creation

by a combination of exchange and crossover

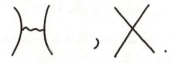

This actually leads to a different state model for the bracket - one that generalizes to give a series of models for infinitely many specializations of the Homfly polynomial! (Compare [JO4].)

In order to see this let's calculate R a bit further:

$$R = (A^{-1} - A^{\text{II}+\text{II}+1})[\,)(\,] + A^{\text{II}+\text{II}+1}[\times].$$

The index set is $\mathcal{I} = \{1,2\}$ with $\underset{12}{\Pi} = +1$, $\underset{21}{\Pi} = -1$, $\underset{11}{\Pi} = \underset{22}{\Pi} = 0$. The lines and cross-lines in the abstract tensors for the diagrams above are Kronecker deltas. Thus for $[\,)(\,]_{12}$ the coefficient is

$$A^{-1} - A^{\overset{12}{\text{II}}+\overset{12}{\Pi}+1} = A^{-1} - A^{2+1} = A^{-1} - A^3.$$

We find:

$$R = (A^{-1} - A^{+3})[\,)<(\,] + A^{-1}[\,)=(\,] + A[\overset{\ast}{\times}].$$

Thus

$$R = A^{+1}\{(A^{-2} - A^2)[\,)<(\,] + A^{-2}[\,)=(\,] + [\overset{\ast}{\times}]\}.$$

From this, we can abstract an R-matrix

$$R_q = (q - q^{-1})[\,)>(\,] + q[\,)=(\,] + [\overset{\ast}{\times}]$$

defined for an **arbitrary ordered index set** $\mathcal{I}$. By the same token, let

$$\overline{R}_q = (q^{-1} - q)[\,)<(\,] + q^{-1}[\,)=(\,] + [\overset{\ast}{\times}].$$

Note that

$$R_q - \overline{R}_q = (q - q^{-1})\{\,[\,\succ\,] + [\,\prec\,] + [\,=\,]\,\}$$
$$\therefore \; R_q - \overline{R}_q = (q - q^{-1})[\,\asymp\,].$$

If $R_q = [\,\times\,]$, $\overline{R}_q = [\,\times\,]$ then this is the analog of the exchange identity for the regular isotopy version of the Homfly polynomial.

As a start in this direction, we observe that

Proposition 9.4. R_q and $\overline{R}_q$ are inverses. Each is a solution to the Yang-Baxter Equation for any ordered index set $\mathcal{I}$.

Remark. We have already verified 9.4 for $\mathcal{I} = \{1, 2\}$. The present form of R_q makes the more general verification easy. From the point of view of particle interactions, these solutions are quite remarkable in that they have a left-right asymmetry - R_q weights only for $\succ\,<\,\prec$ while $\overline{R}_q$ weights for $\succ\,>\,\prec$. In the case of $\mathcal{I} = \{1, 2\}$, this **asymmetry** is not at first apparent from the expansion formula for R **but we see that it arises from the specific choice of time's arrow** that converts one abstract tensor to an exchange, and the other to a creation/annihilation. The loop-adjustment $d = -A^2 - A^{-2}$ assures us that the calculations will be independent of the choice of time/space split. One asymmetry (left/right) compensates for another (time/space).

Where is the physics in all of this? The mathematics is richly motivated by physical ideas, and this part of the story will continue to spin its web. It is also possible that the concept of a topological interaction pattern, such as the knot/link diagram, may contain a clue for modeling observed processes. The link diagram (or the link in three-space) could as a **topological whole** represent a particle or process. This speculation moves in the direction of **embedded** (knotted and linked) **strings**. We have the **choice** to interpret a link diagram as an abstract network with expectation values computable from a mixture of quantum mechanical and topological notions **or** as an embedding in three-space. The three-space auxiliary to a link diagram need not be regarded as the common space of physical observation. The interpretive problem is to understand when these contexts (mathematical versus observational) of three dimensions come together.

Proof of Proposition 9.4. It is easy to verify directly that R_q and $\overline{R}_q$ are inverses. Let's write

$$R = z \big) \!\!<\!\! \big(+ q \big) = \big(+ \text{\Large ⧖}$$

and examine the conditions under which R yields a solution to YBE.

Case I.

Assume that the input and output spins are as indicated in the diagram above. Then we have:

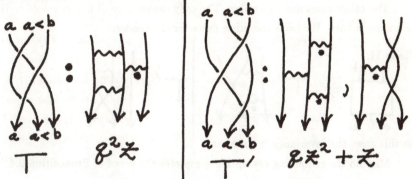

Here I have indicated the possible interaction patterns for the two sides of the Yang-Baxter ledger. I have also adopted the following **notation**:

$$\text{⋎} \;\leftrightarrow\; \big) = \big(\qquad \leftrightarrow \quad \text{equal spin labels}$$

$$\text{⋎} \;\leftrightarrow\; \big) < \big(\qquad \leftrightarrow \quad \text{left label less than right label}$$

$$\text{⋎} \;\leftrightarrow\; \big) > \big(\qquad \leftrightarrow \quad \text{left label greater than right label}$$

$$\text{✕} \;\leftrightarrow\; \text{⧖} \qquad \leftrightarrow \quad \text{crossover with unequal labels}$$

Returning to Case I, we see that for T (the first triangle condition) there is only one admissible interaction consisting in three exchanges (no crossover) and a product of vertex weights of $q^2 z$. For T' we have two possibilities - exchanges, or a pair of canceling crossovers. Here the sum of products of vertex weights is $qz^2 + z$.

In order for the YBE to hold, it is necessary that the contributions from T and T' match. Therefore, we need $q^2 z = qz^2 + z$. Hence

$$q^2 = qz + 1$$
$$z = q - q^{-1}.$$

This completes Case I.

The other cases are all similar. The only demand on R is $z = q - q^{-1}$. Here is a sample case. We leave the other cases for the reader.

Case II.

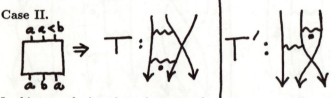

In this case, the invariance is automatic.

Modulo the remaining cases, this completes the proof of Proposition 9.4. //

Discussion. We have seen already that for the index set $\mathcal{I} = \{-1, +1\}$ this solution to the YBE is closely related to the group $SL(2)$.

The general solution shown here for any ordered index set is related to the group $SL(n)$ and these solutions can be used to construct knot polynomial models. Before creating these models, I wish to return to our discussion of $SL(2)$ and show how an extension of its symmetry leads to the idea of a "quantum group."

Bracket, Spin Nets and the Quantum Group for $SL(2)$.

In the discussion surrounding Lemma 9.2 we have seen solutions to the Yang-Baxter Equation and models for the Jones polynomial (a. k. a. bracket) emerge from the deformed epsilon, $\tilde{\epsilon} = \begin{bmatrix} 0 & A \\ -A^{-1} & 0 \end{bmatrix}$. Since $\epsilon = \begin{bmatrix} 0 & 1 \\ -1 & 0 \end{bmatrix}$ is the fundamental defining invariant for $SL(2)$, it is natural to ask the question: **For what algebraic structure is $\tilde{\epsilon}$ the basic invariant?**

Therefore suppose that a 2×2 matrix $P = \begin{pmatrix} a & b \\ c & d \end{pmatrix}$ is given and that the entries of P belong to an associative but not necessarily commutative algebra. Let us demand the invariances:

$$\left. \begin{array}{c} P\tilde{\epsilon}P^T = \tilde{\epsilon} \\ P^T\tilde{\epsilon}P = \tilde{\epsilon} \end{array} \right\} \quad (*)$$

where it is assumed that A commutes with the entries of P. A bit of calculation will reveal the relations demanded by $(*)$.

Proposition 9.5. With $\tilde{\epsilon} = \begin{pmatrix} 0 & A \\ -A^{-1} & 0 \end{pmatrix}$, $P = \begin{pmatrix} a & b \\ c & d \end{pmatrix}$, the equations

$$\left. \begin{array}{lll} 1) & P\tilde{\epsilon}P^T & = \tilde{\epsilon} \\ 1') & P^T\tilde{\epsilon}P & = \tilde{\epsilon} \end{array} \right\} \quad (*)$$

are equivalent to the set of relations $(**)$ shown boxed below: (with $\sqrt{q} = A$)

$$\boxed{\begin{array}{ll} ba = qab & dc = qcd \\ ca = qac & db = qbd \\ bc = cb & \\ ad - da = (q^{-1} - q)bc & \\ ad - q^{-1}bc = 1 & \end{array}} \quad (**)$$

Proof. $P = \begin{pmatrix} a & b \\ c & d \end{pmatrix}$, $\tilde{\epsilon} = \begin{pmatrix} 0 & A \\ -A^{-1} & 0 \end{pmatrix}$.

$$P\tilde{\epsilon}P^T = \begin{pmatrix} a & b \\ c & d \end{pmatrix}\begin{pmatrix} 0 & A \\ -A^{-1} & 0 \end{pmatrix}\begin{pmatrix} a & c \\ b & d \end{pmatrix}$$

$$= \begin{pmatrix} -A^{-1}b & Aa \\ -A^{-1}d & Ac \end{pmatrix}\begin{pmatrix} a & c \\ b & d \end{pmatrix}$$

$$P\tilde{\epsilon}P^T = \begin{pmatrix} -A^{-1}ba + Aab & -A^{-1}bc + Aad \\ -A^{-1}da + Acb & -A^{-1}dc + Acd \end{pmatrix}$$

$$P^T\tilde{\epsilon}P = \begin{pmatrix} -A^{-1}ca + Aac & -A^{-1}cb + Aad \\ -A^{-1}da + Abc & -A^{-1}db + Abd \end{pmatrix}$$

Therefore, the relations that follow directly from $P\tilde{\epsilon}P^T = \tilde{\epsilon}$ and $P^T\tilde{\epsilon}P = \tilde{\epsilon}$ are as follows:

$$A^{-1}ba = Aab \qquad\qquad A^{-1}dc = Acd$$
$$A^{-1}ca = Aac \qquad\qquad A^{-1}db = Abd$$
$$-A^{-1}bc + Aad = A \qquad -A^{-1}cb + Aad = A$$
$$-A^{-1}da + Acb = -A^{-1} \qquad -A^{-1}da + Abc = A^{-1}.$$

It follows that $bc = cb$, and that an equivalent set of relations is given by:

$$\left\{\begin{array}{ll}
ba = qab & dc = qcd \\
ca = qac & db = qbd \\
bc = cb & \\
ad - da = (a^{-1} - q)bc & \\
ad - q^{-1}bc = 1 &
\end{array}\right.$$

where $q = A^2$. Note that the last relation, $ad - q^{-1}bc = 1$, becomes the condition $ad - bc = \mathrm{DET}(P) = 1$ when $q = 1$. Also, when $q = 1$ these relations tell us that the elements of the matrix P commute among themselves.

This completes the proof of Proposition 9.5. $\qquad\qquad //$

Remark. The non-commutativity for elements of P is essential. The only nontrivial case where these elements all commute is when $q = 1$. But one, perhaps unfortunate, consequence of non-commutativity is that if P and Q are matrices satisfying (∗∗), then PQ does not necessarily satisfy (∗∗) (Since $(PQ)^T \neq Q^T P^T$ in the non-commutative case.). Thus Proposition 9.5 does **not** give rise to a natural generalization of the **group** $SL(2)$ to a **group** $SL(2)_q$ leaving $\tilde{\epsilon}$ invariant. Instead, we are left with the universal associative algebra defined by the relations (∗∗). **This algebra will be denoted by** $SL(2)_q$. It is called "the quantum group $SL(2)_q$" in the literature [DRIN1].

While $SL(2)_q$ is not itself a group, it does have a very interesting structure that is directly related to the Lie algebra of $SL(2)$. Furthermore, there is a natural comultiplication $\Delta : SL(2)_q \to SL(2)_q \otimes SL(2)_q$ that is itself a map of algebras. I shall first discuss the comultiplication, and then the relation with the Lie algebra.

The algebraic structure of the quantum group is intimately tied with the properties of the R-matrix and hence with associated topological invariants. The rest of this section constitutes a first pass through this region.

Bialgebra Structure and Abstract Tensors.

Let's begin with a generalization of our construction for $SL(2)_q$. Suppose that we are given a matrix $E = (E^{ij})$ (the analog of $\tilde{\epsilon}$) and a matrix $P = (P^i_j)$. **The elements of E commute with the elements of P.** The indices i, j are assumed to belong to a specified finite index set $\mathcal{I}$. Then the equation $PEP^T = E$ reads in indices as:

$$P^a_i E^{ij} P^b_j = E^{ab}$$

with summation on the repeated indices. In terms of abstract tensors we can write

$$\phi \leftrightarrow P^i_j, \quad \sqcap \leftrightarrow E^{ij}$$

and

$$PEP^T = \leftrightarrow \quad \text{(diagram)}$$

This last equation should be considered carefully since the order of terms in the relations matters. We might write colloquially $\boxed{} = \boxed{}$, preserving the order of the products of elements of P. This latter form makes sense **if** the elements of E commute with the elements of P. Then we can say

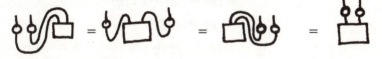

The transposed relationship

$$P^T E P = E$$

is best diagrammed by using E with lowered indices: $E_{ij} = E^{ij}$. Then

$$P^T E P = E \leftrightarrow P^i_a E_{ij} P^j_b = E_{ab}$$

$$\leftrightarrow \quad \text{(diagram)} = \text{(diagram)} .$$

Now, given the relations $PEP^T = E$ and $P^T EP = E$, let $\mathcal{A}(E)$ denote the resulting universal algebra. Thus $\mathcal{A}(\tilde{\epsilon}) = SL(2)_q$ ($\sqrt{q} = A$). **Define** $\Delta : \mathcal{A}(E) \to \mathcal{A}(E) \otimes \mathcal{A}(E)$ by the formula $\Delta(P_j^i) = \sum_{k \in \mathcal{I}} P_k^i \otimes P_j^k$ on the generators - extending linearly over the ground ring [The ground ring is $\mathbf{C}[A, A^{-1}]$ in the case of $SL(2)_q$.].

Proposition 9.6. $\Delta : \mathcal{A}(E) \to \mathcal{A}(E) \otimes \mathcal{A}(E)$ is a map of algebras.

Proof. We must prove that $\Delta(x)\Delta(y) = \Delta(xy)$ for elements $x, y \in \mathcal{A}(E)$. It will suffice to check this on the relations $PEP^T = E$ and $P^T EP = E$. We check the first relation; the second follows by symmetry. Now $PEP^T = E$ corresponds to $P_i^a E^{ij} P_j^b = E^{ab}$. Thus we must show that $\Delta(P_i^a) E^{ij} \Delta(P_j^b) = E^{ab}\iota$ where $\iota = 1 \otimes 1$. (Note that the elements E^{ij} are scalars - commuting with everyone - and that $\Delta(1) = \iota$ by definition, since ι is the identity element in $\mathcal{A}(E) \otimes \mathcal{A}(E)$.) Since ι is the identity element in $\mathcal{A}(E) \otimes \mathcal{A}(E)$ we can regard E as a matrix on any tensor power of $\mathcal{A}(E)$ via $E^{ab} \leftrightarrow E^{ab}(1 \otimes 1 \otimes \ldots \otimes 1)$. With this identification, we write

$$\Delta(P_i^a) E^{ij} \Delta(P_j^b) = E^{ab}$$

as the desired relation. Computing, we find

$$\Delta(P_i^a) E^{ij} \Delta(P_j^b) = (P_k^a \otimes P_i^k) E^{ij} (P_\ell^b \otimes P_j^\ell)$$
$$= (P_k^a P_\ell^b) \otimes (P_i^k P_j^\ell) E^{ij}$$
$$= (P_k^a P_\ell^b) \otimes (P_i^k E^{ij} P_j^\ell)$$
$$= (P_k^a P_\ell^b) \otimes E^{k\ell}$$
$$= (P_k^a E^{k\ell} P_\ell^b) \otimes 1$$
$$= E^{ab} \otimes 1$$
$$= E^{ab}(1 \otimes 1)$$
$$= E^{ab} \qquad \text{(sic.)}$$

This completes the proof of Proposition 9.6. //

Remark. The upshot of Proposition 9.6 is that $\Delta(\mathcal{A}(E)) \subset \mathcal{A}(E) \otimes \mathcal{A}(E)$ is also an algebra leaving the form E invariant. Furthermore, the structure of this proof is well illustrated via the abstract tensor diagrams. We have

$$P_b^a \longleftrightarrow \quad , \quad E^{ab} \longleftrightarrow$$

$$\Delta(P_b^a) \longleftrightarrow \Delta(\;) = \quad .$$

So that

$$\Delta(\;) \;\Delta(\;) = \quad$$

$$= \quad$$

$$= \quad$$

$$= \quad \otimes 1$$

$$= \quad \otimes 1$$

$$= \quad (1 \otimes 1)$$

$$\equiv \quad . \qquad\qquad //$$

The tensor diagrams give a direct view of the repeated indices (via tied lines), laying bare the structure of this calculation. In particular, we see at once that if

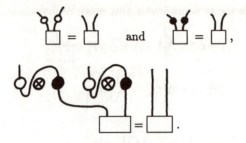

then

In other words, if $PEP^T = E$ and $QEQ^T = E$ and we **define**

$$(P * Q)_b^a = \sum_k P_k^a \otimes Q_b^k \in \mathcal{A}_P(E) \otimes \mathcal{A}_Q(E)$$

then

$$(P * Q)E(P * Q)^T = E$$

in this extended sense of $E \equiv E(1 \otimes 1)$ in the tensor product. The tensor product formalism provides the simplest structure in which we can partially reconstruct something like a group of matrices leaving the form E invariant. (Since $P * P^{-1}$ is distinct from the identity, we do not get a group.)

To return to our specific algebra $\mathcal{A} = \mathcal{A}(\tilde{\epsilon})$ we have: $P = \begin{pmatrix} a & b \\ c & d \end{pmatrix}$

$$\left\{ \begin{array}{ll} ba = qab & dc = qcd \\ ca = qac & db = qbd \\ bc = cb & \\ ad - da = (q^{-1} - q)bc & \\ ad - q^{-1}bc = 1 & \end{array} \right\}$$

$$\left\{ \begin{array}{lll} \Delta(a) & = & a \otimes a + b \otimes c \\ \Delta(b) & = & a \otimes b + b \otimes d \\ \Delta(c) & = & c \otimes a + d \otimes c \\ \Delta(d) & = & c \otimes b + d \otimes d \end{array} \right\}$$

So far, we have shown that $\Delta : \mathcal{A}(\tilde{\epsilon}) \to \mathcal{A}(\tilde{\epsilon}) \otimes \mathcal{A}(\tilde{\epsilon})$ is an algebra homomorphism. In fact, Δ is **coassociative**. This means that the following diagram commutes ($A = \mathcal{A}(\tilde{\epsilon})$):

$$
\begin{array}{ccc}
A \otimes A & \xrightarrow{\;1 \otimes \Delta\;} & A \otimes A \otimes A \\
\Delta \big\downarrow & & \big\uparrow \Delta \otimes 1 \\
A & \xrightarrow{\quad \Delta \quad} & A \otimes A
\end{array}
$$

The coassociativity property is obvious in this construction, since

$$(1 \otimes \Delta)\Delta(P^i_j) = (1 \otimes \Delta)(P^i_k \otimes P^k_j)$$
$$= P^i_k \otimes P^k_\ell \otimes P^\ell_j$$
$$= (\Delta \otimes 1)(P^i_\ell \otimes P^\ell_j)$$
$$= (\Delta \otimes 1)\Delta(P^i_j).$$

An associative algebra A with unit 1 and multiplication $m : A \otimes A \to A$ is a **bialgebra** if there is a homomorphism of algebras $\Delta : A \to A \otimes A$ that satisfies coassociativity, and so that A has a **co-unit**. A **co-unit** is an algebra homomorphism $\epsilon : A \to \mathbf{C}$ ($\mathbf{C}$ denotes the ground ring) such that the following diagrams commute:

$$
\begin{array}{ccc}
A & \xrightarrow{\Delta} & A \otimes A \\
1\downarrow & & \downarrow \epsilon \otimes 1 \\
A & \xrightarrow{\quad 1 \quad} & A
\end{array}
\qquad
\begin{array}{ccc}
A & \xrightarrow{\Delta} & A \otimes A \\
1\downarrow & & \downarrow 1 \otimes \epsilon \\
A & \xrightarrow{\quad 1 \quad} & A
\end{array}
$$

In the case of our construction with $\Delta(P^i_j) = P^i_k \otimes P^k_j$, let $\epsilon(P^i_j) = \delta^i_j$ (Kronecker delta).

Then

$$(\epsilon \otimes 1)\Delta(P^i_j) = \epsilon(P^i_k) \otimes P^k_j$$
$$= \delta^i_k \otimes P^k_j$$
$$= 1 \otimes P^i_j$$
$$= P^i_j.$$

Thus we have verified that $\mathcal{A}(E)$ is a bialgebra.

A bialgebra with certain extra structure (an antipode) is called a **Hopf algebra**.

The algebra $\mathcal{A}(\tilde{\epsilon})$ is an example of a Hopf algebra. In this case the antipode is a mapping $\gamma : \mathcal{A}(\tilde{\epsilon}) \to \mathcal{A}(\tilde{\epsilon})$ defined by $\gamma(a) = d$, $\gamma(d) = a$, $\gamma(b) = -qb$, $\gamma(c) = -q^{-1}c$. Extend γ linearly on sums and define it on products so that $\gamma(xy) = \gamma(y)\gamma(x)$ (an anti-homomorphism).

Note here that

$$P' = \gamma(P) = \begin{pmatrix} \gamma(a) & \gamma(b) \\ \gamma(c) & \gamma(d) \end{pmatrix}$$
$$= \begin{pmatrix} d & -qb \\ -q^{-1}c & a \end{pmatrix}$$

is actually the **inverse matrix of** P in this non-commutative setting:

$$PP' = \begin{pmatrix} a & b \\ c & d \end{pmatrix} \begin{pmatrix} d & -qb \\ -q^{-1}c & a \end{pmatrix}$$
$$= \begin{pmatrix} ad - q^{-1}bc & -qab + ba \\ cd - q^{-1}dc & -qcb + da \end{pmatrix}$$
$$= \begin{pmatrix} 1 & 0 \\ 0 & 1 \end{pmatrix}.$$

It is also true that $P'P = I$. In general, the antipode $\gamma : A \to A$ is intended to be the analog of an inverse. Its abstract definition is that the following diagram should commute:

$$
\begin{array}{ccccc}
A & \xrightarrow{\Delta} & A \otimes A & \underset{1 \otimes \gamma}{\overset{\gamma \otimes 1}{\rightrightarrows}} & A \otimes A \\
{\scriptstyle \epsilon}\downarrow & & & & \downarrow{\scriptstyle m} \\
\mathbf{C} & & \xrightarrow{\qquad \eta \qquad} & & A
\end{array}
$$

Here ϵ is the co-unit, and η is the unit. In the case of $\mathcal{O}(E)$ we have

$$\epsilon(P^i_j) = \delta^i_j \text{ and } \eta(\delta^i_j) = \delta^i_j.$$

Thus the condition that $\mathcal{A}(E)$ be a Hopf algebra is that

$$\gamma(P^i_k)P^k_j = \delta^i_j$$
$$P^i_k\gamma(P^k_j) = \delta^i_j.$$

This is the same as saying that P and $\gamma(P)$ are inverse matrices.

Remark. The algebra $\mathcal{A}(\tilde{\epsilon})$ is a deformation of the algebra of functions (co-ordinate functions) on the group $SL(2, \mathbf{C})$. In the classical case of an arbitrary group G, let $F(G)$ denote the collection of functions $f : G \to \mathbf{C}$. Define $m : F(G) \otimes F(G) \to F(G)$ by

$$m(f \otimes g)(x) = f(x)g(x)$$

and $\Delta : F(G) \to F(G) \otimes F(G)$ by

$$\Delta(f)(x \otimes y) = f(x)f(y).$$

Let $\gamma : F(G) \to F(G)$ be defined by $\gamma(f)(x) = f(x^{-1})$, and $\epsilon : F(G) \to \mathbf{C}$, $\epsilon(f) = f(e)$, e = identity element in G. $\eta : \mathbf{C} \to G$ by $\eta(z) = e$ for all $z \in \mathbf{C}$. This gives $F(G)$ the structure of a Hopf algebra.

Now $F(SL(2, \mathbf{C}))$ (or $F(G)$ for any Lie group G) contains (by taking derivatives) the functions on the Lie algebra of $SL(2, \mathbf{C})$. Thus we should **expect** that $F_q(SL(2)) \underset{\text{def.}}{=} SL(2)_q$ should be related to **functions on a deformation of the Lie algebra** of $SL(2)$. We can draw a connection between $\mathcal{A}(\tilde{\epsilon}) = F_q(SL(2))$ and the Lie algebra of $SL(2)$.

Deforming the Lie algebra of $SL(2, \mathbf{C})$.

The Lie algebra of $SL(2, \mathbf{C})$ consists in those matrices M such that

$$e^M \in SL(2, \mathbf{C}).$$

Since $\mathrm{DET}(e^M) = e^{\mathrm{tr}(M)}$ where $\mathrm{tr}(M)$ denotes the trace of the matrix M

$$\left[\text{e.g., } \mathrm{DET}\exp\left(\begin{bmatrix} \lambda & 0 \\ 0 & \mu \end{bmatrix} \right) = \mathrm{DET} \begin{bmatrix} e^\lambda & 0 \\ 0 & e^\mu \end{bmatrix} = e^\lambda e^\mu = e^{\lambda+\mu} \right]$$

we see that the condition $\mathrm{DET}(e^M) = 1$ corresponds to the condition $\mathrm{tr}(M) = 0$. Let $s\ell_2 = \{M \mid \mathrm{tr}(M) = 0\}$. This is the Lie algebra of $SL(2)$. It is generated by the matrices:

$$H = \begin{pmatrix} 1 & 0 \\ 0 & -1 \end{pmatrix}, \quad X^+ = \begin{pmatrix} 0 & 1 \\ 0 & 0 \end{pmatrix}, \quad X^- = \begin{pmatrix} 0 & 0 \\ 1 & 0 \end{pmatrix}$$

$$s\ell_2 = \left\{ \begin{pmatrix} r & s \\ t & -r \end{pmatrix} = rH + sX^+ + tX^- \right\}.$$

With $[A, B] = AB - BA$ we have the fundamental relations:

$$[H, X^+] = 2X^+$$
$$[H, X^-] = -2X^-$$
$$[X^+, X^-] = H.$$

Now define a **deformation** of this Lie Algebra (and its tensor powers) and call it $\mathcal{U}_q(s\ell_2)$ - **the quantum universal enveloping algebra of** $s\ell_2$. It is an abstract algebra generated by symbols H, X^+, X^- and the relations:

$$[H, X^+] = 2X^+$$

$$[H, X^-] = -2X^-$$

$$[X^+, X^-] = \sinh((\hbar/2)H)/\sinh(\hbar/2)$$

with $q = e^\hbar$ our familiar deformation parameter. Define a co-multiplication $\Delta : \mathcal{U}_q \to \mathcal{U}_q \otimes \mathcal{U}_q \ (\mathcal{U}_q \equiv \mathcal{U}(s\ell(2))_q)$ by the formulas:

$$\Delta(H) = H \otimes 1 + 1 \otimes H$$

$$\Delta(X^\pm) = X^\pm \otimes e^{(\hbar/4)H} + e^{(-\hbar/4)H} \otimes X^\pm.$$

One can verify directly that $\mathcal{U}_q$ is a bialgebra (in fact a Hopf algebra). However, we can see at once that $\mathcal{U}_q$ is really $\mathcal{A}(\widetilde{\epsilon}) = SL(2)_q$ in disguise! The mask of this disguise is a process of dualization: Consider the algebra dual of $\mathcal{U}_q$. This is

$$\mathcal{U}_q^* = \{\rho : \mathcal{U} \to \mathbf{C}[[\hbar, \hbar^{-1}]]\}$$

(We'll work over the complex numbers.)

Note that $\mathcal{U}_q^*$ inherits a multiplication from $\mathcal{U}_q$'s co-multiplication and a co-multiplication from $\mathcal{U}_q$'s multiplication. This occurs thus:

$$\rho, \rho' : \mathcal{U} \to \mathbf{C}[\hbar, \hbar^{-1}].$$

Define $\rho\rho'(\alpha) = \rho \otimes \rho'(\Delta(\alpha))$. Define $\Delta(\rho) : \mathcal{U} \otimes \mathcal{U} \to \mathbf{C}[\hbar, \hbar^{-1}]$ by

$$\Delta(\rho)(\alpha \otimes \beta) = \rho(\alpha\beta),$$

where $\alpha\beta$ denotes the product of $\mathcal{U}$.

Consider the following **representation of** $\mathcal{U}_q$: $\widetilde{\rho} : \mathcal{U}_q \to s\ell_2$

$$\widetilde{\rho}(H) = \begin{pmatrix} 1 & 0 \\ 0 & -1 \end{pmatrix}$$

$$\widetilde{\rho}(X^+) = \begin{pmatrix} 0 & 1 \\ 0 & 0 \end{pmatrix}$$

$$\widetilde{\rho}(X^-) = \begin{pmatrix} 0 & 0 \\ 1 & 0 \end{pmatrix}.$$

In other words, we map each of the generators of $\mathcal{U}_q$ to its "matrix of origin" in sl_2.

It is easy to verify that $\tilde{\rho}$ is a representation of $\mathcal{U}_q$ as algebra. Note that

$$\sinh\left(\frac{\hbar}{2}\tilde{\rho}(H)\right) = \sinh\left(\frac{\hbar}{2}\begin{pmatrix} 1 & 0 \\ 0 & -1 \end{pmatrix}\right)$$

$$= \frac{1}{2}\left[\begin{pmatrix} e^{\hbar/2} & 0 \\ 0 & e^{-\hbar/2} \end{pmatrix} - \begin{pmatrix} e^{-\hbar/2} & 0 \\ 0 & e^{\hbar/2} \end{pmatrix}\right]$$

$$= \begin{pmatrix} \sinh(\hbar/2) & 0 \\ 0 & -\sinh(\hbar/2) \end{pmatrix}.$$

Hence

$$\frac{\sinh((\hbar/2)\tilde{\rho}(H))}{\sinh(\hbar/2)} = \tilde{\rho}(H).$$

We can write $\tilde{\rho}$ as a matrix of functions on $\mathcal{U}_q$:

$$\tilde{\rho} = \begin{pmatrix} a & b \\ c & d \end{pmatrix}.$$

Thus

$$
\begin{array}{lll}
a(H) = 1, & a(X^+) = 0, & a(X^-) = 0 \\
b(H) = 0, & b(X^+) = 1, & b(X^-) = 0 \\
c(H) = 0, & c(X^+) = 0, & c(X^-) = 1 \\
d(H) = -1, & d(X^+) = 0, & d(X^-) = 0.
\end{array}
$$

Lemma 9.7. The (Hopf) bialgebra generated by a, b, c, d above is isomorphic to the bialgebra $\mathcal{A}(\tilde{\epsilon}) = F_q(SL(2))$. Thus $F_q(SL(2)) \simeq \mathcal{U}_q^*$ with $q = e^{\hbar}$.

We have already described the multiplication and comultiplication on $\mathcal{U}_q^*$. The counit for $\mathcal{U}_q$ is given by $\epsilon(1) = 1$, $\epsilon(H) = \epsilon(X^\pm) = 0$. The antipode is defined via $S(H) = -H$, $S(X^\pm) = -e^{-\frac{\hbar}{4}H}X^\pm e^{\frac{\hbar}{4}H}$. In the proof of the Lemma, I will just check bialgebra structure, leaving the Hopf algebra verifications to the reader.

Proof. We will do some representative checking. Consider $ab \in \mathcal{U}_q^*$. By definition, $ab(x) = a \otimes b(\Delta(x))$ for $x \in \mathcal{U}_q$. Thus

$$ab(H) = a \otimes b(H \otimes 1 + 1 \otimes H)$$
$$\Rightarrow ab(H) = 0.$$
$$ab(X^+) = a \otimes b(X^+ \otimes e^{(\hbar/4)H} + e^{-(\hbar/4)H} \otimes X^+)$$
$$= 0 + a(e^{-(\hbar/4)H})$$
$$= e^{-(\hbar/4)}$$

and $ba(X^+) = e^{(\hbar/4)}$.

Thus $ba = e^{+(\hbar/2)}ab = qab$.

Now look at $\Delta(a)$: $\Delta(a)(x \otimes y) = a(xy)$. Thus $\Delta(a)(X^+ \otimes X^-) = a(X^+X^-)$. But in general, we see that since $a(xy)$ is the $1-1$ coordinate of the matrix product $\rho(x)\rho(y)$.

$$\rho(x) = \begin{pmatrix} a(x) & b(x) \\ c(x) & d(x) \end{pmatrix}$$
$$\rho(x)\rho(y) = \begin{pmatrix} a(x) & b(x) \\ c(x) & d(x) \end{pmatrix} \begin{pmatrix} a(y) & b(y) \\ c(y) & d(y) \end{pmatrix}$$
$$= \begin{pmatrix} a(x)a(y) + b(x)c(y) & a(x)b(y) + b(x)d(y) \\ c(x)a(y) + d(x)c(y) & c(x)b(y) + d(x)d(y) \end{pmatrix}.$$

Whence

$$\Delta(a)(x \otimes y) = a(x)a(y) + b(x)c(y) = (a \otimes a + b \otimes c)(x \otimes y)$$
$$\Rightarrow \Delta(a) = a \otimes a + b \otimes d$$
$$\Delta(b) = a \otimes b + b \otimes d$$
$$\Delta(c) = c \otimes a + d \otimes c$$
$$\Delta(d) = c \otimes b + d \otimes d.$$

The other facts about products are easy to verify, and we leave them for the reader. This completes the proof of the Lemma. //

The dual view of $SL(2)_q^*$ as

$$\mathcal{U}_q(s\ell_2) : \begin{cases} [H, X^+] = 2X^+ \\ [H, X^-] = -2X^- \\ [X^+, X^-] = \sinh(\hbar/2\ H)/\sinh(\hbar/2) \\ (\text{and } \Delta \text{ as above.}) \end{cases}$$

is important because it establishes this algebra as a deformation of the Lie algebra of $SL(2)$.

Just as one can study the representation theory of $SL(2)$ (and $s\ell_2$) one can study the representation theory of these quantum groups. Furthermore, the structure of a **universal R-matrix** (a solution to the Yang-Baxter Equation) emerges from the Lie algebra formalism. This universal solution then specializes to a number of different specific solutions-depending upon the representation that we take. In the fundamental representation $(\tilde{\rho})$, we recover the R-matrix that we have obtained in this section for the bracket.

Comment. This section has lead from the bracket as vacuum-vacuum expectation through the internal symmetries of this model that give rise to the quantum group $SL(2)_q$ and its (dual) universal quantum enveloping algebra $\mathcal{U}_q(s\ell_2)$. This algebraic structure arises naturally from the idea of the bracket $[\langle \asymp \rangle = A\langle \approx \rangle + B\langle \supset \subset \rangle]$ and the conditions of topological invariance - coupled with elementary quantum mechanical ideas. The bare bones of a discrete topological quantum field theory and generalizations of classical symmetry are as inevitable as the construction of elementary arithmetic.

10^0. The Form of the Universal R-matrix.

First of all, there is a convenient way to regard an R-matrix algebraically. Suppose that we are given a bialgebra $\mathcal{U}$. Let $R \in \mathcal{U} \otimes \mathcal{U}$. Then we can define **three** elements of $\mathcal{U} \otimes \mathcal{U} \otimes \mathcal{U}$:

$$R_{12}, R_{13}, R_{23} \in \mathcal{U} \otimes \mathcal{U} \otimes \mathcal{U},$$

by placing the tensor factors of R in the indicated factors of $\mathcal{U} \otimes \mathcal{U} \otimes \mathcal{U}$, with a 1 in the remaining factor. Thus, if

$$R = \sum_s e_s \otimes e^s \in \mathcal{U} \otimes \mathcal{U},$$

$$R_{12} = \sum_s e_s \otimes e^s \otimes 1$$

$$R_{13} = \sum_s e_s \otimes 1 \otimes e^s$$

$$R_{23} = \sum_s 1 \otimes e_s \otimes e^s.$$

The appropriate version of the Yang-Baxter Equation for this formalism is

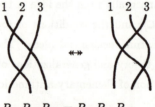

$$R_{12} R_{13} R_{23} = R_{23} R_{13} R_{12}.$$

Here the vector spaces (or modules) $\mathcal{U}$ label the lines in the diagram above the equation.

In order to relate this algebraic form of the YBE to our knot theorist's version, **we must compose with a permutation.** To see this, note that the knot theory uses a matrix representation of $R \in \mathcal{U} \otimes \mathcal{U}$. Let $\rho(R)$ denote such a representation. Thus $\rho(R) = (\rho_{cd}^{ab})$ where a and c are **indices corresponding to the first** $\mathcal{U}$ **factor and** b **and** d **are indices corresponding to the representation of the**

second $\mathcal{U}$ factor. This is the natural convention for algebra. Now **define R^{ab}_{cd}** by the equation

$$R^{ab}_{cd} = \rho^{ab}_{dc}$$

(permuting the bottom indices). We then have the diagram

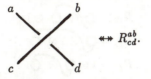

$$\leftrightarrow R^{ab}_{cd}.$$

Now, as desired, the crossing lines each hold the appropriate indices, and the equation $R_{12}R_{13}R_{23} = R_{23}R_{13}R_{12}$ will translate into our familiar index form of the Yang-Baxter Equation.

Now the extraordinary thing about this algebraic form of the Yang-Baxter equation is that it is possible to give general algebraic conditions that lead to solutions. To illustrate this, suppose that we know products for $\{e_s\}$ and $\{e^s\}$ in the algebra:

$$e_s e_t = m^i_{st} e_i$$

and

$$e^s e^t = \mu^{st}_i e^i \leftrightarrow$$

Here the abstract tensors and correspond to e^s and e_s respectively. We cannot assume that and commute, but the coefficient tensors , commute with everybody.

Finally, assume the **commutation relation**

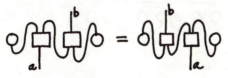

$$e_s \mu^{si}_a m^b_{it} e^t = e^t m^b_{ti} \mu^{is}_a e_s$$

Call this equation the **Yang-Baxter commutation relation.**

150

Theorem 10.1. Let ⧫ and ⧫ be abstract tensors satisfying a Yang-Baxter

commutation relation. Let $R = $ ⊗ and

$$R_{12} = \text{⊗} \otimes 1$$
$$R_{13} = \text{⊗} 1 \otimes$$
$$R_{23} = 1 \otimes \text{⊗} .$$

Then $R_{12}R_{13}R_{23} = R_{23}R_{13}R_{12}$.

Proof.

$$R_{12}R_{13}R_{23} = [\text{⊗} \otimes 1][\text{⊗} 1 \otimes][1 \otimes \text{⊗}]$$

(diagram)

(diagram)

(diagram)

(diagram)   (Y-B commutation)

(diagram)

(diagram)

$$= R_{23}R_{13}R_{12}.$$

//

The simple combinatorial pattern of this theorem controls the relationship of Lie algebras and Hopf algebras to knot theory and the YBE.

Suppose that $\mathcal{U}$ is a bialgebra with comultiplication $\Delta : \mathcal{U} \to \mathcal{U} \otimes \mathcal{U}$. Further suppose that $\{e_s\}$, $\{e^s\} \subset \mathcal{U}$ with

$$e_s e_t = m^i_{st} e_i$$
$$e^s e^t = \mu^{st}_i e^i$$

and that the subalgebras A and $\widehat{A}$ generated by $\{e_s\}$ and $\{e^s\}$ are **dual** in the sense that the multiplicative and comultiplicative structures in these algebras are interchanged. Thus

$$\Delta(e_i) = \mu^{st}_i e_s \otimes e_t$$

and

$$\Delta(e^i) = m^i_{st} e^s \otimes e^t.$$

Now suppose that $R = \sum_s e_s \otimes e^s$ satisfies the equation

$$R\Delta = \Delta' R \tag{$*$}$$

where $\Delta' = \Delta \circ \sigma$ denotes the composition of Δ with the map $\sigma : \mathcal{U} \otimes \mathcal{U} \to \mathcal{U} \otimes \mathcal{U}$ that interchanges factors.

Proposition 10.2. Under the above circumstances the equation $R\Delta = \Delta' R$ is equivalent to the Y-B commutation relation.

Proof.

$$
\begin{aligned}
R\Delta(e_t) &= (e_s \otimes e^s)(\mu^{ij}_t e_i \otimes e_j) \\
&= \mu^{ij}_t e_s e_i \otimes e^s e_j \\
&= \mu^{ij}_t m^\ell_{si} e_\ell \otimes e^s e_j \\
&= e_\ell \otimes e^s m^\ell_{si} \mu^{ij}_t e_j.
\end{aligned}
$$

$$
\begin{aligned}
\Delta'(e_t)R &= (\mu^{ij}_t e_j \otimes e_i)(e_s \otimes e^s) \\
&= \mu^{ij}_t e_j e_s \otimes e_i e^s \\
&= \mu^{ij}_t m^\ell_{js} e_\ell \otimes e_i e^s \\
&= e_\ell \otimes e_i \mu^{ij}_t m^\ell_{js} e^s.
\end{aligned}
$$

Since $\{e_\ell\}$ is a basis for A, we conclude that

$$e^s m^\ell_{si} \mu^{ij}_t e_j = e_i \mu^{ij}_t m^\ell_{js} e^s.$$

This is the Y-B commutation relation, completing the proof of the claim. $\qquad$ //

Remark. The circumstance of $\mathcal{U} \supset A, \widehat{A}$ with these duality patterns is characteristic of the properties of quantum groups associated with Lie algebras.

In the case of $\mathcal{U}_q(s\ell_2)$ the algebras are the Borel subalgebras generated by H, X^+ and H, X^-. $R = \sum_s e_s \otimes e^s$ becomes a specific formula involving power series in these elements. The construction of Theorem 10.1 and its relation to the equation $R\Delta = \Delta' R$ is due to Drinfeld [DRIN1], and is called the **quantum double construction** for Hopf algebras.

While we shall not exhibit the details of the derivation of the universal R-matrix for $\mathcal{U}_q(s\ell(2))$, it is worth showing the end result! Here is Drinfeld's formula:

$$R = \sum_{k=0}^{\infty} \hbar^k Q_k(\hbar) \left\{ \exp \frac{\hbar}{4}[H \otimes H + k(H \otimes 1 - 1 \otimes H)] \right\} (X^+)^k \otimes (X^-)^k$$

where $Q_k(\hbar) = e^{-k\hbar/2} \prod_{r=1}^{k} (e^{\hbar} - 1)/(e^{r\hbar} - 1)$. Note that in this situation the algebras in question are infinite dimensional. Hence the sum $R = \sum_s e_s \otimes e^s$ becomes a formal power series (and the correct theory involves completions of these algebras.)

The Faddeev-Reshetikhin-Takhtajan Construction.

The equation $R\Delta = \Delta' R$ is worth contemplating on its own grounds, for it leads to a generalization of the construction of the $SL(2)$ quantum group via invariants. To see this generalization, suppose that $\mathcal{U}$ contains an R with $R\Delta = \Delta' R$, and let $T : \mathcal{U} \to \mathcal{M}$ be a matrix representation of $\mathcal{U}$. If $T(u) = (t^i_j(u))$, then $t^i_j : \mathcal{U} \to \mathbf{C}$ whence $t^i_j \in \mathcal{U}^*$. [Compare with the construction in section 9^0.] The coproduct in $\mathcal{U}^*$ is given by

$$\Delta(t^i_j)(u \otimes v) = t^i_j(uv)$$
$$= \sum_k t^i_k(u) t^k_j(v)$$

since $T(uv) = T(u)T(v)$. Therefore $\Delta(t^i_j) = t^i_k \otimes t^k_j$.

With this in mind, we can translate $R\Delta = \Delta' R$ into an equation about $\mathcal{U}^*$, letting

$$R^{ij}_{k\ell} = t^i_k \otimes t^j_\ell(R). \qquad \text{(algebraist's convention)}$$

Lemma. $R\Delta = \Delta'R$ corresponds to the equation $\boxed{RT_1T_2 = T_2T_1R}$ where

$$R = (R_{k\ell}^{ij}), \ T = (t_j^i), \ T_1 = T \otimes 1, \ T_2 = 1 \otimes T.$$

Proof.

$$R\Delta(X): \qquad t_k^i \otimes t_\ell^j(R\Delta(X))$$

$$= \sum_{m,n} t_m^i \otimes t_n^j(R)t_k^m \otimes t_\ell^n(\Delta(X))$$

$$= \sum_{m,n} R_{mn}^{ij} t_k^m t_\ell^n(X)$$

(definition of product in $\mathcal{U}^*$)

$$\Delta'(X)R: \qquad t_k^i \otimes t_\ell^j(\Delta'(X)R)$$

$$= t_n^j \otimes t_m^i(\Delta(X))t_k^m \otimes t_\ell^n(R)$$

$$= t_n^j t_m^i(X)R_{k\ell}^{mn}.$$

Thus $R\Delta(X) = \Delta'(X)R$ implies that

$$\sum_{m,n} R_{mn}^{ij} t_k^m t_\ell^n = \sum_{m,n} t_n^j t_m^i R_{k\ell}^{mn}.$$

Letting $T = (t_j^i)$ and $T_1 = T \otimes 1$, $T_2 = 1 \otimes T$. This matrix equation reads

$$\boxed{RT_1T_2 = T_2T_1R}.$$

Faddeev, Reshetikhin and Takhtajan [FRT] take the equation $RT_1T_2 = T_2T_1R$ as a starting place for constructing a bialgebra **from** a solution to the Yang-Baxter Equation. We shall call this the **FRT construction**. In the FRT construction

154

one starts with a given solution R to YBE and associates to it the bialgebra $A(R)$ with generators $\{t_j^i\}$ and relations $T_2 T_1 R = R T_1 T_2$. The coproduct is given by $\Delta(t_j^i) = \sum_k t_k^i \otimes t_j^k$.

Now in order to perform explicit computations with the FRT construction, I shall assume that **the R-matrix is in knot-theory format.** That is, we replace R_{cd}^{ab} by R_{dc}^{ab} in the above discussion. Then it is clear that the defining equation for the FRT construction becomes

$$\boxed{R T_1 T_2 = T_1 T_2 R}\ .$$

In index form, the FRT relations become:

$$\boxed{T_i^a T_j^b R_{cd}^{ij} = R_{ij}^{ab} T_c^i T_d^j}\ .$$

If ⊗ denotes T_j^i, and ⊞ denotes R, then these relations read:

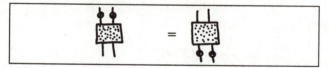

Since we assume that the elements $R_{k\ell}^{ij}$ commute with the algebra generators T_b^a, we can write the Faddeev-Reshetikhin-Takhtajan construction as:

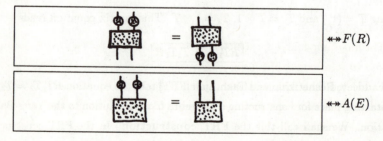

Given an R matrix, we shall denote this algebra by $F(R)$. It is a bialgebra for exactly the same reasons as the related case of invariance (the algebra $A(E)$) discussed in section 9^0. Diagrammatically, the comparison of $A(E)$ and $F(R)$ is quite clear:

A clue regarding the relationship of these constructions is provided by a direct look at the R-matrix for $SL(2)_q$. Recall from section 9^0 that we have $SL(2)_q = A(E)$ where

$$E = \widetilde{\epsilon} = \sqrt{-1}\epsilon_{ab}q^{\epsilon_{ab}/2}$$

and that the knot theory provided us with the associated R-matrix

$$R = \sqrt{q}\,\cup\!\cap + \frac{1}{\sqrt{q}}\,)\,($$

Thus we see at once that any matrix $\otimes$ satisfying $\boxtimes = \boxtimes$ will automatically satisfy the relations for $F(R)$:

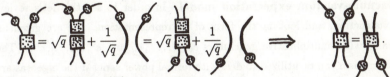

Therefore $F(R)$ has $A(E)$ as a quotient.

In fact, a direct check reveals that $F(R)$ is isomorphic to the algebra with generators a, b, c, d and relations:

$$
\begin{array}{ll}
ba = qab & da = qad \\
db = qbd & ca = qac \\
bc = cb & \\
\multicolumn{2}{c}{ad - da = (q^{-1} - q)bc}
\end{array}
$$

The center of $F(R)$ is generated by the element $(ad - q^{-1}bc)$, and, as we know, $A(E) \simeq F(R)/\langle ad - q^{-1}bc\rangle$ where the bracket denotes the ideal generated by this relation. In any case, the Faddeev-Reshetikhin-Takhtajan construction is very significant in that it shows how to directly associate a Hopf algebra or at least a bialgebra to any solution of the Yang-Baxter Equation.

The quantum double construction shows that certain bialgebras will naturally give rise to solutions to the Yang-Baxter Equation from their internal structure. The associated bialgebra (quantum group) provides a rich structure that can be utilized to understand the structure of these solutions and the structure of allied topological invariants.

Hopf Algebras and Link Invariants.

It may not yet be clear where the Hopf algebra structure is implicated in the construction of a link invariant. We have shown that the Drinfeld double construction yields formal solutions to the Yang-Baxter Equation, but this is accomplished without actually using the comultiplication if one just accepts the "Yang-Baxter commutation relation"

As for the antipode $\gamma : \mathcal{A} \to \mathcal{A}$, it doesn't seem to play any role at all. Not so! **Seen rightly, the Drinfeld construction is an algebraic version of the vacuum-vacuum expectation model.** In order to see this let's review the construction, and look at the form of its representation. We are given an algebra $\mathcal{A}$ with multiplicative generators $e_0, e_1, \dots, e_n$ and $e^0, e^1, \dots, e^n$. (The same formalism can be utilized to deal with formal power series if the algebras are not finite dimensional.) We can assume that $e_0 = e^0 = 1$ (a multiplicative unit) in $\mathcal{A}$, and we assume rules for multiplication:

$$e_i e_j = m_{ij}^k e_k$$
$$e^i e^j = \mu_k^{ij} e^k$$

(summation convention operative). We further assume a comultiplication $\Delta : \mathcal{A} \to \mathcal{A} \otimes \mathcal{A}$ such that

$$\Delta(e_k) = \mu_k^{ij} e_i \otimes e_j$$
$$\Delta(e^k) = m_{ij}^k e^i \otimes e^j.$$

That is, lower and upper indexed algebras are dual as bialgebras, the comultiplication of one being the multiplication of the other. Letting $\rho = e_s \otimes e^s$, we assume that ρ satisfies the Yang-Baxter commutation relation in the form $\rho \Delta = \Delta' \rho$ where Δ' is the composition of Δ with the permutation map on $\mathcal{A} \otimes \mathcal{A}$. With these assumptions it follows that ρ satisfies the algebraic form of the Yang-Baxter Equation: $\rho_{12} \rho_{13} \rho_{23} = \rho_{23} \rho_{13} \rho_{12}$, (Theorem 10.1). Finally we assume that there is an anti-automorphism $\gamma : \mathcal{A} \to \mathcal{A}$ that is an **antipode** in the (direct) sense that

$$m((1 \otimes \gamma)\Delta(e_k)) = m((\gamma \otimes 1)\Delta(e^k)) = \begin{cases} 1 & \text{if } k = 0 \\ 0 & \text{otherwise.} \end{cases}$$

(Same statement for e^k and e_k in these positions.).

Now suppose that **rep** $: \mathcal{A} \to \mathbf{M}$ where $\mathbf{M}$ is an algebra of matrices (with commutative entries). If we let $\bullet$ denote an element of $\mathcal{A}$, then $\mathbf{rep}(\bullet) = $ ⬧. That is, the representation of $\bullet$ becomes an object with upper and lower matrix indices. For thinking about the representation, it is convenient to suppress the indices k and ℓ in e_k, e^ℓ. Thus, we can let ●～● denote $\rho = e_k \otimes e^k$ with the wavy line indicating the summation over product of elements that constitute ρ. Then we have:

$$\mathbf{rep}(\bullet \!\!\sim\!\! \bullet) = \text{⫯⫯}$$

and we define an R-matrix via

$$R = \text{⟨⟩}$$

Note once again how the permutation of the lower lines makes R into a knot-theoretic solution to the Yang-Baxter Equation:

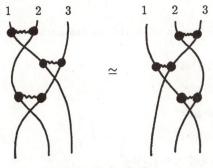

$$\rho_{12}\rho_{13}\rho_{23} = \rho_{23}\rho_{13}\rho_{12}.$$

For the purpose of the knot theory, we can let R correspond to a crossing of the "A" form:

(with respect to a vertical arrow of time). Then we know that R^{-1} corresponds to the opposite crossing, and **by the twist relation** R^{-1} is given by the formula

$$R^{-1} = \quad \raisebox{-1em}{} \quad = \quad \raisebox{-1em}{}$$

where it is understood that the raising and lowering of indices on R to form R^{-1} is accomplished through the creation and annihilation matrices

$$M^{ab} \longleftrightarrow \raisebox{-0.5em}{} \quad \text{and} \quad M_{ab} \longleftrightarrow \raisebox{-0.5em}{}$$

Keeping this in mind, we can deform the diagrammatic for R^{-1} so long as we keep track of the maxima and minima. Hence

$$R^{-1} = \quad \raisebox{-1em}{}$$

This suggests that there should be a mapping $\gamma : \mathcal{A} \to \mathcal{A}$ such that

$$\mathbf{rep}(\gamma)\left(\raisebox{-0.3em}{}\right) = \raisebox{-0.3em}{}$$

Therefore, **we shall assume that the antipode** $\gamma : \mathcal{A} \to \mathcal{A}$ **is so represented.** Then we see that the equation $RR^{-1} = I$ becomes:

$$\raisebox{-1em}{} = \raisebox{-1em}{} = \raisebox{-1em}{}$$

In other words, we must have $\sum_s (e_s \otimes e^s)(\gamma(e_t) \otimes e^t) = 1 \otimes 1$ in the algebra.

With γ the antipode of $\mathcal{A}$, this equation is true:

$$
\begin{aligned}
&(e_s \otimes e^s)(\gamma(e_t) \otimes e^t) \\
&= e_s \gamma(e_t) \otimes e^s e^t \\
&= e_s \gamma(e_t) \otimes \mu_\ell^{st} e^\ell \\
&= \mu_\ell^{st} e_s \gamma(e_t) \otimes e^\ell \\
&= m(1 \otimes \gamma)(\mu_\ell^{st} e_s \otimes e_t) \otimes e^\ell \\
&= m((1 \otimes \gamma)\Delta(e_\ell)) \otimes e^\ell \qquad \text{(by assumption about comultiplication)} \\
&= 1 \otimes 1 \qquad\qquad\qquad\qquad\quad \text{(by definition of antipode)}.
\end{aligned}
$$

Thus we have shown:

Theorem 10.3. If $\mathcal{A}$ is a Hopf algebra satisfying the assumptions of the Drinfeld construction, and if $\mathcal{A}$ is represented so that the antipode is given by matrices M_{ab}, M^{ab} (inverse to one another) so that

$$\text{rep}(\gamma(\text{\large\char"21})) = \text{\large\char"52}$$

Then the matrices R, R^{-1}, M_{ab}, M^{ab} (specified above) satisfy the conditions for the construction of a regular isotopy invariant of links via the vacuum-vacuum expectation model detailed in this section.

The formalism of the Hopf algebra provides an algebraic model for the category of abstract tensors that provide regular isotopy invariants. In this sense the Drinfeld construction creates a universal link invariant (compare [LK26] and [LAW1]).

A Short Guide to the Definition of Quasi-Triangular Hopf Algebras.

Drinfeld [DRIN1] generalized the formalism of the double construction to the notion of **quasi-triangular Hopf algebra**. A Hopf algebra $\mathcal{A}$ is said to be quasi-triangular if it contains an element R satisfying the following identities:

(1) $R\Delta = \Delta'R$

(2) $R_{13}R_{23} = (\Delta \otimes 1)(R)$

(3) $R_{12}R_{13} = (1 \otimes \Delta)(R)$. We have already discussed the motivation behind (1). To understand (2) and (3) let's work in the double formalism; then

$$R_{13}R_{23} = e_s \otimes e_t \otimes e^s e^t$$
$$= e_s \otimes e_t \otimes \mu_i^{st} e^i$$
$$= \mu_i^{st} e_s \otimes e_t \otimes e^i$$
$$= \Delta(e_i) \otimes e^i$$
$$\therefore \ R_{13}R_{23} = (\Delta \otimes 1)(R).$$

Similarly, $R_{12}R_{13} = (1 \otimes \Delta)(R)$.

The notion of quasi-triangularity encapsulates the duality structure of the double construction indirectly and axiomatically. We obtain index-free proofs of all the standard properties. For example, if $\mathcal{A}$ is quasi-triangular, then $(\Delta' \otimes 1)(R) =$

$R_{23}R_{13}$ since this formula is obtained by switching two factors in $(\Delta \otimes 1)R = R_{13}R_{23}$ (interchange 1 and 2). From this we calculate

$$R_{12}R_{13}R_{23} = R_{12}(\Delta \otimes 1)(R)$$
$$= (\Delta' \otimes 1)(R)R_{12} \qquad (R\Delta = \Delta'R)$$
$$= R_{23}R_{13}R_{12}.$$

Hence R satisfies the Yang-Baxter Equation. It is also the case that $R^{-1} = (1 \otimes \gamma)R$ when $\mathcal{A}$ is quasi-triangular.

Statistical Mechanics and the FRT Construction.

The Yang-Baxter Equation originated in a statistical mechanics problem that demanded that an R-matrix associated with a 4-valent vertex commute with the row-to-row transfer matrix for the lattice [BA1]. Diagrammatically, this has the form

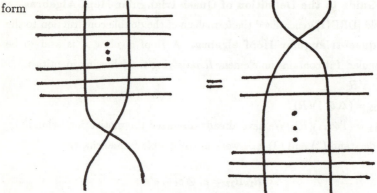

In these terms, we can regard this equation as a representation of the equation $T_1T_2R = RT_1T_2$ with $T = \rlap{$\;$}\text{\hspace{0.5em}}$ and T_1T_2 denoting matrix multiplication as shown in the diagram. In particular, this means that, for a knot-theoretic R solving the Yang-Baxter Equation, $(t_b^a)_j^i = R_{bj}^{ia}$ gives a solution to the FRT equations $t_i^a t_j^b R_{cd}^{ij} = R_{ij}^{ab} t_c^i t_d^j$. Thus, for R a solution to the Yang-Baxter Equation, the FRT construction comes equipped with a matrix representation that is constructed from submatrices of the R-matrix itself.

11^0. Yang-Baxter Models for Specializations of the Homfly Polynomial.

We have seen in section 9^0 that a vacuum-vacuum expectation model for the bracket leads directly to a solution to the YBE and that this solution generalizes to the R-matrices

$$R = (q - q^{-1}) \rangle\hspace{-3pt}\bullet\hspace{-3pt}\langle + q \rangle\hspace{-3pt}\langle + \times$$

$$\overline{R} = (q^{-1} - q) \rangle\hspace{-3pt}\bullet\hspace{-3pt}\langle + q^{-1} \rangle\hspace{-3pt}\langle + \times$$

The edges of these small diagrams are labelled with "spins" from an arbitrary index set $\mathcal{I}$ and

$$a\rangle\hspace{-3pt}\bullet\hspace{-3pt}\langle b \leftrightarrow a < b, \qquad a\rangle\hspace{-3pt}\bullet\hspace{-3pt}\langle b \leftrightarrow a > b$$

$$a\rangle\hspace{-3pt}\langle b \leftrightarrow a = b$$

$$a\times b \leftrightarrow a \neq b$$

so that (e.g.) $\begin{smallmatrix}a & b \\ \rangle\hspace{-3pt}\bullet\hspace{-3pt}\langle \\ c & d\end{smallmatrix} = [a < b]\delta_c^a \delta_d^b$ where $[a < b] = \begin{cases} 1 & \text{if } a < b \\ 0 & \text{if } a \not< b \end{cases}$

$\left(\text{In general, } [P] = \begin{cases} 1 & \text{if } P \text{ is true} \\ 0 & \text{if } P \text{ is false} \end{cases} \right)$. Thus

$$R_{cd}^{ab} = (q - q^{-1})[a < b]\delta_c^a \delta_d^b + q[a = b]\delta_c^a \delta_d^b + [a \neq b]\delta_d^a \delta_c^b.$$

The structure of these diagrammatic forms fits directly into corresponding state models for knot polynomials - as we shall see.

The models that we shall consider in this section are defined for oriented knot and link diagrams. I shall continue to use the bracket formalism:

$$\langle K \rangle = \sum_\sigma \langle K|\sigma\rangle \delta^{\|\sigma\|}$$

where the states σ, products of vertex weights $\langle K|\sigma\rangle$, and state norm $\|\sigma\|$ will be defined below. We shall then find that $\langle K \rangle$ can be adjusted to give invariants of regular isotopy for an infinite class of specializations of the Homfly polynomial. This gives a good class of models, enough to establish the existence of the full 2-variable Homfly polynomial. Recall that we have discussed this polynomial and a combinatorial (not Yang-Baxter) model in section 5^0.

How should this state model be defined? Since the R-matrix has a nice expansion, it makes sense to posit:

$$\left\langle \cancel{\,}\right\rangle = (q - q^{-1})\left\langle \,\right\rangle + q\left\langle \,\right\rangle + \left\langle \,\right\rangle$$

$$\left\langle \cancel{\,}\right\rangle = (q^{-1} - q)\left\langle \,\right\rangle + q^{-1}\left\langle \,\right\rangle + \left\langle \,\right\rangle$$

for the state model. This is equivalent to saying that the R-matrix determines the local vertex weights.

Note that, we then have

$$\left\langle \,\right\rangle - \left\langle \,\right\rangle = (q - q^{-1})\left[\left\langle \,\right\rangle + \left\langle \,\right\rangle + \left\langle \,\right\rangle\right]$$

and, since given two spins a, b we have either $a < b$, $a > b$ or $a = b$, we have

$$\left\langle \,\right\rangle = \left\langle \,\right\rangle + \left\langle \,\right\rangle + \left\langle \,\right\rangle.$$

Therefore,

$$\left\langle \,\right\rangle - \left\langle \,\right\rangle = (q - q^{-1})\left\langle \,\right\rangle.$$

Hence the model satisfies the exchange identity for the regular isotopy version of the Homfly polynomial. Since the edges are labelled with spins from the index set $\mathcal{I}$, we see that the **states** σ of this model are obtained by

1. Replace each crossing by either a decorated splice:

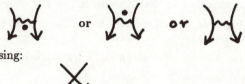

   or by a graphical crossing:

2. Label the resulting diagram σ with spins from $\mathcal{I}$ so that each loop in σ has

constant spin, and so that the spins obey the rules:

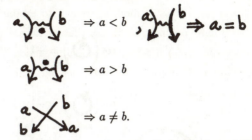

[A diagram may have no labelling - in such a case it does not contribute to $\langle K \rangle$.]

Example. If K is a trefoil diagram

then a typical state σ is

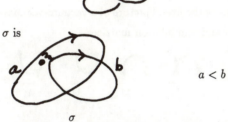

$a < b$

σ

Note that the loops in a state have no self-crossings (since such a crossing would force a label to be unequal to itself).

In conformity with the notion that $\langle K \rangle$ is a vacuum-vacuum expectation (generalizing from the case of the unoriented bracket of section 3^0), we see that each individual loop of constant spin a must receive a value that depends only on a and the rotational direction of the loop. We choose this evaluation by the formula

$$\delta^{\|\sigma\|} = \langle \sigma \rangle$$

where δ is a variable whose relation with q is as-yet to be elucidated, and $\|\sigma\|$ is **the sum of the spins of the loops in σ multiplied by ± 1 according to the rotational sense of the loop.** That is,

$$\|\sigma\| = \sum_{\ell \in \text{components}(\sigma)} \text{rot}(\ell) \cdot \text{label}(\ell)$$

where label(ℓ) is the spin assigned to the loop ℓ and rot(ℓ) = ± 1 with the convention:

$$\mathrm{rot}\left(\bigcirc\right) = +1$$

$$\mathrm{rot}\left(\bigcirc\right) = -1.$$

Thus for

$$= \sigma$$

we have $\|\sigma\| = -a - b$ and $\langle\sigma\rangle = \delta^{-a-b}$.

Remark. In terms of the literal picture of a vacuum-vacuum expectation we need to specify creation and annihilation matrices

$$\overrightarrow{M}_{ab} \qquad \overleftarrow{M}_{ab} \qquad \overrightarrow{M}^{ab} \qquad \overleftarrow{M}^{ab}$$

so that

$$\left\langle\bigcirc\right\rangle = \sum_{a \in \mathcal{I}} q^{-a}$$

and

$$\left\langle\bigcirc\right\rangle = \sum_{a \in \mathcal{I}} q^{a}$$

since a given loop has states with arbitrary labels from $\mathcal{I}$.

There is one case where these matrices appear naturally. Suppose that $a \in \mathcal{I} \Leftrightarrow -a \in \mathcal{I}$. (Positive spin is theoretically accompanied by corresponding negative spin.) Then interpret

$$a \overparen{\qquad} b = q^{a/2}\Delta_{a,b}$$

as a deformation of a reversed Kronecker delta:

$$\Delta_{a,b} = \delta_{a,-b} = \begin{cases} 1 & a = -b \\ 0 & \text{otherwise.} \end{cases}$$

Here we use the vertical direction (of time for the vacuum-vacuum process) so that a constant labelling of the arc $\cap$ is interpreted as

$$\uparrow +a \qquad \text{and} \qquad \downarrow -a$$

depending on the arrow's ascent or descent.

(A particle travelling backwards in time reverses its spin and charge.) The matrices become:

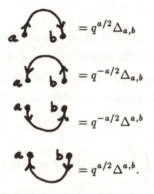

$$= q^{a/2}\Delta_{a,b}$$

$$= q^{-a/2}\Delta_{a,b}$$

$$= q^{-a/2}\Delta^{a,b}$$

$$= q^{a/2}\Delta^{a,b}.$$

For the symmetrical index set ($a \in \mathcal{I} \Leftrightarrow -a \in \mathcal{I}$) these matrices produce the given expectations for the trivial loops. The matrices simply state probabilities for the creation or annihilation of particles/antiparticles with opposite spin or the continued trajectory of a given particle "turning around in time". The whole point about a topological theory of this type is that the vacuum-vacuum expectations are independent of time and of the direction of time's arrow.

Returning now to the model itself we have $\langle K \rangle = \sum_{\sigma} \langle K|\sigma \rangle \langle \sigma \rangle$ where the σ can be regarded as diagrammatic "state-holders" (ready to be labelled with spins) or as labelled spin-states. A state-holder such as

is the diagrammatic form of a particular process whose expectation involves the summation over all spin assignments that satisfy the process. Note that we are free

166

to visualize particles moving around the loops, or a sequence of creations, interactions and annihilations. The latter involves a choice for the arrow of time. The
former involves a choice of **meta-time** for the processes running in these trajectories. The topological theory maintains independence of meta-time conventions
just as it is aloof from time's arrow.

In order for this theory to really be topological, we must examine the behavior
of $\langle K \rangle$ under Reidemeister moves. We shall adjust δ to make $\langle K \rangle$ an invariant of
regular isotopy.

For arbitrary q, δ and index set $\mathcal{I}$, it is easy to see that $\langle K \rangle$ is an invariant
of moves II(A) and III(A) (non-cyclic with all crossings of some type). II(A) is
easy since R and $\overline{R}$ are inverses and since the state evaluations $\delta^{\|\sigma\|}$ do not involve
crossing types. Nevertheless, the technique of checking II(A) by state-expansion
is worth illustrating. Therefore

Lemma 11.1. $\left\langle \vcenter{\hbox{\includegraphics{a}}} \right\rangle = \left\langle \vcenter{\hbox{\includegraphics{b}}} \right\rangle$ for any index set $\mathcal{I}$ and variables q and δ.

Proof.

$$\left\langle \vcenter{\hbox{}} \right\rangle = z\left\langle \vcenter{\hbox{}} \right\rangle + q\left\langle \vcenter{\hbox{}} \right\rangle + \left\langle \vcenter{\hbox{}} \right\rangle$$
$$(z = q - q^{-1})$$
$$\left\langle \vcenter{\hbox{}} \right\rangle = z\left\{ -z\left\langle \vcenter{\hbox{}} \right\rangle + q^{-1}\left\langle \vcenter{\hbox{}} \right\rangle + \left\langle \vcenter{\hbox{}} \right\rangle \right\}$$
$$+ q\left\{ -z\left\langle \vcenter{\hbox{}} \right\rangle + q^{-1}\left\langle \vcenter{\hbox{}} \right\rangle + \left\langle \vcenter{\hbox{}} \right\rangle \right\}$$
$$+ \left\{ -z\left\langle \vcenter{\hbox{}} \right\rangle + q^{-1}\left\langle \vcenter{\hbox{}} \right\rangle + \left\langle \vcenter{\hbox{}} \right\rangle \right\}.$$

Now certain combinations are incompatible. Thus $\left\langle \vcenter{\hbox{}} \right\rangle = 0 = \left\langle \vcenter{\hbox{}} \right\rangle$.

Therefore

$$\left\langle \text{⧖} \right\rangle = z\left\{0 + 0 + \left\langle \text{⧖} \right\rangle\right\}$$
$$+ q\left\{0 + q^{-1}\left\langle \text{⧗} \right\rangle + 0\right\}$$
$$+ \left\{-z\left\langle \text{⧓} \right\rangle + \left\langle \text{⧖} \right\rangle\right\}.$$

Now note that $\left\langle \text{⧖} \right\rangle = \left\langle \text{)(} \right\rangle - \left\langle \text{⧗} \right\rangle$ and that $\left\langle \text{⧓} \right\rangle = \left\langle \text{⧓} \right\rangle$.

Thus $\left\langle \text{⧖} \right\rangle = \left\langle \text{)(} \right\rangle$ and this proves the lemma. $\qquad //$

Lemma 11.2. The state model $\langle K \rangle$, as described above, is invariant under the oriented move III(A) for any ordered set $\mathcal{I}$ and any choice of the variables q and δ.

Proof. Note that the model has the form

$$\langle K \rangle = \sum_{\sigma} \langle K | \sigma \rangle \delta^{\|\sigma\|}$$

where σ runs over all spin-labellings of the diagram K, $\langle K | \sigma \rangle$ is the product of vertex weights corresponding to the matrices R and $\overline{R}$ and $\|\sigma\|$ is the loop count as explained above. Since R and $\overline{R}$ satisfy the YBE, it suffices to check that for given inputs and outputs to a triangle (for the III - move) $\|\sigma\|$ is the same for all contributing internal states. A comparison with the verification in Proposition 9.4 shows that this is indeed the case. For example

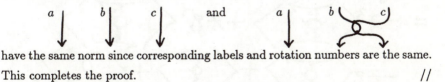

have the same norm since corresponding labels and rotation numbers are the same. This completes the proof. $\qquad //$

The next lemma is the key to bracket behavior under the reverse type II move.

Lemma 11.3. $\langle K \rangle$ is invariant under the reversed type II move, and hence is an invariant of regular isotopy, if and only if the following conditions are satisfied:

(i) $q^{-1}\left\langle \vcenter{\hbox{}} \right\rangle = q\left\langle \vcenter{\hbox{}} \right\rangle$

(ii) $q^{-1}\left\langle \vcenter{\hbox{}} \right\rangle = q\left\langle \vcenter{\hbox{}} \right\rangle$

Call these the **cycle conditions** for the model $\langle K \rangle$.

Remark. Note that we have reverted to the direct notation - indicating by inequality signs the spin relationships between neighboring lines.

Proof. The proof is given for one type of orientation for the type IIB move. The other case is left for the reader. Let $z = q - q^{-1}$. Expanding the form of the II move, we find:

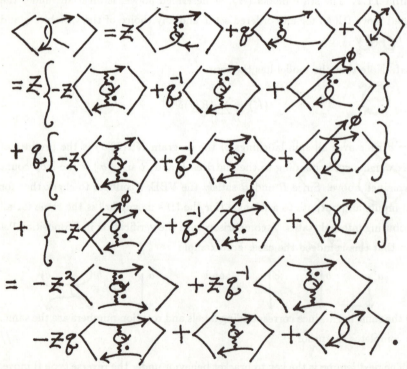

In order to simplify this expression, note that

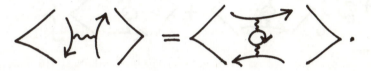

since crossed lines have unequal indices.

Also, it is easy to check that

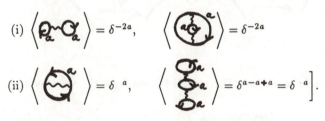

$\Big[$This is an identity about rotation numbers:

(i) $\left\langle \text{⬭} \right\rangle = \delta^{-2a}, \qquad \left\langle \text{⬭}^{a} \right\rangle = \delta^{-2a}$

(ii) $\left\langle \text{⬭}^{a} \right\rangle = \delta^{\ a}, \qquad \left\langle \text{⬭} \right\rangle = \delta^{a-a+a} = \delta^{\ a} \Big].$

Therefore,

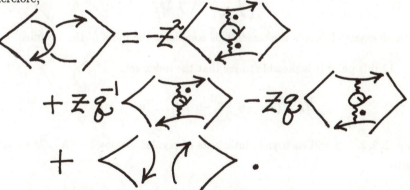

Thus, a necessary condition for invariance under the move IIB is that the sum of the first three terms on the right hand side of this equation should be zero.

170

Dividing by z, we have:

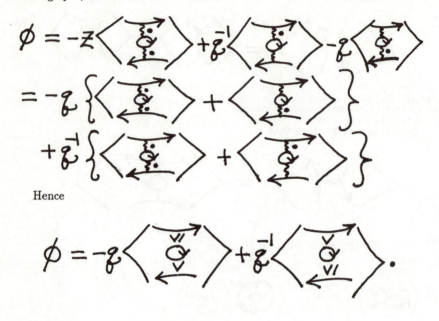

Hence

This proves the first cycle condition. The second follows from the need for the invariance

$$\left\langle \text{⟨knot diagram⟩} \right\rangle = \left\langle \text{⟨knot diagram⟩} \right\rangle$$

with all essential details the same. This completes the proof of the Lemma.   //

At this point it is possible to see that the index sets

$$\mathcal{I}_n = \{-n, -n+2, -n+4, \ldots, n-2, n\}$$

($n = 1, 2, 3, \ldots$) will each yield invariants of regular isotopy for $\langle K \rangle$. More precisely:

Proposition 11.4. Let $\mathcal{I}_n = \{-n, -n+2, \ldots, n-2, n\}$ ($n \in \{1, 2, 3, \ldots\}$) and let $\delta = q$, $z = q - q^{-1}$ so that $\langle K \rangle = \sum_\sigma \langle K | \sigma \rangle q^{\|\sigma\|}$. Then $\langle K \rangle$ is an invariant of regular isotopy, and it has the following behavior under type I moves:

$$\left\langle \vphantom{} \right\rangle = q^{n+1} \left\langle \vphantom{} \right\rangle$$

$$\left\langle \vphantom{} \right\rangle = q^{-n-1} \left\langle \vphantom{} \right\rangle .$$

Proof. To check regular isotopy, it is sufficient to show that the cycle conditions (Lemma 11.3) are satisfied. Here is the calculation for the first condition.

$$q^{-1} \left\langle \vphantom{} \right\rangle - q \left\langle \vphantom{} \right\rangle$$

$$= \sum_{a>c} \left[q^{-1} \sum_{a>b\geq c} q^{-b} - q \sum_{a\geq b>c} q^{-b} \right] \left\langle \vphantom{} \right\rangle .$$

With the choice of index set as given above, it is easy to see that each coefficient $\left[q^{-1} \sum_{a>b\geq c} q^{-b} - q \sum_{a\geq b>c} q^{-b} \right]$ vanishes identically. The second cycle condition is satisfied in exactly the same manner.

Direct calculation now shows the behavior under the type I moves. Since we shall generalize this calculation in the next proposition, I omit it here. $/\!/$

Remark. With index set $\mathcal{I}_n$, and $\delta = q$ we can define

$$P_K^{(n)}(q) = (q^{n+1})^{-w(K)} \langle K \rangle / \langle 0 \rangle$$

where $w(k)$ denotes the writhe of K. Then $P_k^{(n)}$ satisfies the conditions

1. K ambient isotopic to $K' \Rightarrow P_K^{(n)}(q) = P_{K'}^{(n)}(q)$.
2. $P_O^{(n)} = 1$
3. $q^{n+1} P^{(n)} \diagdown - q^{-n-1} P^{(n)} \diagdown = (q - q^{-1}) P^{(n)} \diagdown$.

Thus we see that $P_K^{(n)}$ is a one-variable specialization of the Homfly polynomial (Compare [JO4]). The entire collection of specializations $\{ P_K^{(n)} \mid n = 1, 2, 3, \ldots \}$ is sufficient to establish the existence of the Homfly polynomial itself (as a 2-variable polynomial invariant of knots and links). Although this route to the 2-variable oriented polynomial is a bit indirect, it is a very good way to understand the nature of the invariant.

Note that in this formulation the original Jones polynomial appears for the index set $\mathcal{I}_1 = \{-1, +1\}$ and $q = \sqrt{t}$. This is an oriented state model for the Jones polynomial based on the Yang-Baxter solution

$$R = (q - q^{-1}) \, \bigg\downarrow\!\!<\!\bigg\{ + q \, \bigg\downarrow\bigg\} = \bigg\{ + \text{\Large\lightning}$$

We have already remarked (9^0.) on the relationship between this R (for a two-element index set) and the Yang-Baxter solution that arises naturally from the bracket model. It is a nice exercise to see that these models translate into one another via the Fierz identity, as explained in section 10^0, for the R-matrices.

A further mystery about this sequence of models is the way they avoid the classical Alexander-Conway polynomial. We may regard the Conway polynomial as an ambient isotopy invariant ∇_K, with $\nabla_O = 1$ and

$$\nabla_{\text{\Large\diagdown\!\!\!\diagup}} - \nabla_{\text{\Large\diagup\!\!\!\diagdown}} = z \nabla_{\text{\Large)(}} \, .$$

Thus ∇_K seems to require $n = -1$, a zero dimensional spin set! In fact ∇_K has no direct interpretation in this series of models, **but** there is a Yang-Baxter model for it via a different R-matrix. We shall take up this matter in the next section.

To conclude this section, I will show that $\langle K \rangle$ really needed an equally spaced spin set in the form $\mathcal{I} = \{-a, -a + \Delta, -a + 2\Delta, \dots, a - \Delta, a\}$ in order to satisfy $\langle \text{\scriptsize◯→} \rangle = \alpha \langle \text{\scriptsize→} \rangle$ for some α. Thus the demand for multiplicativity under the type I move focuses these models into the specific sequence of regular isotopy invariants that we have just elucidated.

Proposition 11.5. Let $\langle K \rangle$ denote the state model of this section, with independent variables z, δ and arbitrary ordered index set $\mathcal{I} = \{N_0 < N_1 < \dots < N_m\}$. Suppose that $\langle K \rangle$ is simple in the sense that there exists an algebraic element α such that

$$\left\langle \text{\scriptsize◯→} \right\rangle = \alpha \left\langle \text{\scriptsize→} \right\rangle \qquad \text{and}$$

$$\left\langle \text{\scriptsize◯→} \right\rangle = \alpha^{-1} \left\langle \text{\scriptsize→} \right\rangle .$$

Then all the gaps $N_{k+1} - N_k$ are equal **and** $N_0 = -N_m$.

Proof.

$$\left\langle \vcenter{\hbox{}} \right\rangle = z\left\langle \vcenter{\hbox{}} \right\rangle + q\left\langle \vcenter{\hbox{}} \right\rangle$$

$$= z\sum_{a>b}\delta^{-b}\left\langle \vcenter{\hbox{}} \right\rangle + q\sum_{a}\delta^{-a}\left\langle \vcenter{\hbox{}} \right\rangle$$

$$= \sum_{a}\left(\sum_{a>b}z\delta^{-b} + q\delta^{-a}\right)\left\langle \vcenter{\hbox{}} \right\rangle.$$

In order for the model to be simple, we need that all of these coefficients are equal:

$$\alpha = q\delta^{-N_0}$$
$$= (q - q^{-1})\delta^{-N_0} + q\delta^{-N_1}$$
$$= \ldots$$
$$= (q - q^{-1})\sum_{i=0}^{m-1}\delta^{-N_i} + q\delta^{-N_m}.$$

Thus

$$q\delta^{-N_0} = q\delta^{-N_0} - q^{-1}\delta^{-N_0} + q\delta^{-N_1}$$
$$\Rightarrow q^{-1}\delta^{-N_0} = q\delta^{-N_1}$$
$$\Rightarrow q^2 = \delta^{N_1 - N_0}.$$

Similarly, $q^2 = \delta^{N_{k+1} - N_k}$, $k = 0, \ldots, m-1$. This completes the proof. //

12^0. The Alexander Polynomial.

The main purpose of this section is to show that a Yang-Baxter model essentially similar to the one discussed in section 11^0 can produce the Alexander-Conway polynomial.

Along with this construction a number of interesting relationships emerge - both about solutions to the Yang-Baxter Equation and about the classical viewpoints for the Alexander polynomial.

The model discussed in this section is the author's re-working [LK23] of a model for the Alexander polynomial, discovered by Francois Jaeger [JA4] (in ice-model language). Other people (Wadati [AW3], H. C. Lee [LEE1]) have seen the relevance of the R-matrix we use below to the Alexander polynomial. In [LK18] and [JA8] it is shown how this approach is related to free-fermion models in statistical mechanics. (See also the exercise at the end of section 13^0.) Also, see [MUK2] for a related treatment of the multi-variable Alexander polynomial. The R-matrix itself first appeared in [KUS], [PS].

The first state model for a knot polynomial was the author's FKT model [LK2] for the Alexander-Conway polynomial. The FKT-model gives a (non Yang-Baxter) state model for the (multi-variable) Alexander-Conway polynomial. Its structure goes back to Alexander's original definition of the polynomial. It came as something of a surprise to realize that there was a simple Yang-Baxter model for the Alexander polynomial, and that this model was also related to the classical topology. This connection will be drawn in section 13^0.

I shall begin by giving the model and stating the corresponding Yang-Baxter solution. We shall then backtrack across a number of the related topics.

The Yang-Baxter Model.

Recall that the Conway version of the one-variable Alexander polynomial is determined by the properties

(i) $\nabla_K = \nabla_{K'}$ whenever K is ambient isotopic to K'.

(ii) $\nabla_O = 1$

(iii) $\nabla_{\overcrossing} - \nabla_{\undercrossing} = z \nabla_{\smoothing}$

One important consequence of these axioms is that ∇_K **vanishes on split links** K. For suppose K is split. Then it can be represented diagrammatically as

 $K.$

(Take this as a definition of **split**.) Then we have

$$\nabla_A - \nabla_B = z\nabla_K$$

where

$$A = \left\{ \vbox{} \right\} \quad \text{and} \quad \left\{ \vbox{} \right\} = B.$$

Since A and B are ambient isotopic (via a 2π twist), we conclude that $z\nabla_K = 0$, whence $\nabla_K = 0$.

This vanishing property seems to raise a problem for state models of the form

$$\langle K \rangle = \sum_\sigma \langle K|\sigma\rangle \delta^{\|\sigma\|}$$

since these models have the property that $\langle\, \bigcirc\ K \,\rangle = \langle\, \bigcirc\, \rangle\langle K \rangle$ and we would need therefore that $\langle\, \bigcirc\, \rangle = 0$ for this equation to hold - sending the whole model to zero!

In principle, there is a way out (and the idea works for Alexander - as we shall see). We go back to the idea of knot theory as the study of knots (links) on a string with end-points:

One is allowed ambient or regular isotopy **keeping the endpoints fixed** and so that no movement is allowed past the endpoints. More precisely, the diagrams all live in strip $\mathbf{R} \times I$ for a finite interval I. A link L is represented as a single-input, single-output tangle with one end attached to $\mathbf{R} \times 0$, and the other end attached to $\mathbf{R} \times 1$. Only the end-points of L touch $(\mathbf{R} \times 0) \cup (\mathbf{R} \times 1)$. All Reidemeister moves are performed inside $\mathbf{R} \times I$ (i.e. in the interior of $\mathbf{R} \times I$).

This two-strand tangle theory of knots and links is equivalent to the usual theory (exercise) - and I shall use it to create this model. For we can now imagine a model where

$$\left\langle \rightsquigarrow \right\rangle = 1$$

but

$$\left\langle \bigcirc \right\rangle = 0 \text{ and } \left\langle \rightsquigarrow \right\rangle = 0.$$

The split link property is possible in this context.

[An n-**strand tangle** is a link diagram with n free ends; usually the free ends are configured on the outside of a **tangle-box** (a rectangle in the plane) with the rest of the diagram confined to the interior of the box.]

To produce the model explicitly, we need a Yang-Baxter solution. Deus ex machina, here it is:

Let $\mathcal{I} = \{-1, +1\}$ be the index set. Let

$$R = (q - q^{-1}) \;\bigg)\bigg(\; + q \;\bigg)\bigg(\; - q^{-1} \;\bigg)\bigg(\; + \;\bigg\times$$

$$\overline{R} = (q^{-1} - q) \;\bigg)\bigg(\; + q^{-1} \;\bigg)\bigg(\; - q \;\bigg)\bigg(\; + \;\bigg\times$$

Here we use the same shorthand as in section 11^0. Thus

$$\left. \begin{matrix} a & & b \\ & \cdot & \\ c & & d \end{matrix} \right. = \begin{cases} 1 & \text{if } a = c = +1 \text{ and} \\ & \quad b = d = -1 \\ 0 & \text{otherwise.} \end{cases}$$

Note that

$$R - \overline{R} = (q - q^{-1}) \left[\;\big)\big(\; + \;\big)\big(\; + \;\big)\big(\; + \;\big)\big(\; \right]$$

$$= (q - q^{-1}) \left[\;\big)\big(\; \right].$$

Thus this tensor is a candidate for a model satisfying the identity

$$\nabla\!\!\!\!\diagdown\!\!\!\diagup - \nabla\!\!\!\!\diagdown\!\!\!\diagup = z\nabla\,)\,($$

if we associate R to positive crossings, and $\overline{R}$ to negative crossings. That R and $\overline{R}$ are indeed solutions to the Yang-Baxter Equation, and inverses of each other can be checked directly. [See also the Appendix where we derive this solution in the course of finding all spin preserving solutions to YBE for $\mathcal{I} = \{-1, +1\}$].

Therefore, suppose that we do set up a model

$$\langle K \rangle \doteq \sum_{\sigma} \langle K|\sigma\rangle \delta^{\|\sigma\|}$$

using this R-matrix and $\mathcal{I} = \{-1, +1\}$ with the norms $\|\sigma\|$ calculated by labelling a rotation exactly as in section 11^0. We want $\langle \curvearrowright \rangle = \langle \curvearrowright \rangle$, but a first pass is to try for **simplicity**: $\langle \curvearrowright \rangle = \alpha \langle \curvearrowright \rangle$ and see if that restriction determines the structure of the model. Therefore, we must calculate:

$$\left\langle \curvearrowright \right\rangle = (q - q^{-1})\left\langle \curvearrowright \right\rangle + q\left\langle \curvearrowright \right\rangle - q^{-1}\left\langle \curvearrowright \right\rangle$$

$$= (q - q^{-1})\delta^{-1}\left\langle \rightarrow \right\rangle + q\delta^{-1}\left\langle \rightrightarrows \right\rangle - q^{-1}\delta\left\langle \rightarrow \right\rangle$$

$$= [(q - q^{-1})\delta^{-1} - q^{-1}\delta]\left\langle \rightarrow \right\rangle + q\delta^{-1}\left\langle \rightrightarrows \right\rangle.$$

Thus, we demand that

$$(q - q^{-1})\delta^{-1} - q^{-1}\delta = q\delta^{-1}.$$

Whence

$$q^{-1}(\delta^{-1} + \delta) = 0.$$

Therefore $\delta^{-1} + \delta = 0$, and we shall take $\delta = i$ where $i^2 = -1$.

It is now easy to verify that the model is multiplicative for all choices of curl. In fact, it also does not depend upon the crossing type:

Lemma 12.1. With $\delta = i$ ($i^2 = -1$) and $\langle K \rangle$ as described above, we have

(i) $$\left\langle \vartheta^{\!\nearrow} \right\rangle = -iq \left\langle \rightarrow \right\rangle$$

(ii) $$\left\langle \neg\mathcal{O}^{\!\nearrow} \right\rangle = -iq \left\langle \rightarrow \right\rangle$$

(iii) $$\left\langle \mathcal{D}^{\!\rightarrow} \right\rangle = iq^{-1} \left\langle \rightarrow \right\rangle$$

(iv) $$\left\langle \neg\mathcal{C}_{\!\rightarrow} \right\rangle = iq^{-1} \left\langle \rightarrow \right\rangle$$

Proof. We verify (iii) and leave the rest for the reader.

$$\left\langle \mathcal{D}^{\!\rightarrow} \right\rangle = (q^{-1} - q) \left\langle \underset{-}{\overset{\mathcal{O}_+}{\rightarrow}} \right\rangle + q^{-1} \left\langle \underset{+}{\overset{\mathcal{O}_+}{\rightarrow}} \right\rangle - q \left\langle \underset{-}{\overset{\mathcal{O}_-}{\rightarrow}} \right\rangle$$

$$= ((q^{-1} - q)i - qi^{-1}) \left\langle \overset{\frown}{\underset{-}{\rightarrow}} \right\rangle + q^{-1}i \left\langle \overset{\frown}{\underset{+}{\rightarrow}} \right\rangle$$

$$= q^{-1}i \left\langle \rightarrow \right\rangle . \qquad\qquad //$$

The lemma implies that if

$$\nabla_K = (iq^{-1})^{-\mathrm{rot}(k)} \langle K \rangle,$$

then $\nabla_{\vartheta^{\!\nearrow}} = \nabla_{\rightarrow}$ for a curl. Certainly $\nabla_{\!\nwarrow\!\!\nearrow} - \nabla_{\!\nearrow\!\!\nwarrow} = (q - q^{-1})\nabla_{\rightarrow}$. It remains to verify that ∇_K is invariant under the type IIB move. Once we verify this, ∇_K becomes an invariant of regular isotopy and hence ambient isotopy - via the curl invariance.

Note that we shall assign rotation number zero to the bare string:

$$\mathrm{rot}(\bullet\!\!\longrightarrow\!\!\bullet) = 0.$$

Also, we assume that $\langle \bullet\!\overset{+}{\longrightarrow}\!\bullet \rangle + \langle \bullet\!\overset{-}{\longrightarrow}\!\bullet \rangle = \langle \bullet\!\longrightarrow\!\bullet \rangle = 1$. The model now has the form

$$\langle K \rangle = \sum_{\sigma} \langle K | \sigma \rangle i^{\|\sigma\|}$$

where $i^2 = -1$, $\langle K | \sigma \rangle$ is the product of vertex weights determined by the R-matrix, and $\|\sigma\|$ is the rotational norm for index set $\mathcal{I} = \{-1, +1\}$.

Proposition 12.2. With $\langle K \rangle$ defined as above,

$$\left\langle \mathbf{\Large\bowtie} \right\rangle = \left\langle \mathbf{\Large)(} \right\rangle.$$

Proof. Consider the terms in the expansion of $\left\langle \mathbf{\Large\bowtie} \right\rangle$. The contributing forms are

and

Hence

$$\left\langle \mathbf{\Large\bowtie} \right\rangle = qq^{-1} \left\langle \right\rangle + (-q^{-1})(-q) \left\langle \right\rangle$$

$$+ zq^{-1} \left\langle \right\rangle + (-q^{-1})(-z) \left\langle \right\rangle$$

$$+ \left\langle \right\rangle + \left\langle \right\rangle$$

$$= i^{-1} \left\langle \right\rangle + i \left\langle \right\rangle + \left\langle \right\rangle$$

$$= i^{-1} \left\langle \right\rangle + i \left\langle \right\rangle$$

$$- \left\langle \right\rangle - \left\langle \right\rangle$$

$$+ \left\langle \mathbf{)(} \right\rangle.$$

The last step follows from the identity

$$\left\langle \right\rangle = \left\langle \right\rangle - \left\langle \right\rangle.$$

I now assert the following two identities:

180

$$i^{-1}\left\langle \overset{+}{\underset{\longleftarrow}{\overset{\longrightarrow}{+}}} \right\rangle - \left\langle \overset{+}{\downarrow}\,\overset{+}{\uparrow} \right\rangle = 0$$

and

$$i\left\langle \overset{\longrightarrow}{\underset{\longleftarrow}{}} \right\rangle - \left\langle \downarrow\,\overset{+}{\uparrow} \right\rangle = 0.$$

$\bigg[$These are trivial to verify by checking cases. Thus

$$\left\langle \begin{array}{c}\text{+}\\ \text{+}\end{array} \right\rangle = i^2 = -1$$

$$\left\langle \,\text{+}\, \right\rangle = i$$

$$\Rightarrow i^{-1}\left\langle \begin{array}{c}\text{+}\\ \text{+}\end{array} \right\rangle - \left\langle \,\text{+}\, \right\rangle = -i^{-1} - i = 0$$

and

$$\left\langle \overset{+}{\longrightarrow}\,\text{+} \right\rangle = i\left\langle \overset{+}{\longrightarrow}\bullet \right\rangle$$

$$\left\langle \text{+}\, \right\rangle = \left\langle \bullet\overset{\longrightarrow}{\underset{+}{}} \right\rangle$$

$$\Rightarrow i^{-1}\left\langle \overset{+}{\longrightarrow}\,\text{+} \right\rangle - \left\langle \,\text{+}\, \right\rangle = 0.\bigg]$$

Note that we must consider separately those cases when one of the local arcs is part of the input/output string.

As a result of these last two identities, we conclude that

$$\left\langle \overset{\longrightarrow}{} \right\rangle = \left\langle \downarrow\,\uparrow \right\rangle.$$

This completes the proof. $\qquad\qquad$ //

We have now completed the proof that $\langle K \rangle = \sum_{\sigma} \langle K | \sigma \rangle i^{\|\sigma\|}$ is a regular isotopy invariant of single input/output tangles. It follows that $\nabla_K = (iq^{-1})^{-\mathrm{rot}K} \langle K \rangle$ is an invariant of ambient isotopy for these tangles, and that

1. $\nabla_{\text{⤢}} - \nabla_{\text{⤡}} = z \nabla_{\text{⇉}}$
2. $\nabla_{\text{◠}} = \nabla_{\text{⤻}} + \nabla_{\text{⤸}} = 1$
   $\nabla_{\text{◯}} = 0$.

Having verified these axioms, we see that $\nabla_K = \nabla_K^{+} \nabla_{\text{⤻}} + \nabla_K^{-} \nabla_{\text{⤸}}$ being free (by using 1. and 2. recursively) of $\nabla_{\text{⤻}}$ or $\nabla_{\text{⤸}}$, it follows that $\nabla_K^{+} = \nabla_K^{-} = \nabla_K$. In other words, the coefficients of $\langle \text{⤻} \rangle$ and $\langle \text{⤸} \rangle$ in the state expansion of $\langle K \rangle$ are identical. For calculation this means that we can assume that all states have (say) positive input and positive output:

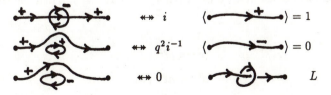

Since we already know that these axioms will compute the Conway polynomial of $\overline{K}$ where $\overline{K}$ is the closure of the tangle K:

$$\text{⟶}\boxed{K}\text{⟶} \qquad \text{⟨}\!\blacksquare\!\text{⟩} \quad \overline{K},$$

we conclude that ∇_K is indeed the Conway polynomial.

Finally, we remark that ∇_K will give the same answer from a knot $\overline{K}$ for any choice of associating a tangle K to $\overline{K}$ by dropping a segment. I assume this segment is dropped to produce a planar tangle in the form $\text{⟶}\blacksquare\text{⟶}$, fitting our constructions. The independence of segment removal again follows from the corresponding skein calculation using properties 1. and 2.

In fact, the model's independence of this choice can be verified directly and combinatorially in the form of the state summation. I leave this as an exercise for the reader.

Example.

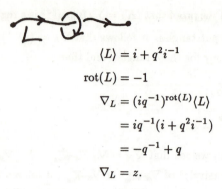

$$\langle L \rangle = i + q^2 i^{-1}$$

$$\mathrm{rot}(L) = -1$$

$$\nabla_L = (iq^{-1})^{\mathrm{rot}(L)}\langle L \rangle$$

$$= iq^{-1}(i + q^2 i^{-1})$$

$$= -q^{-1} + q$$

$$\nabla_L = z.$$

There are many mysteries about this model for the Conway polynomial. Not the least of these is the genesis of its R-matrix. In the Appendix we classify all spin preserving solutions of the Yang-Baxter Equation for the spin set $\mathcal{I} = \{-1, +1\}$. Such solutions must take the form

$$R = \ell \; \Bigr)\!\!\Bigl(\; + r \; \Bigr)\!\!\Bigl(\; + p \; \Bigr)\!\!\Bigl(\; + n \; \Bigr)\!\!\Bigl($$

$$+ d \; \times \; + s \; \times$$

for some coefficients ℓ, r, p, n, d, and s. We show in the appendix that the following relations are necessary and sufficient for R to satisfy the YBE:

$$
\begin{array}{c}
r\ell d = 0 \\
r\ell s = 0 \\
r\ell(\ell - r) = 0 \\
p^2 \ell = p\ell^2 + \ell ds \\
n^2 \ell = n\ell^2 + \ell ds \\
p^2 r = pr^2 + rds \\
n^2 r = nr^2 + rds
\end{array}
$$

One natural class of solutions arises from $d = s = 1$, $r = 0$. Then the conditions reduce to:

$$p^2 \ell = p\ell^2 + \ell$$

$$n^2 \ell = n\ell^2 + \ell.$$

Assuming n and p invertible, we get $\ell = p - p^{-1}$ and $\ell = n - n^{-1}$. Thus $p - p^{-1} = n - n^{-1}$.

$$\Leftrightarrow np^2 - n = n^2 p - p$$

$$\Leftrightarrow \boxed{np(p-n) = n - p}.$$

Thus we see that there are **two** basic solutions:

1. $n = p$
2. $np = -1$.

The first gives the R-matrix

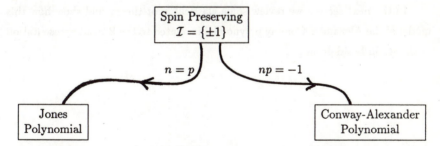

$$R = (p - p^{-1}) \left(\cdots + p \cdots = \cdots + \cdots \right)$$

and gives the Jones polynomial as in section 11^0.

The second gives the R-matrix

$$R = (p - p^{-1}) \left(\cdots + p \cdots - p^{-1} \cdots + \cdots \right)$$

and yields the Conway-Alexander polynomial as we have seen in this chapter. Thus from the viewpoint of the YBE we have the diagram

Spin Preserving
$\mathcal{I} = \{\pm 1\}$

$n = p$

$np = -1$

Jones
Polynomial

Conway-Alexander
Polynomial

In this sense the Jones and Conway polynomials have equal footing. But in another sense they are very different and the difference is apparent in the state models.

Let $J\langle K \rangle$ denote the oriented state model for the Jones polynomial as constructed in section 11^0. Then

$$J\langle K \rangle = \sum_{\sigma} J\langle K | \sigma \rangle t^{\|\sigma\|}$$

where t is an algebraic variable. Let $A\langle K \rangle$ denote the model in this chapter. Then

$$A\langle K \rangle = \sum_{\sigma} A\langle K | \sigma \rangle i^{\|\sigma\|}$$

where $i^2 = -1$.

That the loop term is numerical ($i^{\|\sigma\|}$) for the Conway polynomial, and algebraic ($t^{\|\sigma\|}$) for the Jones polynomial, is the first visible difference between these models.

There are many results known about the Alexander polynomial via its classical definitions. The classical Alexander polynomial is denoted $\Delta_K(t)$ and it is related to the Conway version as $\Delta_K(t) \doteq \nabla_K(t^{1/2} - t^{-1/2})$ where $\doteq$ denotes equality up to sign and powers of t. Since the model $A\langle K \rangle$ of this section is expressed via $z = q - q^{-1}$ we have that $\Delta_K(q^2) \doteq A\langle K \rangle(q)$. One problem about this model is particularly interesting to me. We know [FOX1] that if K is a **ribbon knot**, then $\Delta_K(t) \doteq f(t)f(t^{-1})$ where $f(t)$ is a polynomial in t. Is there a proof of this fact using the state model of this chapter? A ribbon knot is a knot that bounds a disk immersed in three space with only **ribbon singularities**. In a ribbon singularity the disk intersects itself transversely, matching two arcs. One arc is an arc interior to the disk; the other goes from one boundary point to another.

In the next section we review some classical knot theory and show how this model of the Alexander-Conway polynomial is related to the Burau representation of the Artin braid group.

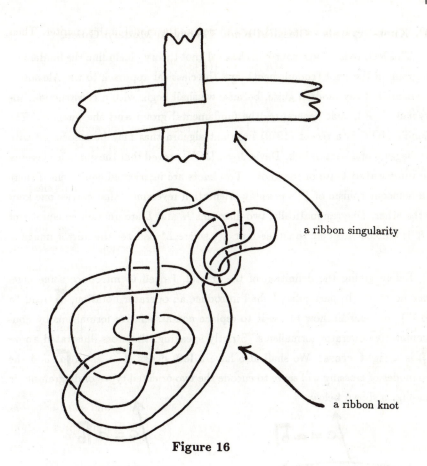

a ribbon singularity

a ribbon knot

Figure 16

13⁰. Knot-Crystals - Classical Knot Theory in Modern Guise.

This lecture is a short course in classical knot theory - including the fundamental group of the knot (complement), and the classical approach to the Alexander polynomial. I say modern guise, because we shall begin with a construction, the **crystal**, $C(K)$, that generalizes the fundamental group **and** the quandle. The quandle [JOY] is a recent (1979) invariant algebra that may be associated with the diagram of a knot or link. David Joyce [JOY] proved that the quandle classifies the **unoriented type** of the knot. (Two knots are unoriented equivalent if there is a homeomorphism of S^3 - possibly orientation reversing - that carries one knot to the other. Diagrammatically, two diagrams K and L are unoriented equivalent if K is ambient isotopic to either L or L^*, where L^* denotes the mirror image of L.)

Before giving the definition of the crystal, I need to introduce some algebraic notation. In particular, I shall introduce an operator notation, $a\rceil$ (read "a cross"), and explain how to use it to replace non-associative formalisms by non-commutative operator formalisms. Strictly speaking, the cross illustrated above ($\lceil$) is a **right cross**. We shall also have a **left cross** ($\lceil$: as in $\lceil a$) and the two modes of crossing will serve to encode the two oriented types of diagrammatic crossing as shown below:

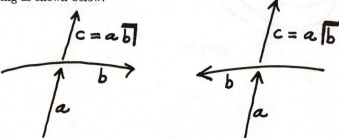

In this formalism, each arc in the oriented link diagram K is labelled with a letter $(a, b, c, \dots)$. In this mode I shall refer to the arc as an **operand**. The arc can also assume the **operator mode**, and will be labelled by a left or right cross ($a\rceil$ or $\lceil a$) according to the context (left or right handed crossing).

As illustrated above, we regard the arc c emanating from an undercrossing as the result of the overcrossing arc b, acting ($b\rceil$ or $\lceil b$) on the incoming undercrossing arc a. Thus we write $c = ab\rceil$ or $c = a\lceil b$ according to whether the crossing is of right or left handed type.

At this stage, these notations do not yet abide in an algebraic system. Rather, they constitute a **code** for the knot. For example, in the case of the trefoil, we obtain three code-equations as shown below.

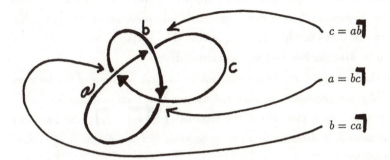

$$c = ab\rceil$$
$$a = bc\rceil$$
$$b = ca\rceil$$

What is the simplest algebra compatible with this notation and giving a regular isotopy invariant of oriented knots and links?

In order to answer this question, let's examine how this formalism behaves in respect to the Reidemeister moves. First, consider the type II move:

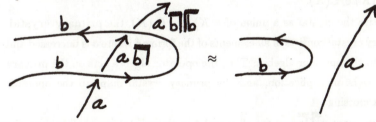

From this we see that type II invariance corresponds to the rules

$$ab\rceil\lceil b = a \qquad \text{and} \qquad a\lceil b\ b\rceil = a$$

for any a and b. We also see that in order to form an algebra it will be necessary to be able to multiply x and $y\rceil$ to form $xy\rceil$ and to multiply x and $\lceil y$ to form $x\lceil y$ for **any** two elements x and y in the algebra.

Example.

The stipulation that we can form $xy\rceil$ for any two elements x and y means that $xy\rceil$ does not necessarily have a direct geometric interpretation on the given diagram. Note also that the multiplication, indicated by juxtaposition of symbols, is assumed to be **associative and non-commutative**.

Since the knot diagrams do not ever ask us to form products of the form xy where x and y are uncrossed, we shall not assume that such products live in the crystal. However, since products of the form $a_0 b_1\rceil\ b_2\rceil\ b_3\rceil \ldots b_k\rceil$ occur naturally with a_0 uncrossed, it **does** make sense to consider isolated products of crossed elements such as

$$a\rceil\ bc\rceil\ d\rceil$$

Call these **operator products**. Let 1 denote the empty word. The subalgebra of the crystal $C(K)$ generated by operator products will be called the **operator algebra** $\pi(K) \subset C(K)$.

We write the crystal as a union of $\pi(K)$ and $C_0(K)$, the **primary crystal**. The primary crystal consists of all elements of the form $a\beta$ where a is uncrossed and β belongs to the operator algebra. Thus the operator algebra acts on the primary crystal by right multiplication, and the primary crystal maps to the operator algebra via crossing.

Since we want $ab\rceil\ \lceil b = a$ for all a and b in the crystal, we shall take as the first axiom (labelled II. for obvious reasons) that

II. $$b\rceil\ \lceil b = \lceil b\ b\rceil = 1$$

for all $b \in C(K)$. Note that this equality refers to $\pi(K)$, and that therefore $\pi(K)$ is a group. We shall see that, with the addition of one more axiom, $\pi(K)$ is isomorphic with the classical fundamental group of the link K.

In order to add the remaining axiom, we must analyze the behavior under the type III move. I shall use the following move, dubbed the **detour**:

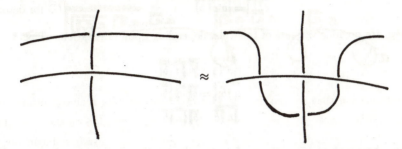

It is easy to see that, **in the presence of the type II move**, the detour is equivalent to the type III move:

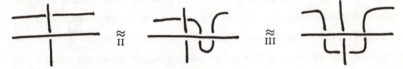

Now examine the algebra of the detour:

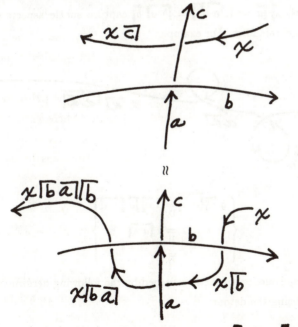

Thus, we see that the axiom should be: **When** $c = a\overline{b}\rceil$, **then** $\overline{c}\rceil = \overline{\overline{b}\ a}\rceil\ \overline{b}\rceil$.

190

Or, more succinctly:

D. $$ab\rceil = \lceil b \; a \rceil \; b\rceil$$ (D for detour)

There are obvious variants of this rule, depending upon orientations. Thus

$$\lceil a \; b\rceil = b\rceil \lceil a \; b\rceil$$
$$\overline{a \lceil b\rceil} = b\rceil \; a\rceil \lceil b$$
$$\lceil a \lceil b = b\rceil \lceil a \lceil b$$

From the point of view of crossing, the detour axiom is a **depth reduction rule**. It takes an expression of (nesting) depth two $(a \; \overline{b}\rceil\,)$ and replaces it by an expression of depth one $(\lceil b \; a\rceil \; b\rceil)$. Note the pattern: The crossed expression $(\overline{b}\,)$ inside is duplicated on the right in the same form as its inside appearance, and duplicated on the left in reversed form. The crossed expression $(\overline{b}\,)$ is removed from the middle, leaving the a with a cross standing over it whose direction corresponds to the original outer cross on the whole expression.

These rules, $a\rceil \; \lceil a = 1$, $a \; \overline{b}\rceil\rceil = \lceil b \; a\rceil \; b\rceil$ capture subtle aspects of regular homotopy in algebraic form. For example:

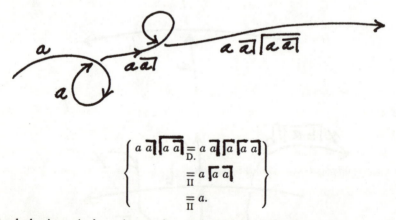

$$\left\{\begin{array}{l} a \; a\rceil\lceil a \; a\rceil \underset{\text{D.}}{=} a \; a\rceil\lceil a \lceil a \; a\rceil \\[4pt] \underset{\text{II}}{=} a \lceil a \; a\rceil \\[4pt] \underset{\text{II}}{=} a. \end{array}\right\}$$

This algebraic equivalence is exactly matched by the following performance of the Whitney trick, using the detour:

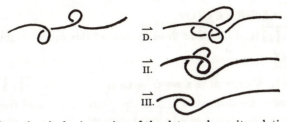

Another look at the algebraic version of the detour shows its relationship with the fundamental group. For in the Wirtinger presentation of $\pi_1(S^3 - K)$ [ST] there is one relation for each crossing and one generator for each arc. The relation is $\gamma = \beta^{-1}\alpha\beta$ when the crossing is positive as shown below:

$$\gamma = \beta^{-1}\alpha\beta.$$

If we are working in the crystal with crossing labels as shown below

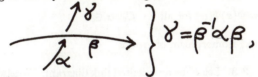

then we have the relation on operators

$$c\rceil = \lceil b \; a\rceil \; b\rceil.$$

This is recognizable as

$$\gamma = \beta^{-1}\alpha\beta$$

with $\gamma = c\rceil$, $\alpha = a\rceil$, $\beta = b\rceil$. This is the standard relation at a crossing in the Wirtinger presentation of the fundamental group of the link complement. **Thus the group of operators, $\pi(K)$, is isomorphic to the fundamental group of the link complement.**

It should be clear by now how to make a **formal definition of a crystal.** We have the **notations $a\rceil$** and $\lceil a$ for maps $f, g : C_0 \to \pi$. $f(x) = x\rceil$, $g(x) = \lceil x$. π is a group with binary operation denoted $x, y \mapsto xy$. π **acts on C_0 on the right** via

$$C_0 \times \pi \to C_0$$

$$a, \lambda \mapsto a\lambda$$

such that $(a\lambda)\mu = a(\lambda\mu)$. And it is given that

II. $\overline{x}\,\lceil x = 1 \in \pi$ for all $x \in C_0 \cup \pi$.

D. $\overline{x\ \overline{y}} = \overline{y}\ \overline{x}\ \overline{y}$ (and the other three variants of this equation) for all $x, y \in C_0 \cup \pi$.

Definition 13.1. A **crystal** is a set $C = C_0 \cup \pi$ with maps $\rceil,\ \lceil\, : C_0 \to \pi$ such that C_0 and π are disjoint, π is a group acting on C_0, and the maps from C_0 to π satisfy the conditions II. and D. above. Note that D. (above) becomes $\overline{a\alpha} = \alpha^{-1}\overline{a}\,\alpha$ and $\lceil a\alpha = \alpha^{-1}\lceil a\ \alpha$ for $\alpha \in \pi$, $a \in C_0$.

Definition 13.2. Two crystals C and C' are **isomorphic** if there is a 1-1 onto map $\Phi : C \to C'$ such that $\Phi(C_0) = C_0'$, $\Phi(\pi) = \pi'$ and

1. $\overline{\Phi(x)} = \Phi(\overline{x})$
   $\lceil \Phi(x) = \Phi(\lceil x)$ for all $x \in C_0$.

2. $\Phi(x\alpha) = \Phi(x)\Phi(\alpha)$ for all $x \in C_0$, $\alpha \in \pi$

3. $\Phi \mid \pi$ is an isomorphism of groups.

Definition 13.3. Let K be an oriented link diagram. We associate the **link crystal** $C(K)$, to K by assigning one generator of $C_0(K)$ for each arc in the diagram, and one relation (of the form $c = a\overline{b}$ or $c = a\lceil b$) to each crossing. The resulting structure is then made into a crystal in the usual way (of universal algebra) by allowing all products (as specified above) and taking equivalence classes after imposing the axioms.

Example.
$$C(K) = (a, b, c \mid c = a\overline{b},\, b = c\overline{a},\, a = b\overline{c}).$$

Here I use the abbreviation

$$C(K) = (a_1, a_2, \ldots, a_n \mid r_1, r_2, \ldots, r_m)$$

to denote the free crystal on $a_1, \ldots, a_n$ modulo the relations $r_1, \ldots, r_m$.

Because it contains information about how the fundamental group $\pi(K)$ acts on the "peripheral" elements $C_0(K)$, the crystal has more information about the link than the fundamental group alone. In fact, if two knots K and K' have isomorphic crystals, then K is ambient isotopic to either K' or the mirror image K'^*. This follows from our discussion below, relating the crystal to David Joyce's Quandle [JOY]. But before we do this, we will finish the proof of the

Theorem 13.4. Let K and K' be diagrams for two oriented links. If K is regularly isotopic to K' then $C(K)$ is isomorphic to $C(K')$. If K is ambient isotopic with K', then $\pi(K)$ and $\pi(K')$ are isomorphic.

Proof. Since $\overline{a\ \overline{a}|} = \lceil\overline{a}\ \overline{a}\rceil\ \overline{a}] = \overline{a}]$ we see

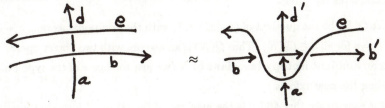

that $\pi(K)$ is invariant under Reidemeister moves of type I. $[\mathcal{O}(K)$ is not invariant under type I, since we do not insist that $a\overline{a}]$ is equal to a.] It remains to check invariance for the "over" version of the detour.

Referring to the diagram above, we have $d = a\overline{b}]\ \lceil\overline{e}$ and $d' = a\lceil e\ b\ \lceil e\ \rceil$, $b' = b\lceil e\ \overline{e}]$. Thus

$$d' = a\lceil e\ \lceil\overline{b}\lceil e = a\lceil e\ e\rceil\ \overline{b}\rceil\lceil e$$
$$= a\overline{b}]\ \lceil\overline{e}$$
$$d' = d.$$

This shows the remaining case (modulo a few variations of orientation) for the invariance of $C(K)$ under the moves of type II and III. Hence $C(K)$ is a regular isotopy invariant. //

The Quandle.

Let K be an oriented link, and let $C(K)$ be its crystal. Define binary operations $*$ and $\overline{*}$ on $C_0(K)$ as follows:

$$a * b = a\overline{b}]$$
$$a \,\overline{*}\, b = a\lceil b.$$

Lemma 13.5. $(a * b) \bar{*} b = a$

$\qquad\qquad (a * b) * c = (a * c) * (b * c)$

(The same identities hold if $*$ and $\bar{*}$ are interchanged.)

Proof.

$$(a * b) \bar{*} b = a\overline{b}\,\overline{b} = a$$

$$(a * b) * c = a\overline{b}\,\overline{c}$$

$$(a * c) * (b * c) = a\overline{c}\,\overline{b\,\overline{c}}$$

$$= a\overline{c}\,\overline{c}\,\overline{b}\,\overline{c}$$

$$= a\overline{b}\,\overline{c}$$

$$= (a * b) * c.$$

This completes the proof. $\qquad\qquad\qquad\qquad\qquad\qquad\qquad //$

Let $Q(K)$ be the quotient of $(C_0(K), *, \bar{*})$ with the additional axioms $a * a = a$, $a \bar{*} a = a$ for all $a \in C_0(K)$. Thus $Q(K)$ is an algebra with two binary operations that is an **ambient isotopy invariant of** K (we just took care of the type I move by adding the new axioms.)

It is easy to see that $Q(K)$ is the quotient of the set of all formal products of arc labels $\{a_1, \ldots, a_n\}$ for K under $*$ and $\bar{*}$ (these are non-associative products) by the **axioms of the quandle** [JOY].

1. $a * a = a$, $a \bar{*} a = a$ for all a.

2. $(a * b) \bar{*} b = a$

   $(a \bar{*} b) * b = a$ for all a and b.

3. $(a * b) * c = (a * c) * (b * c)$

   $(a \bar{*} b) \bar{*} c = (a \bar{*} c) \bar{*} (b \bar{*} c)$.

Thus $Q(K)$ is identical with Joyce's quandle. This shows that the crystal $C(K)$ classifies oriented knots up to mirror images.

The quandle is a non-associative algebra. In the crystal, we have avoided non-associativity via the operator notation:

$$(a * b) * c = a b\,\overline{}\,\overline{c}$$

$$a * (b * c) = a b\,\overline{c}\,\overline{}$$

A fully left associated product in the quandle (such as $((a*b)*c)*d$) corresponds to an operator product of depth 1:

$$((a*b)*c)*d = a\overline{b}\,\overline{c}\,\overline{d}$$

while other associations can lead to a nesting of operators such as

$$a*(b*(c*d)) = a\ b\ c\ \overline{\overline{\overline{d}}}$$

It might seem that the quandle is very difficult to deal with because of this multiplicity of associated products. However, we see that the second basic crystal axiom $(\overline{a\alpha} = \alpha^{-1}\overline{a}\ \alpha,\ a \in \mathcal{C},\ \alpha \in \pi)$ shows at once that

Theorem 13.6. (Winker [WIN]) Any quandle product can be rewritten as a left-associated product.

This theorem is the core of Winker's Thesis. It provides canonical representations of quandle elements and permits one to do extensive analysis of the structure of knot and link quandles. Winker did not have the crystal formalism, and so proved this result directly in the quandle. The crystalline proof is given below.

Proof of Winker's Theorem. It is clear that the rules $\overline{a\alpha} = \alpha^{-1}\overline{a}\ \alpha$ and $\underline{a\alpha} = \alpha^{-1}\ \underline{a\alpha}$ for $a \in \mathcal{C}_0,\ \alpha \in \pi$ give depth reduction rules for any formal product in $\mathcal{C}_0$. When a product has been reduced to depth 0 or depth 1 then it corresponds to a left-associated product in the quandle notation. This completes the proof.//

Example.

$$a*(b*c) = a\ b\ \overline{\overline{c}}$$
$$= a\ \underline{c}\ \overline{b}\ \overline{c}$$
$$= ((a*c)*b)*c$$
$$a*(b*(c*d)) = a\ b\ c\ \overline{\overline{\overline{d}}}$$
$$= a\ b\ \overline{\underline{d}\ \overline{c}\ \overline{d}}$$
$$= a\ \underline{d}\ \underline{c}\ \overline{d}\ \overline{b}\ \underline{d}\ \overline{c}\ \overline{d}$$
$$= ((((((a \mp d) \mp c) * d) * b) \mp d) * c) * d.$$

Remark. The quandle axioms are also images of the Reidemeister moves. For example,

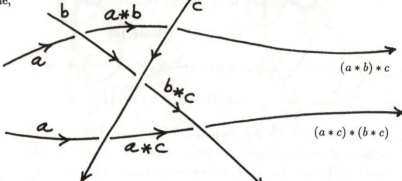

Thus the distributivity $(a * b) * c = (a * c) * (b * c)$ expresses this form of the type III move.

Remark. The quandle has a nice homotopy theoretic interpretation (see [JOY] or [WIN]). The idea is that each element of $Q(K)$ is represented by a **lasso** consisting of a disk transverse to the knot that is connected by a string attached to the boundary of the disk to the base point p:

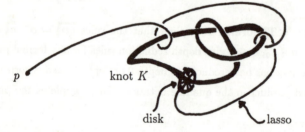

Two lassos are multiplied $(a * b)$ by **adding extra string to a via a path along b** [Go from base point to the disk, around the disk **in the direction** of its orientation for $\bar{*}$ (**against** for $*$) and back to p along this string, **then** down a's string to attach the disk!].

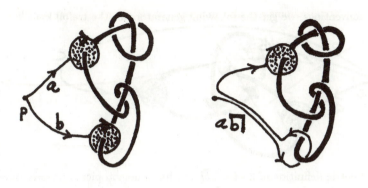

$$a * b \equiv a\overline{b\,|}$$

Note that in this interpretation we can think of $a * b = a\overline{b\,|}$ where $\overline{b\,|}$ is the element in the fundamental group of $S^3 - K$ corresponding to b via [go down string, around disk and back to base point].

The string of a lasso, b, can entangle the link in a very complex way, but the element $\overline{b\,|}$ (or $\overline{|b}$) in the group π is a conjugate of one of the peripheral generators of π. Such a generator is obtained by going from base point to an arc of the link, encircling the arc once, and returning to base point. For diagrams, we can standardize these peripheral generators by

1) choosing a base point in the unbounded region of the diagram,

and

2) stipulating that the string of the lasso goes over all strands of the diagram that it interacts with.

With these conventions, we get the following generators for the trefoil knot T:

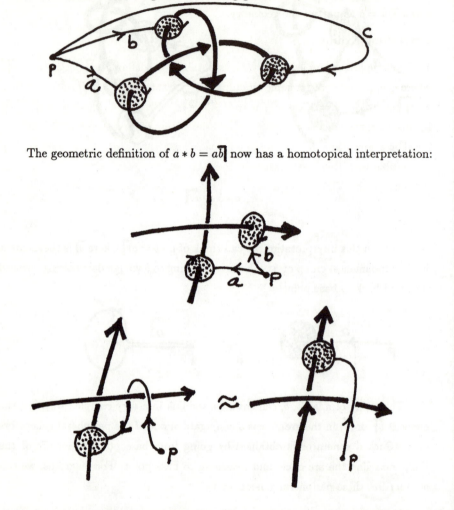

The geometric definition of $a * b = a\overline{b}$ now has a homotopical interpretation:

We see that the relation at a crossing ($c = a * b$) is actually a description of the existence of a homotopy from the lasso $a * b$ to the lasso c. In terms of the fundamental group, this says that $\overline{c} = \overline{b}^{-1}\,\overline{a}\,\overline{b} = \overline{b}\,\overline{a}\,\overline{b}$ (as we have seen algebraically in the crystal). Sliding the lasso gives a neat way to picture this relation in the fundamental group.

The homotopy interpretation of the quandle includes the relations $a * a = a$ and $a \mathbin{\bar{*}} a = a$ as part of the geometry.

Crystalline Examples.

1. The simplest example of a crystal is obtained by defining $a\overline{b\rceil} = 2b - a = a\lceil\overline{b}$ where $a, b, \ldots \in \mathbf{Z}/m\mathbf{Z}$ for some modulus m. Note that $a\overline{b\rceil}\,\overline{b\rceil} = a\overline{b\rceil}\lceil\overline{b} = a$ and that $xa\,\overline{b\rceil\rceil} = 2(a\overline{b\rceil}) - x = 2(2b - a) - x = 4b - 2a - x = 2b - (2a - (2b - x)) = x\lceil\overline{b}\,\overline{a\rceil}\,\overline{b\rceil}$. In this case, the associated group structure π is the additive group of $\mathbf{Z}/m\mathbf{Z}$.

For a given link K, we can ask for the **least modulus** m $(m \neq 1)$ that is compatible with this crystal structure **and** compatible with crossing relations for the diagram K. Thus [using $\bar{a} = a\rceil = \lceil a$ in this system] for the trefoil we have

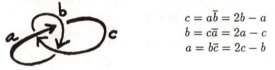

$$c = a\bar{b} = 2b - a$$
$$b = c\bar{a} = 2a - c$$
$$a = b\bar{c} = 2c - b$$

$$\Rightarrow \quad \begin{aligned} -a + 2b - c &= 0 \\ 2a - b - c &= 0 \\ -a - b + 2c &= 0 \end{aligned}$$

$$\begin{bmatrix} -1 & 2 & -1 \\ 2 & -1 & -1 \\ -1 & -1 & 2 \end{bmatrix} \xrightarrow[\text{row}]{} \begin{bmatrix} -1 & 2 & -1 \\ 0 & -3 & 0 \\ 0 & 0 & 0 \end{bmatrix} \xrightarrow[\text{col}]{} \begin{bmatrix} 1 & 0 & 0 \\ 0 & 3 & 0 \\ 0 & 0 & 0 \end{bmatrix}$$

(doing integer-invertible row and column operations on this system). Hence $m(K) = 3$.

The least modulus for the trefoil is three. This corresponds to the labelling

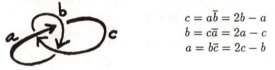

$$(2 \cdot 0 - 2 \equiv 1 (\mathrm{mod}\ 3)).$$

Thus the three-coloration of the trefoil corresponds to the $(\mathbf{Z}/3\mathbf{Z})$-crystal. $m(K) = 3$ implies that the trefoil is non-trivial. (Note that in the $(\mathbf{Z}/m\mathbf{Z})$-crystal, $a\bar{a} = 2a - a = a$ hence $m(K)$ is an ambient isotopy invariant of K.)

2. One way to generalize Example 1 is to consider the most general crystal such that $a\rceil = \lceil a$. Thus we require that $a\bar{b}\,\bar{b} = a$ and that $\overline{a\bar{b}} = \bar{b}\,\bar{a}\,\bar{b}$ for all $a, b \in \mathcal{C}_0$. Call such a crystal a **light crystal** (it corresponds to the **involuntary quandle**

[JOY]). Let $\mathcal{CL}(K)$ denote the light crystal corresponding to a given knot or link K. Note that $\mathcal{CL}(K)$ does not depend upon the orientation of K.

In the case of the trefoil, we have

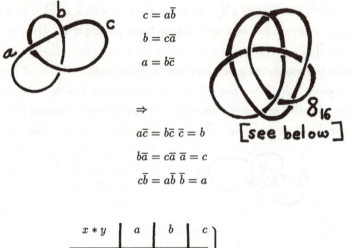

$$c = a\bar{b}$$
$$b = c\bar{a}$$
$$a = b\bar{c}$$

$$\Rightarrow$$

$$a\bar{c} = b\bar{c}\,\bar{c} = b$$
$$b\bar{a} = c\bar{a}\,\bar{a} = c$$
$$c\bar{b} = a\bar{b}\,\bar{b} = a$$

8_{16}

[see below]

$x * y$	a	b	c
a	a	c	b
b	c	a	a
c	b	a	a

$$\Big\} \; x * y = x\bar{y}$$

This calculation shows that $\mathcal{CL}(K)$ is finite (for the trefoil) with three elements in $C_0\mathcal{L}(K) = \{a, b, c\}$, and $\pi\mathcal{L}(K) \simeq \mathbf{Z}/3\mathbf{Z}$. In general, the involuntary quandle is not finite. Winker [WIN] shows that the first example of an infinite involuntary quandle occurs for the knot 8_{16}. Relatively simple links, such as ⬭⬭ have infinite involuntary quandles.

It is instructive to compute the light crystal (involuntary quandle) for the figure eight knot E. It is finite, but has five elements in $C_0\mathcal{L}(E)$, one more element than there are arcs in the diagram. This phenomenon (more algebra elements than diagrammatic arcs) shows how **the crystal (or quandle) is a generalization of the idea of coloring the arcs of the diagram.** In general there is a given supply of colors, but a specific configuration of the knot or link uses only a subset of these colors.

3. One of the most important crystal structures is the **Alexander Crystal** (after Alexander [ALEX2]). Let $\mathcal{M}$ be any module over the ring $\mathbf{Z}[t, t^{-1}]$. Given $a, b \in \mathcal{M}$, define $ab\rceil$ and $a\lceil b$ by the equations

$$ab\rceil = ta + (1 - t)b$$
$$a\lceil b = t^{-1}a + (1 - t^{-1})b.$$

It is easy to verify the crystal axioms for these equations:

$$ab\rceil\lceil b = t^{-1}(ta + (1 - t)b) + (1 - t^{-1})b$$
$$= a + (t^{-1} - 1)b + (1 - t^{-1})b$$
$$= a.$$

$$ab\ \overline{c\rceil}\rceil = ta + (1 - t)bc\rceil$$
$$= ta + (1 - t)(tb + (1 - t)c)$$
$$= ta + (t - t^2)b + (1 - t)^2 c.$$

$$a\lceil c\ \overline{b}\rceil c\rceil = t(a\lceil c\ \overline{b}\rceil) + (1 - t)c$$
$$= t(ta\lceil c + (1 - t)b) + (1 - t)c$$
$$= t(t(t^{-1}a + (1 - t^{-1})c) + (1 - t)b) + (1 - t)c$$
$$= ta + (t^2 - t)c + (t - t^2)b + (1 - t)c$$
$$= ta + (t - t^2)b + (1 - t)^2 c.$$

Thus $b\rceil\lceil b = 1$ and $b\ \overline{c\rceil}\rceil = \lceil c\ \overline{b}\rceil c\rceil$. Note that if $\mathcal{A}(\mathcal{M})$ denotes the Alexander crystal of the module $\mathcal{M}$, then $C_0(\mathcal{M}) = \mathcal{M}$ and $\pi(\mathcal{M})$ is the group of automorphisms of $\mathcal{M}$ generated by the maps $b\rceil, \lceil b : \mathcal{M} \to \mathcal{M}$, $ab\rceil = \overline{b}\rceil(a)$, $a\lceil b = \lceil b(a)$.

Note also that

$$aa\rceil = ta + (1 - t)a = a$$

and

$$a\lceil a = t^{-1}a + (1 - t^{-1})a = a.$$

Thus each element $a \in \mathcal{M}$ corresponds to an automorphism $a\rceil : \mathcal{M} \to \mathcal{M}$ leaving a fixed. From the point of view of the knot theory, $\mathcal{A}(\mathcal{M})$ will give rise to an ambient isotopy invariant module $\mathcal{M}(K)$ (called the Alexander Module [CF], [FOX2]) if we define $\mathcal{M}(K)$ to be the module over $\mathbf{Z}[t, t^{-1}]$ generated by the edge labels of the diagram K, modulo the relations

$$\uparrow \quad c = ta + (1-t)b = a\overline{b}\rceil \qquad\qquad \uparrow \quad c = t^{-1}a + (1-t^{-1})b = a\rceil b$$

For a given knot diagram, any one of the crossing relations is a consequence of all the others. As a result, we can calculate $\mathcal{M}(K)$ for a knot diagram of n crossings by taking $(n-1)$ crossing relations.

Example.

$$c \;\; = a\overline{b}\rceil = ta + (1-t)b$$

$$b \;\; = c\overline{a}\rceil = tc + (1-t)b$$

$$a \;\; = b\overline{c}\rceil = tb + (1-t)c$$

Here $\mathcal{M}(T)$ is generated by the relations

$$ta + (1-t)b - c = 0$$

$$(1-t)a - b + tc = 0$$

$$-a + tb + (1-t)c = 0$$

$$\Rightarrow \begin{bmatrix} t & 1-t & -1 \\ 1-t & -1 & t \\ -1 & t & 1-t \end{bmatrix} \begin{bmatrix} a \\ b \\ c \end{bmatrix} = \begin{bmatrix} 0 \\ 0 \\ 0 \end{bmatrix}$$

and we see that the first relation is the negative of the sum of the second two relations.

A closer analysis reveals that **the determinant of any $(n-1) \times (n-1)$ minor of this relation matrix is a generator of the ideal in $\mathbf{Z}[t, t^{-1}]$ of Laurent polynomials $f(t)$ such that $f(t) \cdot m = 0$ for any $m \in \mathcal{M}$.** This determinant is determined up to sign and power of t and it is called the **Alexander polynomial** $\Delta(t)$. The Alexander polynomial is an ambient isotopy invariant of the knot.

In the case of the trefoil, we have

$$\Delta_T(t) \doteq \text{Det} \begin{pmatrix} -1 & t \\ t & 1-t \end{pmatrix} = (t-1) - t^2$$
$$\doteq t^2 - t + 1.$$

($\doteq$ denotes equality up to sign and power of t).

Think of the Alexander polynomial as a generalization of the modulus of a knot or link - as described in the first crystalline example.

For small examples the Alexander polynomial can be calculated by putting the relations directly on the diagram - first splitting the diagram as a 2-strand tangle and assigning an input and an output value of 0:

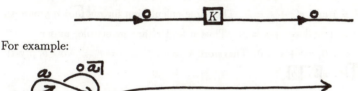

For example:

$$a0\overline{a}] = a\overline{a} \; 0] \; a] = a0] \; a]$$

Note

$$x0] = tx + (1-t)0 = tx$$
$$x\overline{0} = t^{-1}x + (1-t^{-1})0 = t^{-1}x.$$

Thus

$$a\overline{0}] \; a] = t(ta) + (1-t)a$$
$$= (t^2 - t + 1)a$$
$$= \Delta(t)a.$$

The requirement that $\Delta(t)a = 0$ shows that $\Delta(t) = t^2 - t + 1$ is the Alexander polynomial.

We produced the Alexander Crystal as a rabbit from a hat. In fact, the following result shows that it arises as the unique linear representation of the crystal axioms.

Theorem 13.7. Let $\mathcal{M}$ be a free module of rank ≥ 3 over a commutative ring R. Suppose that $\mathcal{M}$ has a crystal structure with operations defined by the formulas

$$ab\rceil = ra + sb$$
$$a\lceil b = r'a + s'b$$

where $r, s, r', s' \in R$ and $a, b \in \mathcal{M}$. If r, s, r', s' are invertible elements of R, then we may write $r = t, s = (1 - t), r' = t^{-1}, s' = (1 - t^{-1})$. Thus, this linear crystal representation necessarily has the form of the Alexander Crystal.

Proof. We may assume (by the hypothesis) that a and b are independent over R. By the crystal structure, $ab\rceil \lceil b = a$. Hence $a = (ra + sb) \lceil b = rr'a + (r's + s')b$. Therefore $rr' = 1$ and $r's + s' = 0$. The same calculation for $a\lceil b\, \rceil b = a$ yields the equations $r'r = 1$ and $rs' + s = 0$. Thus r and r' are invertible, and $r' = r^{-1}$ while $r^{-1}s + s' = 0$, $rs' + s = 0$. Therefore $s' = -r^{-1}s$, $r' = r^{-1}$. Now apply the equation $xab\rceil\rceil = x\lceil b\, \rceil a\, \rceil b$:

$$xa\,\overline{b\rceil}\rceil = x\overline{ra + sb}\rceil = rx + s(ra + sb)$$
$$x\lceil b\, \rceil a\, \rceil b = r(r(r'x + s'b) + sa) + sb$$
$$= rx + r^2 s'b + rsa + sb$$
$$= rx + (r^2 s' + s)b + rsa$$
$$= rx + (-rs + s)b + rsa. \qquad (s' = -r^{-1}s)$$

Therefore, we obtain the further condition:

$$s^2 = -rs + s$$
$$\Rightarrow s = 1 - r.$$

Thus, if $r = t$, then $s = 1 - t$, $r' = t^{-1}$, $s' = 1 - t^{-1}$. This completes the proof.//

Remark. This theorem shows that the structure of the classical Alexander module arises naturally and directly from basic combinatorial considerations about colored states for link diagrams. The reader may enjoy comparing this development with the classical treatments using covering spaces ([FOX], [LK3], [ROLF], [B2]).

The Burau Representation.

We have already encountered the Alexander polynomial in the framework of the Yang-Baxter models. There is a remarkable relationship between the Alexander module and the R-matrix for the statistical mechanics model of the Alexander polynomial.

In order to see this relationship, I need to first extract the **Burau representation of the braid group** from the Alexander crystal:

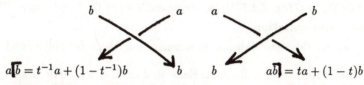

$$a\overline{b} = t^{-1}a + (1 - t^{-1})b \qquad\qquad a\overline{b} = ta + (1 - t)b$$

Let a and b now denote two basis elements for a vector space V. Define $T : V \to V$ by $T(b) = a\overline{b} = ta + (1 - t)b$ and $T(a) = b$. Note that, with respect

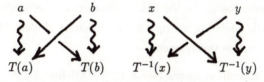

to this basis, T has the matrix (also denoted by T)

$$T = \begin{bmatrix} 0 & t \\ 1 & (1 - t) \end{bmatrix}$$

and that

$$T^{-1} = \begin{bmatrix} 1 - t^{-1} & 1 \\ t^{-1} & 0 \end{bmatrix}.$$

To obtain a representation of the n-strand braid group B_n, let

$$\rho(\sigma_1) = \begin{bmatrix} [T] & & & & \bigcirc \\ & 1 & & & \\ & & 1 & & \\ & & & \ddots & \\ \bigcirc & & & & 1 \end{bmatrix}$$

$$\rho(\sigma_2) = \begin{bmatrix} 1 & & & & \\ & [T] & & \bigcirc & \\ & & 1 & & \\ & \bigcirc & & \ddots & \\ & & & & 1 \end{bmatrix}$$

$$\rho(\sigma_{n-1}) = \begin{bmatrix} 1 & & & \bigcirc \\ & \ddots & & \\ \bigcirc & & 1 & \\ & & & [T] \end{bmatrix}$$

where the large matrices are $n \times n$, and $[T]$ takes up a 2×2 block. $(\rho(\sigma_i^{-1})$ is obtained by replacing T by T^{-1} in the blocks.)

It is easy to see (from our discussion of invariance for the crystal) that the mapping $\rho : B_n \to GL(n; \mathbf{Z}[t, t^{-1}])$ is a representation of the braid group. This is the **Burau representation**.

One can use this representation to calculate the Alexander polynomial.

Proposition 13.8. Let $w \in B_n$ be an element of the n-strand Artin braid group. Let $\rho(w) \in GL(n; \mathbf{Z}[t, t^{-1}])$ denote the matrix representing w under the Burau representation. Let $A(t) = \rho(w) - I$ where I is an $n \times n$ identity matrix. Then $A(t)$ is an Alexander matrix for the link $\overline{w}$ obtained by closing the braid w. (An Alexander matrix is a relation matrix for the Alexander Module $\mathcal{M}(\overline{w})$ as defined in Example 3 just before this remark.)

Remark. This proposition means (see e.g. [FOX]) that for $K = \overline{w}$, $\Delta_K(t)$ is the generator of the ideal generated by all $(n-1) \times (n-1)$ minors of $A(t)$. In the case under consideration, $\Delta_K(t)$ is equal (up to an indeterminacy of sign and power of t) to any $(n-1) \times (n-1)$ minor of $A(t)$. We write $\Delta_K(t) \doteq \mathrm{Det}(A'(t))$ where $A'(t)$ denotes such a minor and $\doteq$ denotes equality up to a factor of the form $\pm t^{\ell} (\ell \in \mathbf{Z})$.

I omit a proof of this proposition, but point out that it is equivalent to stating that $A(t) - I$ is a form of writing the relations in $\mathcal{M}$ (as described for the crystal) that ensue from closing the braid. For example:

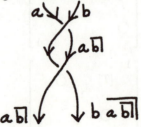

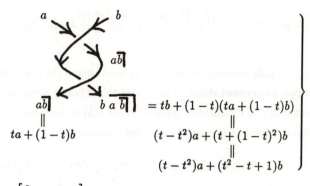

$$\rho(\sigma_1) = \begin{bmatrix} 0 & t \\ 1 & (1-t) \end{bmatrix}$$

$$\rho(\sigma_1^2) = \rho(\sigma_1)^2 = \begin{bmatrix} 0 & t \\ 1 & (1-t) \end{bmatrix} \begin{bmatrix} 0 & t \\ 1 & (1-t) \end{bmatrix} = \begin{bmatrix} t & t-t^2 \\ (1-t) & t^2-t+1 \end{bmatrix}.$$

We see that $\rho(\sigma_1^2)$ catalogues the mapping from the top to the bottom of the braid. On closing the braid, we have

$$\Rightarrow a = a\,\overline{b} = ta + (1-t)b$$
$$b = b\,\overline{a\,\overline{b}} = (t-t^2)a + (t^2-t+1)b$$

These relations become the matrix $A(t) = \rho(w) - I$. $\left(w = \sigma_1^2 \right)$

In the example we have

$$A(t) = \begin{bmatrix} t-1 & t-t^2 \\ 1-t & t^2-t \end{bmatrix}.$$

Hence (taking 1×1 minors) we have

$$\Delta \overline{w} \doteq t - 1.$$

To continue this example, let's take $K = \overline{\sigma_1^3}$. Then

$$\rho(\sigma_1^3) = \rho(\sigma_1)^3 = \begin{bmatrix} t & t-t^2 \\ (1-t) & t^2-t+1 \end{bmatrix} \begin{bmatrix} 0 & t \\ 1 & 1-t \end{bmatrix}$$

$$= \begin{bmatrix} t-t^2 & t^2+(1-t)(t-t^2) \\ t^2-t+1 & t(1-t)+(t^2-t+1)(1-t) \end{bmatrix}$$

$$= \begin{bmatrix} t-t^2 & t-t^2+t^3 \\ t^2-t+1 & t^2-t^3-t+1 \end{bmatrix}.$$

Thus $A_K(t) = \rho(\sigma_1)^3 - I = \begin{bmatrix} t - t^2 - 1 & t - t^2 + t^3 \\ t^2 - t + 1 & t^2 - t^3 - t \end{bmatrix}$. Hence $\Delta_K(t) \doteq t^2 - t + 1$

as before.

Exercise. It has only recently (Spring 1990) been shown (by John Moody [MD]) that **the Burau representation is not faithful**. That is, there is a non-trivial braid b such that $\rho(b)$ is the identity matrix, where $\rho(b)$ is the Burau representation described above. Moody's example is the commutator:

$$[765432'7'7'667765'4'32'1'23'456'7'7'6'6'7723'4'5'6'7', 7'7'87788778]$$

Here 7 denotes σ_7, $7'$ denotes σ_7^{-1} and $[a, b] = aba^{-1}b^{-1}$. Verify that ρ is the identity matrix on Moody's nine-strand commutator, and show that the braid is non-trivial by closing it to a non-trivial link.

Deriving an R-matrix from the Burau Representation.

This last section of the lecture is devoted to showing a remarkable relationship between the Burau representation and the R-matrix that we used in section 12^0 to create a model for the Alexander-Conway polynomial. We regard the Burau representation as associating to each braid generator σ_i, a mapping $\rho(\sigma_i) : V \to V$ where V is a module over $\mathbf{Z}[t, t^{-1}]$. Let $\wedge^*(V)$ denote the exterior algebra on V. This means that $\wedge^*(V) = \wedge^0(V) \oplus \wedge^1(V) \oplus \ldots \oplus \wedge^d(V)$ where $d = \dim(V/\mathbf{Z}[t, t^{-1}])$ and $\wedge^0(V) = \mathbf{Z}[t, t^{-1}]$ while $\wedge^k(V)$ consists in formal products $v_{i_1} \wedge \ldots \wedge v_{i_k}$ of basis elements $\{v_1, \ldots, v_d\}$ of V and their sums with coefficients in $\mathbf{Z}[t, t^{-1}]$. These wedge products are associative, anti-commutative ($v_i \wedge v_j = -v_j \wedge v_i$) and linear ($(v + w) \wedge z = (v \wedge z) + (w \wedge z)$). We shall see that the R-matrix emerges when we prolong ρ to $\hat{\rho}$ where $\hat{\rho} : B_n \to \text{Aut}(\wedge^*(V))$. I am indebted to Vaughan Jones for this observation [JO8]. The relationship leads to a number of interesting and (I believe) important questions about the interplay between classical topology and state models arising from solutions to the Yang-Baxter Equation.

Here is the construction: First just consider the Burau representation for 2-strand braids. Then, if V has basis $\{a, b\}$ $(a = v_1, b = v_2)$ then we have

$$\rho(\sigma)(a) = b$$
$$\rho(\sigma)(b) = ta + (1 - t)b.$$

In $\wedge^*(V)$, we have the basis $\{1, a, b, a \wedge b\}$ and $\widehat{\rho}(\sigma)$ acts via the formulas:

$$\widehat{\rho}(\sigma)(1) = 1$$
$$\widehat{\rho}(\sigma)(a) = b$$
$$\widehat{\rho}(\sigma)(b) = ta + (1 - t)b$$
$$\widehat{\rho}(\sigma)(a \wedge b) = (\rho(\sigma)(a)) \wedge (\rho(\sigma)(b)) = (-t)a \wedge b.$$

Let R denote the matrix of $\widehat{\rho}(\sigma)$ with respect to the basis $\{1, a, b, a \wedge b\}$. Then

$$R = \begin{bmatrix} 1 & 0 & 0 & 0 \\ 0 & 0 & t & 0 \\ 0 & 1 & (1-t) & 0 \\ 0 & 0 & 0 & (-t) \end{bmatrix}.$$

We can see at once, using the criteria of the Appendix that R is a solution to the Yang-Baxter Equation. To see how it fits into 2-spin ($\pm$) spin-preserving models, let

$$++ \longleftrightarrow 1$$
$$+- \longleftrightarrow a$$
$$-+ \longleftrightarrow b$$
$$-- \longleftrightarrow a \wedge b$$

so that

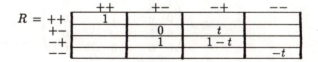

Now refer to the Appendix to see that R satisfies the Yang-Baxter Equation.

On the other hand, we can change basis by first replacing t by t^{-2} and then multiplying all rows by t:

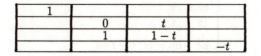

$$\longmapsto$$

1			
	0	t^{-2}	
	1	$1 - t^{-2}$	
			$-t^{-2}$

$$\longmapsto$$

t			
	0	t^{-1}	
	t	$t - t^{-1}$	
			$-t^{-1}$

Now a straightforward basis change on the middle 2×2 matrix

$$\left(\begin{pmatrix} 1 & 0 \\ 0 & t^{-1} \end{pmatrix} \begin{pmatrix} 0 & t^{-1} \\ t & x \end{pmatrix} \begin{pmatrix} 1 & 0 \\ 0 & t \end{pmatrix} = \begin{pmatrix} 0 & 1 \\ 1 & x \end{pmatrix} \right)$$

yields the matrix

t			
	0	1	
	1	$t - t^{-1}$	
			$-t^{-1}$

This is exactly the Yang-Baxter matrix that we used in section 12^0 to create a model for the Alexander-Conway polynomial.

Exercise. Using the FRT construction (section 10^0), show that the bialgebra associated with the R-matrix

$$R = \begin{pmatrix} t & 0 & 0 & 0 \\ 0 & 0 & 1 & 0 \\ 0 & 1 & (t - t^{-1}) & 0 \\ 0 & 0 & 0 & -t^{-1} \end{pmatrix}$$

for $T = \begin{pmatrix} a & b \\ c & d \end{pmatrix}$ has relations

$$\begin{cases} ba = -t^{-1}ab & db = tbd \\ ca = -t^{-1}ac & dc = tcd \\ bc = cb & \\ b^2 = c^2 = 0 & \\ ad - da = (t^{-1} - t)bc \end{cases}$$

and comultiplication

$$\Delta(a) = a \otimes a + b \otimes c$$
$$\Delta(b) = a \otimes b + b \otimes d$$
$$\Delta(c) = c \otimes a + d \otimes c$$
$$\Delta(d) = c \otimes b + d \otimes d.$$

(See [LK23], p.314 and also [GE]. Compare with [LEE1] and [LK18], [MUK2].)

Remarks and Questions. I leave it to the reader to explore the full representation $\widehat{\rho} : B_n \to \text{Aut}(\wedge^* V)$ and how it is supported by the matrix R ($\widehat{\rho}(\sigma_i) = I \otimes I \otimes \ldots \otimes I \otimes R \otimes I \otimes \ldots \otimes I$ where I is the identity matrix - for appropriate bases.)

The state model of section 12^0 and the determinant definitions of the Alexander polynomial (related to the Burau representation) are, in fact, intertwined by these remarks. See [LK18] and [JA8] for a complete account. The key to the relationship is the following theorem.

Theorem 13.9. Let A be a linear transformation of a finite dimensional vector space V. Let $\widehat{A} : \wedge^*(V) \to \wedge^*(V)$ be the extension of A to the exterior algebra of V. Let $\mathcal{S} : \wedge^*(V) \to \wedge^*(V)$ be defined by $\mathcal{S} \mid \wedge^k(V)(x) = (-\lambda)^k x$. Then

$$\text{Det}(A - \lambda I) = \text{Tr}(\mathcal{S}\widehat{A}).$$

In other words, the characteristic polynomial of a given linear transformation can be expressed as the trace of an associated transformation on the exterior algebra of the vector space for A. This trace can then be related to our statistical mechanics model for the Alexander polynomial.

In [LK18] H. Saleur and I show that this interpretation of the state model for the Alexander polynomial fits directly into the pattern of the free-fermion model in statistical mechanics. This means that the state summation can be rewritten as a discrete Berezin integral with respect to the link diagram. This expresses the Alexander polynomial as a determinant in a new way. Murakami [MUK2] has observed that these same methods yield Yang-Baxter models for the multi-variable Alexander-Conway polynomials for colored links. (Compare [MUK3].)

In general, knot polynomials defined as Yang-Baxter state models do not have interpretations as annihilators of modules (such as the Alexander module). A reformulation of the state models of section 12^0 in this direction could pave the way for an analysis of classical theorems about the Alexander polynomial in a statistical mechanics context, and to generalizations of these theorems for the Jones polynomial and its relatives.

Exercise. (The Berezin Integral). The Berezin integral [RY] is another way to work with Grassman variables. If θ is a single Grassman variable, then by definition

$$\int d\theta 1 = 0$$

while

$$\int d\theta\, \theta = 1$$

and $\theta d\theta = -d\theta\, \theta$ if it comes up. Since $\theta^2 = 0$, the most general function of θ, $F(\theta)$ has the form $F(\theta) = a + b\theta$. Therefore,

$$\int d\theta F(\theta) = b = dF/d\theta,$$

a strange joke.

Let $x^1, x^2, \ldots, x^n; y^1, y^2, \ldots, y^n$ be distinct anti-commuting Grassman variables. Thus $(x^i)^2 = (y^i)^2 = 0$ while $x^i x^j = -x^j x^i$, $y^i y^j = -y^j y^i$ for $i \neq j$, and $x^i y^j = -y^j x^i$. Let M be an $n \times n$ matrix with elements in a commutative ring. Let $x^t = (x^1, \ldots, x^n)$ and $y^t = (y^1, \ldots, y^n)$ so that y is a column matrix. Show that

$$\mathrm{DET}(M) = (-1)^{(n(n+1)/2)} \int dx^1 \ldots dx^n dy^1 \ldots dy^2 e^{x^t M y}$$

where this is a Berezin integral.

Exercise. This exercise gives the bare bones of the relationship [LK18] between the free fermion model [SAM], [FW] and the Alexander-Conway state model of section 12^0. Attach Grassman variables to the edges of a universe (shadow of a link diagram) as follows:

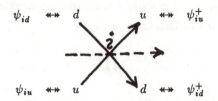

Thus ψ^+ goes **out**, ψ goes **in**. The u and d designations refer to **up** and **down** with respect to the crossing direction

A **state** of the diagram consists in assigning $+1$ or -1 to each edge of the diagram. An edge labelled -1 is said to be **fermionic**. An edge labelled $+1$ is said to be **bosonic**. Assume that vertex weights are assigned as shown below

Here the dotted line denotes $+1$ and the wavy line denotes -1. If K is the diagram let $\langle K \rangle$ denote the state summation

$$\langle K \rangle = \sum_{\sigma} \langle K|\sigma \rangle (-1)^{\mathcal{L}(\sigma)}$$

where σ runs over the states described above, $\langle K|\sigma \rangle$ denotes the product of vertex weights in a given state, an $\mathcal{L}(\sigma)$ is the number of fermionic loops in a given state. Loops are counted by the usual separation convention:

Show that, up to a constant factor, the Alexander polynomial state model of section 12^0 can be put in this form (with, of course, different vertex weights corresponding to the two types of crossing.). **Check** that for the Alexander (Burau) R-matrix the vertex weights satisfy

$$w_1 w_2 = w_5 w_6 - w_3 w_4.$$

Now suppose that the weights in the model are just known to satisfy the "free-fermion condition" $w_1 w_2 = w_5 w_6 - w_3 w_4$ **and** suppose that $w_2 = -1$ (as can always be arranged). Define Grassman forms

$$d\psi d\psi^+ = \prod_i d\psi_i d\psi_i^+$$

and $A = A_I + A_P$. The forms A_I (interaction) and A_P (propagator) are explained below:

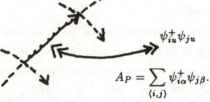

$$\psi_{iu}^+ \psi_{ju}$$

$$A_P = \sum_{\langle i,j \rangle} \psi_{i\alpha}^+ \psi_{j\beta}.$$

The term A_P is the sum over all oriented edges in K of the products of fermionic creation and annihilation operators for that edge.

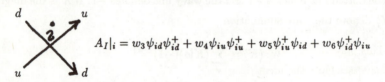

$$A_I|_i = w_3 \psi_{id} \psi_{id}^+ + w_4 \psi_{iu} \psi_{iu}^+ + w_5 \psi_{iu}^+ \psi_{id} + w_6 \psi_{id}^+ \psi_{iu}$$

A_I is the sum, over all vertices, of the interaction terms described above.

Now assume that K is a one-input, one-output tangle universe such. as

$$K$$

with fermionic input and output. **Show** that

$$\langle K \rangle = \int d\psi d\psi^+ e^A$$

where the integral denotes a Berezin integral. Show that this implies that there exists a matrix **A** such that

$$\langle K \rangle = \mathrm{Det}(\mathbf{A}).$$

14^0. The Kauffman Polynomial.

The Kauffman polynomial ([LK8], [JO5], [TH3], [TH4]) (Dubrovnik version) is defined as a regular isotopy invariant D_K of unoriented links K. $D_K = D_K(z, a)$ is a 2-variable polynomial satisfying:

1. $K \approx K' \Rightarrow D_K = D_{K'}$.

2. $D_{\asymp} - D_{\times} = z(D_{\asymp} - D_{\supset\subset})$

3. $D_{\bigcirc} = \mu = ((a - a^{-1})/z) + 1$

   $D_{\curlyvee} = aD$

   $D_{\curlywedge} = a^{-1}D$

The corresponding normalized invariant, Y_K, of ambient isotopy (for oriented links) is given by the formula

$$Y_K = a^{-w(K)} D_K$$

where $w(K)$ denotes the writhe of K.

This polynomial has a companion that I denote L_K. L_K is a regular isotopy invariant satisfying the identities

$$L_{\asymp} + L_{\times} = z(L_{\asymp} + L_{\supset\subset})$$

$$L_{\curlyvee} = aL \quad , \quad L_{\bigcirc} = \lambda = \left(\frac{a + a^{-1}}{z}\right) - 1$$

$$L_{\curlywedge} = a^{-1}L$$

The L-polynomial specializes at $a = 1$ to the Q-polynomial of [BR] a 1-variable invariant of ambient isotopy. As usual, we can normalize L to obtain a 2-variable invariant of ambient isotopy:

$$F_K(z, a) = a^{-w(K)} L_K.$$

In fact, the F and Y polynomials are equivalent by a change of variables. The result is (compare [LICK2]):

Proposition 14.1.

$$L_K(z,a) = (-1)^{c(K)} i^{w(K)} D_K(-iz, ia)$$

and

$$F_K(z,a) = (-1)^{c(K)} Y_K(-iz, ia)$$

where $c(K)$ denotes the number of components of K, and $w(K)$ is the writhe of K. (The first formula is valid for unoriented links K and any choice of orientation for K to compute $w(K)$ - since $i^{w(K)}$ is independent of the choice of orientation. The second formula depends upon the choice of orientation.)

We shall leave the proof of this proposition to the reader. However, some discussion and examples will be useful.

Discussion of D and L.

First note that we have taken the convention that $D(O) = \mu$, $L(O) = \lambda$ where these loop values are fundamental to the polynomials in the sense that the value of the polynomial on the disjoint union $O \amalg K$ of a loop and a given link K is given by the product of the loop value and the value of K;

$$D(O \amalg K) = \mu D(K)$$
$$L(O \amalg K) = \lambda L(K).$$

It is easy to see that the bracket polynomial (see section 3^0):

$$\langle \asymp \rangle = A \langle \asymp \rangle + A^{-1} \langle \supset\subset \rangle$$
$$\langle O \rangle = -A^2 - A^{-2}$$

is a special case of both L and D. Hence the *Jones polynomial is a special case of both F and Y*. In fact, we have

$$\langle \asymp \rangle - \langle \asymp \rangle = (A - A^{-1})[\langle \asymp \rangle - \langle \supset\subset \rangle]$$
$$\langle \mathbf{\sigma} \rangle = (-A^3)\langle \sim \rangle$$
$$\langle \mathbf{\sigma} \rangle = (-A^{-3})\langle \sim \rangle.$$

Hence $\langle K \rangle(A) = D_K(A - A^{-1}, -A^3)$. Similarly,

$$\langle \asymp \rangle + \langle \succ\!\!\prec \rangle = (A + A^{-1})[\langle \,\succeq\, \rangle + \langle \supset \subset \rangle].$$

Hence $\langle K \rangle(A) = L_K((A + A^{-1}), -A^3)$. From this, it is easy to verify direct formulas for the Jones polynomial as a special case of F and of Y. Thus we know that

$$V_K(t) = f_K(t^{-1/4})/f_0(t^{-1/4})$$

where

$$f_K(A) = (-A^3)^{-w(K)}\langle K \rangle(A).$$

Thus

$$f_K(A) = Y_K(A - A^{-1}, -A^3)$$

and

$$f_K(A) = F_K(A + A^{-1}, -A^3).$$

Therefore

$$V_K(t) = \frac{Y_K(t^{-1/4} - t^{1/4}, \, -t^{-3/4})}{(-t^{-1/2} - t^{1/2})}$$

and

$$V_K(t) = \frac{F_K(t^{-1/4} + t^{+1/4}, \, -t^{-3/4})}{(-t^{-1/2} - t^{1/2})}.$$

The polynomials F and Y are pretty good at detecting chirality. They do somewhat better than the Homfly polynomial on this score. See [LK8] for more information along these lines. The book [LK3] contains a table of L-polynomials (normalized so that the unknot has value 1) for knots up to nine crossings. Thistlethwaite [TH5] has computed these polynomials for all knots up to twelve crossings.

Here are two examples for thinking about chirality: The knot 9_{42}

9_{42}

is chiral. Its chirality is *not* detected by either F or by the Homfly polynomial. The knot 10_{79}

10_{79}

is chiral. Its chirality is detected by F (and Y) but not by the Homfly polynomial.

Birman-Wenzl Algebra.

We have mentioned that there is an algebra (the Hecke algebra) related to the Homfly polynomial. The Hecke algebra $\mathcal{H}_n$ is a quotient of the $\mathbf{Z}[z, z^{-1}, a, a^{-1}]$-group algebra of the braid group B_n by the relations $\sigma_i - \sigma_i^{-1} = z$ (or $\sigma_i^2 = 1 + z\sigma_i$). These relations are compatible with the corresponding polynomial relation, $H\!\!\times - H\!\!\times = zH\!\!\,)(\,$, and so we get "traces" defined on the Hecke algebra by evaluating the H-polynomial (the un-normalized Homfly polynomial) on braids lifting elements in the Hecke algebra. Another route to the construction of the Homfly polynomial was to directly construct these traces at the algebra level (see [JO3]).

An analogous idea leads to algebras associated with the versions of the Kauffman polynomial. This is the Birman-Wenzl Algebra [B1]. For example, if we are using the Dubrovnik polynomial then the basic relation is $D\!\!\times - D\!\!\times = z(D\!\!\asymp - D\,)(\,)$ and this corresponds formally to a relation in the braid monoid algebra:

$$\sigma_i - \sigma_i^{-1} = z(U_i - 1)$$

where U_i is a cup-cap combination in the i, $i+1$-th place as we discussed in section 7^0. This algebra involves understanding how the braiding and Temperley-Lieb generators U_i are interrelated. (See [LK8].)

Jaeger's Theorem.

While the Homfly polynomial and the Kauffman polynomial are distinct - with different topological properties, there is a very beautiful relationship between them - due to Francois Jaeger [JA9], and also observed in a special case by Reshetikhin [RES].

Jaeger shows that the Y-polynomial can be obtained as a certain weighted sum of Homfly polynomials on links associated with a given link K. This is a kind of state expansion for Y_K over states that are evaluated via the Homfly polynomial P_K!

The idea behind Jaeger's construction is very simple. Consider the following formalism:

$$\left[\begin{array}{c}\times\end{array}\right] = z\left[\left[\begin{array}{c}\leftrightarrows\end{array}\right] - \left[\begin{array}{c}\uparrow\downarrow\end{array}\right]\right] + \left[\begin{array}{c}\times\end{array}\right] + \left[\begin{array}{c}\times\end{array}\right] + \left[\begin{array}{c}\times\end{array}\right] + \left[\begin{array}{c}\times\end{array}\right] \quad (*)$$

The small diagrams in this formula are intended to stand for parts of a larger diagram (differing only at the indicated site) *and* for the corresponding polynomial evaluation. The oriented diagrams will be evaluated by an oriented polynomial (Homfly polynomial) coupled with data about rotation numbers - we shall specify the evaluations below. Viewed as a state expansion, $(*)$ demands states whose arrow configurations are globally compatible as oriented link diagrams. For example,

$$\left[\begin{array}{c}\infty\end{array}\right] = z\left[\left[\begin{array}{c}\infty\end{array}\right] - \left[\begin{array}{c}\infty\end{array}\right]\right] + \left[\begin{array}{c}\infty\end{array}\right] + \left[\begin{array}{c}\infty\end{array}\right].$$

A *state* for the expansion $[K]$ is obtained by splicing some subset of the crossings of K *and* choosing an orientation of the resulting link. The formula $(*)$ gives vertex weights for each state. Thus we have,

$$[K] = \sum_{\sigma} [K|\sigma][\sigma]$$

where $[K|\sigma]$ is the product of these vertex weights ($\pm z$ or 1), and $[\sigma]$ is an as-yet-to-be-specified evaluation of an oriented link σ.

Define $[\sigma]$ by the formula

$$[\sigma] = (ta^{-1})^{\mathrm{rot}(\sigma)} R_\sigma(t - t^{-1}, a)$$

where $\mathrm{rot}(\sigma)$ denotes the rotation number (Whitney degree) of the oriented link σ, and $R_\sigma(z, a)$ is the regular isotopy version of the Homfly polynomial defined by:

$$\begin{cases} R\!\!\nearrow - R\!\!\nwarrow = zR\!\!\rightrightarrows \\[2mm] R\!\!\curvearrowright = aR\!\!\rightarrow \\[2mm] R\!\!\curvearrowleft = a^{-1}R\!\!\rightarrow \\[2mm] R\!\!\bigcirc = \left(\dfrac{a - a^{-1}}{z}\right), \quad \left(R_{O \amalg K} = \left(\dfrac{a - a^{-1}}{z}\right)R_K\right). \end{cases}$$

Theorem 14.2. (Jaeger). With the above conventions, $[K]$ is a Homfly polynomial expansion of the Dubrovnik version of the Kauffman polynomial:

$$[K](t - t^{-1}, a) = D_K(t - t^{-1}, t^{-1}a^2).$$

Proof (Sketch). First note that this model satisfies the Dubrovnik polynomial exchange identity:

$$\left[\times\right] = z\left(\left[\rightleftarrows\right] - \left[\updownarrow\right]\right) + \left[\times\right] + \left[\times\right] + \left[\times\right] + \left[\times\right]$$

$$\left[\times\right] = z\left(\left[\upuparrows\right] - \left[\rightleftarrows\right]\right) + \left[\times\right] + \left[\times\right] + \left[\times\right] + \left[\times\right].$$

Hence

$$[\times] - [\times] = z\left(\left[\rightrightarrows\right] - \left[\uparrow\downarrow\right] - \left[\downarrow\uparrow\right] + \left[\rightleftarrows\right]\right)$$
$$+ \left[\nearrow\right] - \left[\nwarrow\right] + \left[\swarrow\right] - \left[\searrow\right]$$
$$+ \left[\nwarrow\right] - \left[\nearrow\right] + \left[\nearrow\right] - \left[\nwarrow\right]$$
$$= z\left(\left[\rightrightarrows\right] - \left[\uparrow\downarrow\right] - \left[\downarrow\uparrow\right] + \left[\rightleftarrows\right]\right)$$
$$+ z\left[\rightrightarrows\right] - z\left[\downarrow\downarrow\right] + z\left[\leftleftarrows\right]$$
$$- z\left[\uparrow\uparrow\right] \quad \text{(by Conway identity for state evaluations)}$$
$$= z\left(\left[\rightrightarrows\right] + \left[\rightleftarrows\right] + \left[\rightrightarrows\right] + \left[\leftleftarrows\right]\right) -$$
$$z\left(\left[\uparrow\downarrow\right] + \left[\downarrow\uparrow\right] + \left[\downarrow\downarrow\right] + \left[\uparrow\uparrow\right]\right)$$
$$= z\left(\left[\asymp\right] - \left[\,)(\,\right]\right).$$

Note that the value of the loop in this model is given by

$$\left[\, O \,\right] = \left[\, \circlearrowright \,\right] + \left[\, \circlearrowleft \,\right]$$
$$= (ta^{-1})^{-1}R_{\circlearrowright} + (ta^{-1})R_{\circlearrowleft}$$
$$= (t^{-1}a + ta^{-1})\left(\frac{a - a^{-1}}{t - t^{-1}}\right)$$
$$= \frac{t - t^{-1} + t^{-1}a^2 - ta^{-2}}{t - t^{-1}}$$
$$\left[\, O \,\right] = 1 + \frac{t^{-1}a^2 - ta^{-2}}{t - t^{-1}}.$$

This is the correct μ-value for $D_K(t - t^{-1}, t^{-1}a^2)$.

It is easy to check that

$$\left[\,\sigma\!\!\rightharpoonup\,\right] = a^2 t^{-1}\left[\,\sim\,\right]$$
$$\left[\,\rightharpoonup\!\!\sigma\,\right] = a^{-2} t\left[\,\sim\,\right]$$

by the same sort of calculation.

Finally, it is routine (but lengthy) to check that $[K]$ is a regular isotopy invariant (on the basis of the regular isotopy invariance of R_σ). I shall omit this verification. This completes the sketch of the proof.   //

Discussion. This model of the Dubrovnik polynomial tells us that there is a certain kind of Yang-Baxter model for D_K that is based on the Yang-Baxter model we already know for specializations of R_K. We have

$$D_K(t - t^{-1}, a^2 t^{-1}) = \sum_\sigma [K|\sigma](t^{-1}a)^{\text{rot}(\sigma)} R_\sigma(t - t^{-1}, a)$$

where the states σ are obtained from the unoriented diagram K by splicing a subset of crossings, and orienting the resulting link. The product of vertex weights $[K|\sigma]$ is a product of $\pm z, 1$ according to the expansion rule (∗):

$$\left[\times\right] = z\left(\left[\overset{\leftarrow}{\underset{\rightarrow}{}}\right] - \left[\mathcal{T}\mathcal{L}\right]\right) + \left[\overset{\nearrow}{\times}\right] + \left[\underset{\nearrow}{\times}\right] + \left[\times\right] + \left[\times\right].$$

If we let $a = t^{n+1}$, then we can replace $R_\sigma(t - t^{-1}, t^{n+1})$ by the Yang-Baxter state model of section 11^0. This gives a rather special set of models of D_K whose properties deserve closer investigation.

Research Problem. Investigate the models for D_K described above!

A Yang-Baxter Model and the Quantum Group for $SO(n)$.

There is another path to Yang-Baxter models for D_K that arises directly from the Yang-Baxter solution that we have used for specializations of the Homfly polynomial. Of course the model we have just indicated (using the Homfly expansion for the Dubrovnik polynomial) can be also seen as based on this Yang-Baxter solution. Here we study a simpler state structure, and we shall - in the process - uncover a significant solution to the Yang-Baxter Equation that is related to the group $SO(n)$.

The idea is as follows. Let K be an unoriented link. Define a *state* σ *of* K to be a labelled diagram σ that is obtained from K by first splicing some subset of crossings of K and projecting the rest so that the resulting diagram σ has components that are Jordan curves in the plane (no self-crossings). **Each** component is then assigned an orientation and a label from a given ordered index set $\mathcal{I}$. Thus if K is a trefoil, then

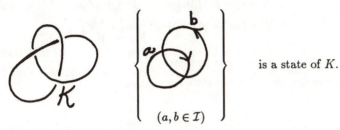

is a state of K.

We shall assume that if $a \in \mathcal{I}$ then $-a \in \mathcal{I}$ ($\mathcal{I} \subset \mathbf{R}$, the real numbers) *and* that a state obtained by switching both orientation and label ($a \mapsto -a$) is equivalent to the original state.

This means that at a splice we can assume that each state has *local* form ⇄ since

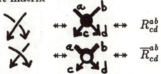

With these assumptions, we can define a loop-count $\|\sigma\|$ just as in the oriented case:

$$\|\sigma\| = \sum_{C \in \text{components}(\sigma)} \text{rot}(C) \cdot \text{label}(C)$$

and, given an oriented R-matrix

we can define $[K|\sigma]$ to be the product of the vertex weights assigned by $R(\overline{R})$ to the crossings in the diagram.

The assumption that the states have either a splice ⇄ or a projection ✕ at each crossing, means that we are here assuming that R and $\overline{R}$ are non-zero only when $a = c$, $b = d$ or $a = d$, $b = c$. In fact, I shall use R, $\overline{R}$ in the form

$$R = (q - q^{-1})\,\big\rangle\!<\!\big\langle\ + q\ \big\rangle\!=\!\big\langle\ +\ \text{⨯}$$

$$\overline{R} = (q^{-1} - q)\,\big\rangle\!>\!\big\langle\ + q^{-1}\,\big\rangle\!=\!\big\langle\ +\ \text{⨯}$$

our familiar solutions from previous sections.

With this choice, we define the state summation

$$[K] = \sum_\sigma [K|\sigma]q^{\|\sigma\|}.$$

We now show that an appropriate choice of the index set $\mathcal{I}$ makes $[K]$ into a series of models for the Dubrovnik polynomial.

Before specializing $\mathcal{I}$, it is easy to see that $[K]$ satisfies the correct exchange identity, for

$$\left[\text{⤬}\right] = (q - q^{-1})\left[\overrightarrow{\mathbf{v}}\right] + q\left[\overrightarrow{\text{II}}\right] + \left[\text{✕}\right]$$
$$+ (q^{-1} - q)\left[\text{↓>↓}\right] + q^{-1}\left[\text{↓=↓}\right]$$

$$\left[\text{✕}\right] = (q^{-1} - q)\left[\overrightarrow{\wedge}\right] + q^{-1}\left[\overrightarrow{\text{II}}\right] + \left[\text{✕}\right]$$
$$+ (q - q^{-1})\left[\text{↓<↓}\right] + q\left[\text{↓=↓}\right].$$

Hence

$$\left[\text{⤬}\right] - \left[\text{✕}\right] = (q - q^{-1})\left(\left[\overrightarrow{\mathbf{v}}\right] + \left[\overrightarrow{\wedge}\right] + \left[\overrightarrow{\text{II}}\right]\right) +$$
$$(q^{-1} - q)\left(\left[\text{↓>↓}\right] + \left[\text{↓<↓}\right] + \left[\text{↓=↓}\right]\right)$$
$$= (q - q^{-1})\left(\left[\asymp\right] - \left[\,\rangle\,\langle\,\right]\right).$$

There are two cases for the index set $\mathcal{I}$. For n *odd* we take
$\mathcal{I}_n = \{-n, -n+2, \ldots, -1, \tilde{0}, +1, 3, \ldots, n-2, n\}$ and for n *even* we take
$\mathcal{I}_n = \{-n, -n+2, \ldots, -2, 0, \tilde{0}, 2, 4, \ldots, n\}$.

Note the presence of an extra zero in each index set. The extra zeros act numerically like ordinary zeros. This $q^{\widetilde{0}} = 1$. Note also that for odd n the index $\widetilde{0}$ is special in that it is spaced by one unit from its neighbors, while all the other indices have spacing equal to two.

These new zeros will be handled by the

Zero Rules

1. If n is odd, then SHIFT the occurrence of $\overset{\widetilde{o}}{\underset{\widetilde{o}}{}}$ to an occurrence of $\times$, and assign it a vertex weight of 1. [For regular indices the crossed line configuration receives weight zero unless the two lines are labelled by distinct labels.]

2. For n even, treat the extra zero as an extra copy of the zero index.

As we shall see, these zero-rules are needed to ensure the invariance of the model under regular isotopy. The rules also make the model multiplicative under the type I move. We shall look at some of these issues shortly, but first it is worthwhile extracting a solution of the Yang-Baxter Equation from this model.

Since the model takes the form

$$\left[\times\right] = z\left[\,\right] - z\left[\,\right] + q\left[\,\right] + q^{-1}\left[\,\right] + \left[\times\right]$$

it is tempting to think that the associated scattering matrix must be

$$z\,\,\, - z\,\,\, + q\,\,\, + q^{-1}\,\,\, + \times$$

with conventions for handing $\times$ corresponding to the zero rules. However, this matrix does **not** satisfy the Yang-Baxter Equation! The correct scattering matrix is

$$M = z\left[\,\,\, - q^{\left(\frac{i-\dot{i}}{2}\right)}\,\,\,\right] + q\,\,\, + q^{-1}\,\,\, + \,\,\, + \,\,\,$$

(for the case n odd). Here I have included the Zero Rules in the diagrammatic notation, and an extra factor of $q^{\frac{i-i}{2}}$ on the horizontal split. **The extra factor is a rotation number compensation.**

In order to see the genesis of this compensation, consider the form of the model:

$$[K] = \sum_{\sigma} [K|\sigma] q^{\|\sigma\|}.$$

In this model it is no longer the case that the factor $q^{\|\sigma\|}$ remains constant for all the states corresponding to a given triangle move (fixed inputs and outputs). The simplest instance (and generic instance) of this change is seen in the difference between rotation numbers for

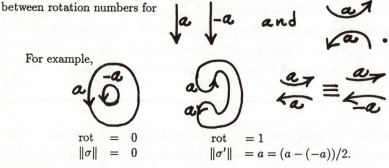

For example,

rot $= 0$ rot $= 1$

$\|\sigma\| = 0$ $\|\sigma'\| = a = (a - (-a))/2.$

The general compensation is given by a power of q as indicated in the diagram below:

$$\xleftrightarrow{} q^{(a-b)/2}.$$

Here is a second example:

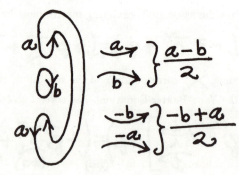

$$\|\sigma''\| = a - b = \left(\frac{a-b}{2}\right) + \left(\frac{-b+a}{2}\right).$$

It follows that **if** M satisfies the QYBE **then** $[K]$ is invariant under the III(A) triangle move. Thus we have,

Proposition 14.3. Let $[K]$ be defined for unoriented diagrams as described above. If the matrix M is a solution to QYBE, then $[K]$ is invariant under the type IIIA move.

Proof. In this model, we have defined

$$[K] = \sum_{\sigma} [K|\sigma] q^{\|\sigma\|}$$

where $[K|\sigma]$ is the given product of vertex weights for oriented states described prior to this proof. Let T denote the tangle corresponding to a configuration for the type III move, and let S denote a tangle representing the rest of K. Thus $K = T\#S$ in the sense of connecting the tangles as shown below:

$$T, \quad \boxed{T} \;\; \boxed{S} \quad\quad T\#S = K.$$

(S actually lives all over the plane exterior to T.)

Any state σ of K can be regarded as the #-connection of states σ_T and σ_S of T and S respectively. In order to prove invariance under the type III move we must consider all states σ_T and show that **for any given state** σ_S, the sum

$$\sum_{\sigma_T \in S(T, \sigma_S)} [K|\sigma] q^{\|\sigma\|}$$

(where $\sigma = \sigma_T \# \sigma_S$ and $S(T, \sigma_S)$ denotes all states of T whose end-conditions match the end-conditions of σ_S) is invariant under a triangle move applied to T. Now we have seen that for any state $\sigma = \sigma_T \# \sigma_S$, the norm, $\|\sigma\|$, is a sum of

contributions of the form $(a-b)/2$ for each occurence of in σ_T (horizontal with respect to T's vertical lines) plus a part that is independent of the choice of σ_T. Thus

$$\|\sigma\| = \left[\sum_{\substack{\text{\tiny in } \sigma_T}} \left(\frac{a-b}{2} \right) \right] + c$$

where c is independent of σ_T. This means that

$$\sum_{\substack{\sigma \in \mathcal{S}(T, \sigma_S) \\ \sigma = \sigma_T \# \sigma_S}} [K|\sigma] q^{\|\sigma\|} = \left[\sum_{\sigma_T} [K|\sigma] \prod q^{(a-b)/2} \right] q^c.$$

Thus, we must verify that the summation

$$\sum_{\sigma_T} [K|\sigma] \prod q^{(a-b)/2}$$

is invariant under the triangle move. If the terms $q^{(a-b)/2}$ are regarded as extra vertex weights, then this invariance follows at once from knowing that the matrix M is a solution to the Yang-Baxter Equation. This completes the proof of the proposition. //

In fact, with the zero-rules in effect, M does satisfy the Yang-Baxter Equation:

Proposition 14.4. Let the index set be taken in the form given in the preceding discussion, with the surrounding conventions for zeros. Then M is a solution to the QYBE, where M is the matrix we have described diagrammatically as

Discussion. Before proving this proposition, it is worthwhile delineating the standard matrix form for M: Let f_j^i denote an elementary matrix: f_j^i has a 1 in the (i, j) place and 0 elsewhere. Here the indices i, j belong to a given index set $\mathcal{I}$. Taking the case n odd, so that $\mathcal{I} = \{-n, -n+2, \ldots, -1, \tilde{0}, 1, 3, \ldots, n\}$, we then have

$$\begin{aligned}
M = {}& z \left[\sum_{i<j} f_i^i \otimes f_j^j \Big| - q^{(i-j)/2} \sum_{i<j} f_{-j}^i \otimes f_j^{-i} \right] \\
& + q \sum_{i \neq 0} f_i^i \otimes f_i^i + q^{-1} \sum_{i \neq 0} f_{-i}^i \otimes f_i^{-i} \\
& + \sum_{i \neq j, -j} f_j^i \otimes f_i^j + f_{\tilde{0}}^{\tilde{0}} \otimes f_{\tilde{0}}^{\tilde{0}} \\
& (z = q - q^{-1}).
\end{aligned}$$

I have arranged the indices on the elementary matrices so that they correspond to our abstract tensor conventions. Thus

$$\equiv f_a^a$$

$$\equiv \sum_{i<j} f_i^i \otimes f_j^j \ , \qquad \equiv \sum_{i\neq 0} f_i^i \otimes f_i^i$$

$$\equiv \sum_{i\neq j,-j} f_j^i \otimes f_i^j .$$

With these conventions, we see that M corresponds directly to the matrices cited in [TU1], [RES], [JI2]. In particular, [RES] gives an independent verification that M satisfies the QYBE via the quantum group for $SO(n)$. It would be interesting to understand how the rotation compensation is related to this representation theory.

In the proof to follow, and the remaining verifications, we return to the abstract tensor form for M. For diagrammatic analysis, this shorthand has clear advantages.

Proof of 14.3. I will consider only the case n odd, and verify two key cases. In the following, the notation $\bar{a}$ will be used for the negative of a: $\bar{a} = -a$.

Case 1. Assume that $0 < a < b$.

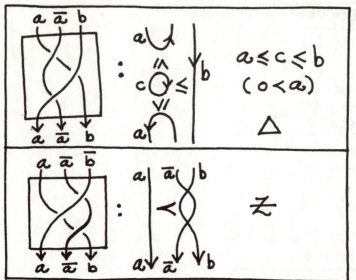

In the top part we have to sum over all the contributions where the center loop has spin c with $a \leq c \leq b$. Call this total contribution Δ. The bottom part contributes a factor of z. Summing up for Δ, we get

$$\Delta = q^{-2}z + (-z)^2 z \sum_{a<c<b} q^{(a-c)} + (-z)^2 q(q^{-2})^{k+1}, (k = (b-a-2)/2)$$

$$= q^{-2}z + z^2 q(1-q^2)[q^{-2} + (q^{-2})^2 + \ldots + (q^{-2})^k] + z^2 q(q^{-2})^{k+1}$$

$$= q^{-2}z + z^2 q(1-q^{-2})[(q^{-2} - (q^{-2})^{k+1})/(1-q^{-2})] + z^2 q(q^{-2})^{k+1}$$

$$= q^{-2}z + z^2 q(q^{-2} - (q^{-2})^{k+1}) + z^2 q(q^{-2})^{k+1}$$

$$= q^{-2}z + z^2 q^{-1}$$

$$= (q^{-2} + zq^{-1})z$$

$$= (q^{-2} + 1 - q^{-2})z$$

$$\Delta = z.$$

Thus M satisfies the Yang-Baxter Equation for this choice of indices.

Case 2.

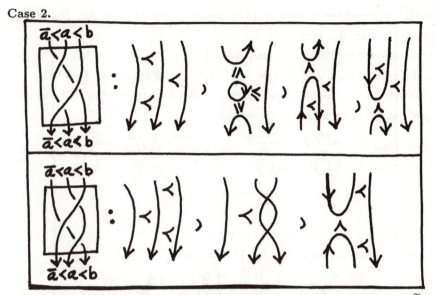

Here it is not hard to see that the extra spin contribution inserted at $\tilde{0}$ by the Zero-Rules is just what is needed to compensate for the apparent extra term in the top part. Thus top and bottom contribute equally.

These two cases are representative of the manner in which this matrix M is a solution to the Yang-Baxter Equation. With the remaining cases left to the reader, this completes the proof. //

Finally, we have

Theorem 14.5. With the index set $\mathcal{I}_n$ taken as in the preceding discussion (and the Zero Rules) the model $[K]$ is an invariant of regular isotopy. Furthermore these models are multiplicative under the type 1 move with

$$\left[\,\rotatebox{0}{σ^{-}}\,\right] = q^{n+1}\left[\,\sim\,\right]$$
$$\left[\,\rotatebox{0}{σ}\,\right] = q^{-n+1}\left[\,\sim\,\right].$$

Since $\left[\,\asymp\,\right] - \left[\,\times\,\right] = z\left(\left[\,\asymp\,\right] - \left[\,\rotatebox{0}{$)($}\,\right]\right)$ (by definition of the model), we conclude that these models give an infinite set of models for the Dubrovnik version of the Kauffman polynomial.

Proof. I will omit the verification of invariance under the move IIB. This leaves us to check multiplicativity under the type I move. Again, we omit the proof for n even, and take the index set to be $\mathcal{I} = \{-n, -n+2, \ldots, -1, \widetilde{0}, 1, 3, 5, \ldots, n\}$ for n odd. In fact, I shall take the case $\mathcal{I} = \{-1, \widetilde{0}, 1\}$ and work it out in detail:

In the diagram below, I have indicated the four types of contribution to $[\,\sigma\!\!\curvearrowright\,]$ and their respective vertex weights.

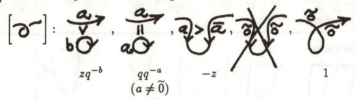

Note that we are crucially using the zero-rules.

The bookkeeping for $\mathcal{I} = \{-1, \widetilde{0}, 1\}$ then gives

$$\left[\sigma\!\!\curvearrowright\right] = \left[\overset{-1}{\sigma\!\!\curvearrowright}\right] + \left[\overset{\widetilde{0}}{\sigma\!\!\curvearrowright}\right] + \left[\overset{1}{\sigma\!\!\curvearrowright}\right].$$

We have

$$\left[\overset{-1}{\sigma\!\!\curvearrowright}\right] = \left[\overset{-1}{\sigma\!\!\!\!-\!\!1}\right] = q^2\left[\overset{-1}{\curvearrowright}\right]$$

$$\left[\overset{\widetilde{0}}{\sigma\!\!\curvearrowright}\right] = \left[\overset{\widetilde{0}}{\sigma\!\!\!-\!\!1}\right] + \left[\overset{}{\sigma\,\widetilde{0}}\right]$$

$$= (zq+1)\left[\overset{\widetilde{0}}{\curvearrowright}\right]$$

$$= (q^2-1+1)\left[\overset{\widetilde{0}}{\curvearrowright}\right]$$

$$= q^2\left[\overset{\widetilde{0}}{\curvearrowright}\right]$$

$$\left[\overset{1}{\sigma\!\!\curvearrowright}\right] = \left[\overset{1}{\sigma\!\!-\!\!1}\right] + \left[\overset{1}{\sigma\,\widetilde{0}}\right] + \left[\overset{1}{\sigma\,1}\right] + \left[\overset{1}{\curvearrowright\!\!-\!\!1}\right]$$

$$= (zq+z+qq^{-1}-z)\left[\overset{1}{\curvearrowright}\right]$$

$$= (q^2-1+q-q^{-1}+1-q+q^{-1})\left[\overset{1}{\curvearrowright}\right]$$

$$= q^2\left[\overset{1}{\curvearrowright}\right].$$

Thus $\left[\,\rotatebox{45}{$\sim$}\,\right] = q^2 \left[\,\sim\,\right]$ for $\mathcal{I} = \{-1, \tilde{0}, 1\}$. A similar calculation shows that for $\mathcal{I} = \mathcal{I}_n$,

$$\left[\,\rotatebox{45}{$\sim$}\,\right] = q^{n+1} \left[\,\sim\,\right]$$

$$\left[\,\rotatebox{-45}{$\sim$}\,\right] = q^{-n-1} \left[\,\sim\,\right].$$

This completes the proof. $//$

Summary.

We started by assuming the existence of a state model $[K]$ for D_K satisfying the expansion

$$\left[\,\times\,\right] = z \left[\,\asymp\,\right] + q \left[\,)=(\,\right] + \left[\,\times\,\right]$$
$$\quad - z \left[\,\rotatebox{45}{$\wedge$}\,\right] + q^{-1} \left[\,\rotatebox{90}{$\asymp$}\,\right]$$

and using the vertex weights for the $SL(n)$ R-matrix

$$\mathrel{\rotatebox{-20}{$\times$}} = z \,)<(\, + q \,)=(\, + \,\times.$$

We found that this led directly to a state model that indeed works (for special index sets)

$$[K] = \sum_{\sigma} [K|\sigma] q^{\|\sigma\|}$$

where the states are obtained by splicing or projecting the crossings of K to obtain a collection of inter-crossing Jordan curves. Orienting these curves, and labelling them from the index set yields the state σ. $[K|\sigma]$ is the product of vertex weights from the oriented R-matrix, and $\|\sigma\|$ denotes the sum of $\mathrm{rot}(\ell)$ $\mathrm{label}(\ell)$ over Jordan curves ℓ in σ. A second set of solutions to the Yang-Baxter Equation emerges from the structure of this model viz:

$$\mathcal{M} = z \left[\,)<(\, + q^{(i-j)/2} \,\rotatebox{45}{$\wedge$}\,\right]$$
$$\quad + q \,)=(\, + q^{-1} \,\rotatebox{90}{$\asymp$}\, + \,\times\!\!\#\, + \,\times\!\!\times$$
$$\qquad (\neq 0) \qquad (\neq 0)$$

(with special index set as specified in the previous discussion).

This identifies these models with the models constructed by Turaev in [TU1]. He begins with the solutions M identified as originating in the work of Jimbo [JI2]. In [RES] Reshetikhin shows that the solutions M emerge from the analysis of the quantum group for $SO(n)$. It would be very nice to see how the $SL(n)$ and $SO(n)$ solutions are related via the quantum groups. There may be a story at this level that is in parallel with the relationship that we have seen in the state models. In particular, there should be a quantum group interpretation for the rotation-compensation factor $q^{(i-j)/2}$. Finally, the matrix M and its inverse $\overline{M}$ can be used to create a specific representation of the Birman-Wenzl algebra [B1]. This is manifest from the structure of the model described here.

15⁰. Oriented Models and Piecewise Linear Models.

Recall from section 9^0 our treatment of unoriented abstract tensor models for link invariants. The invariant is viewed as a generalized vacuum-vacuum amplitude with creation, annihilation and braiding operators:

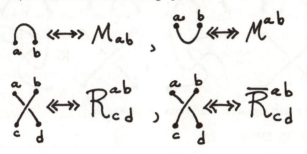

The topology underlying this viewpoint has as its underpinning the generalized Reidemeister moves that take into account maxima and minima in diagrams written with respect to a height function. These moves are:

1.

2.

3.

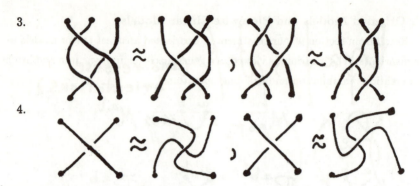

4.

They translate into algebraic conditions on M and R:

1. $M_{ai}M^{ib} = \delta_a^b$, $M^{ai}M_{ib} = \delta_b^a$

2. R and $\overline{R}$ are inverse matrices.

3. Yang-Baxter Equation for R and $\overline{R}$.

4. $\overline{R}_{cd}^{ab} = M_{ci}R_{dj}^{ia}M^{jb}$
   $\overline{R}_{cd}^{ab} = M^{ai}R_{ic}^{bj}M_{jd}$.

We have also remarked that in the case of the Drinfeld double construction, condition 4. translates to a condition about the antipode (and its representations) for this Hopf algebra. (Theorem 10.3).

A similar set of conditions holds for regular isotopy invariants of oriented link diagrams. It is worth recording these conditions here and comparing them both with Hopf algebra structures and with other forms of these models.

The generalization of the Reidemeister moves for the oriented case reads as follows:

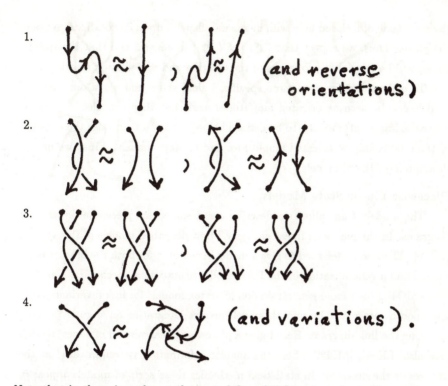

1.

2.

3.

4.

(and reverse orientations)

(and variations).

Note that in the oriented case there are left and right oriented creations and annihilations:  and a medley of braiding operators

The twist moves (4.) coupled with the two types of type 2. moves, let us retain the simple Yang-Baxter relations 3. (The other versions of the triangle move then follow from these relations.) The algebra relations are then the direct (abstract tensor) transcriptions of 1. → 4.

In this way, we can consider oriented models that take the form of vacuum-vacuum amplitudes.

Some features of the oriented amplitudes serve to motivate models that we have already discussed. For example, it is clear from this general framework that the amplitude of a closed loop will, in general, depend upon its rotational sense in the plane. Thus, we expect that $\langle \, \circlearrowright \, \rangle \neq \langle \, \circlearrowleft \, \rangle$ in general and that the specific values will be related to properties of the creation and annihilation matrices.

To make the story even more specific, I shall show that all of our previous models can be seen as oriented amplitudes **and** that they can also be seen as piecewise linear (pl) models (to be defined below). We can then raise the question of the relationship of oriented amplitudes and pl state models. First, we have a description of the pl models.

Piecewise Linear State Models.

The models I am about to describe are designed for piecewise linear knot diagrams. In the piecewise linear category I first described models of this type in [LK10]. These were designed to give a model in common for both the bracket polynomial and a generalization of the Potts model in statistical mechanics. Vaughan Jones [JO4] gave a more general version of vertex models for link invariants using the smooth category. The smooth models involve integrating an angular parameter along the link diagram. Here I give a piecewise linear version of Jones' models (see also [LK14], [LK23]) where the angular information is concentrated at the vertices of the diagram. In statistical mechanics these sorts of models appear in the work of Perk and Wu [PW] and Zamolodchikov [Z3] (and perhaps others).

A piecewise linear link diagram is composed of straight line segments so that the vertices are either **crossings** (locally 4-valent) or **corners** (2-valent). The diagram is oriented, and we assume the usual two types of local crossing. A matrix $S_{cd}^{ab}(\theta)$ is associated with each crossing as indicated below:

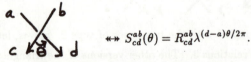

$$\longleftrightarrow \quad S_{cd}^{ab}(\theta) = R_{cd}^{ab}\lambda^{(d-a)\theta/2\pi}.$$

Here θ is the angle in radians between the two crossing segments measured in the counter-clockwise direction. It is assumed that R_{cd}^{ab} is a solution to the Yang-Baxter Equation for $a, b, c, d \in \mathcal{I}$ ($\mathcal{I}$ a specified index set) where $\mathcal{I} \subseteq \mathbf{R}$. Furthermore, we assume that R is **spin-preserving** in the sense that $R_{cd}^{ab} = 0$ unless $a + b = c + d$.

To the reverse crossing we associate $\overline{S}_{cd}^{ab}(\theta)$ as shown below:

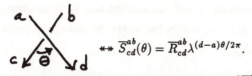

$$\leftrightarrow\rightarrow \overline{S}_{cd}^{ab}(\theta) = \overline{R}_{cd}^{ab}\lambda^{(d-a)\theta/2\pi}.$$

Here $\overline{R}$ is the inverse matrix for R in the sense that $R_{ij}^{ab}\overline{R}_{cd}^{ij} = \delta_c^a\delta_d^b$. Note that the angular part of the contribution is independent of the type of crossing.

Since the model is to be piecewise linear, we also have vertices of valence two. These acquire vertex weights according to the angle between the segments incident to the vertex:

$$\leftrightarrow\rightarrow \lambda^{\theta a/2\pi}$$

The **partition function** (or **amplitude**) for a given piecewise linear link diagram K is then defined by the summation $\langle K\rangle = \sum_\sigma [K|\sigma]$ where $[K|\sigma]$ denotes the product of these vertex weights and a **state** (**configuration**) σ is an assignment of spins to the edges of the link diagram (each edge extends from one vertex to another) such that spins are constant at the corners (two-valent vertices) and preserved ($a + b = c + d$) at the crossings.

We can rewrite this product of vertex weights as $[K|\sigma] = \langle K|\sigma\rangle\lambda^{\|\sigma\|}$ where $\langle K|\sigma\rangle$ is the product of R values from the crossings, and $\|\sigma\|$ is the sum of angle exponents from all the crossings and corners. Thus

$$\langle K\rangle = \sum_\sigma \langle K|\sigma\rangle\lambda^{\|\sigma\|}.$$

A basic theorem [JO4] gives a sufficient condition for $\langle K\rangle$ to be an invariant of regular isotopy:

Theorem 15.1. If the matrix R_{cd}^{ab} satisfies the Yang-Baxter Equation and if R also satisfies the cross-channel inversion

$$\sum_{i,j} R_{jc}^{ia}R_{id}^{jb}\lambda^{((c-i)/2+(d-j)/2)} = \delta_d^a\delta_c^b$$

then the model $\langle K\rangle$, as described above, is an invariant of regular isotopy.

Remark. The cross-channel condition corresponds to invariance under the reversed type II move. The diagram below shows the pattern of index contractions corresponding to $R^{ia}_{jc} R^{jb}_{id}$.

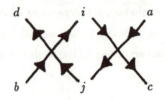

Before proving this theorem some discussion is in order.

Piecewise Linear Reidemeister Moves. These moves are shown in **Figure 17**. (As usual the moves in the figure are representatives, and there are a few more cases - such as switching a crossing in the type III move - that can also be illustrated.) In order to work with the invariance of the model under these moves, we have to keep track of the angles.

For example, consider move zero:

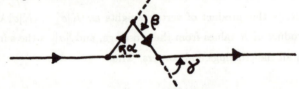

We have that the sum of the angles $\alpha + \beta + \gamma$, equals zero. (Since, in the Euclidean plane, the sum of the angles of a triangle is π.) Thus the product of vertex weights for the above configuration is the same as the product for a straight segment.

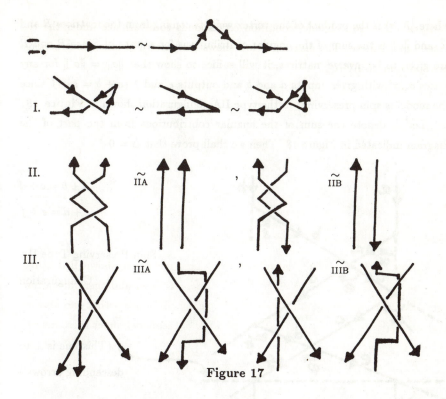

Figure 17

It is not hard to see that it is sufficient to assume that the main (input and output) lines for the type II move are parallel. The angular contributions involve some intricacy. To illustrate this point, and to progress towards the proof of the theorem, let's prove invariance of $\langle K \rangle$ under the move IIA.

Lemma 15.2. If the matrices R_{cd}^{ab} and $\overline{R}_{cd}^{ab}$ are inverses in the sense that $R_{ij}^{ab}\overline{R}_{cd}^{ij} = \delta_c^a \delta_d^b$, then the model $\langle K \rangle$ (using angular vertex weights as described above) satisfies invariance under the pl move IIA.

Proof. We know that $\langle K \rangle$ has a summation of the form

$$\langle K \rangle = \sum_\sigma \langle K|\sigma\rangle \lambda^{\|\sigma\|}$$

242

where $\langle K|\sigma \rangle$ is the product of the vertex weights coming from the matrices R and $\overline{R}$, and $\|\sigma\|$ is the sum of the angular contributions for a state σ. Since R and $\overline{R}$ are given to be inverse matrices, it will suffice to show that $\|\sigma\| = \|\sigma'\|$ for any states σ, σ' with given inputs a and b and outputs e and f $(a+b=e+f$ since the model is spin preserving) in the type IIA configuration shown in **Figure 18**.

Let Δ denote the sum of the angular contributions from the part of the diagram indicated in Figure 18. Then we shall prove that $\Delta = 0$

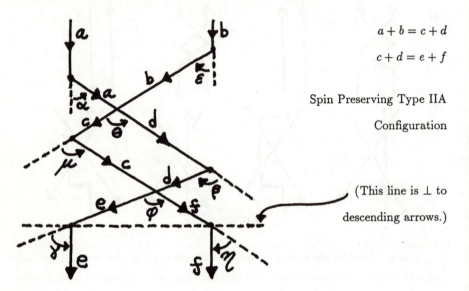

$$a+b = c+d$$

$$c+d = e+f$$

Spin Preserving Type IIA

Configuration

(This line is $\perp$ to

descending arrows.)

Figure 18

From Figure 18 it follows that

$$(*)\begin{cases} \alpha + \beta + \gamma & = 0 \\ \epsilon + \mu + \eta & = 0 \\ \theta + \phi & = \mu - \beta \\ a+b=c+d & = e+f \end{cases}$$

Thus

$$\Delta = a\alpha + b\epsilon + (d-a)\theta + \mu c + d\beta + (f-c)\varphi + \gamma e + \eta f.$$

To show that $\Delta = 0$, our strategy is to use the equations $(*)$ to simplify and reduce this formula for Δ. Thus

$$\Delta = a\alpha + d\beta + e\gamma + (d-a)\theta + b\epsilon + c\mu + f\eta + (f-c)\phi$$
$$= a\alpha + (a+b-c)\beta + (a+b-f)\gamma + (f-c)\phi + b\epsilon$$
$$\quad + (a+b-d)\mu + (a+b-e)\eta + (d-a)\theta$$
$$= (b-c)\beta + (b-f)\gamma + (f-c)\phi + (a-d)\mu + (a-e)\eta + (d-a)\theta$$
$$= (d-a)\beta + (b-f)\gamma + (f-c)\phi + (a-d)\mu + (a-e)\eta + (d-a)\theta$$
$$= (d-a)(\beta + \theta - \mu) + (b-f)\gamma + (f-c)\phi + (a-e)\eta$$
$$= (a-d)\phi + (b-f)\gamma + (f-c)\phi + (a-e)\eta$$
$$= (c-b)\phi + (f-c)\phi + (b-f)\gamma + (a-e)\eta$$
$$= (f-b)\phi + (b-f)\gamma + (f-b)\eta$$
$$= (f-b)(\phi - \gamma - \eta)$$
$$= (f-b)(0) \qquad \text{(By Figure 18)}$$
$$\therefore \Delta = 0.$$

//

An exactly similar calculation shows that in the case of the IIB move we have $\Delta/2\pi = [(c-b) + (f-d)]/2$ where the spins are labelled as in Figure 19

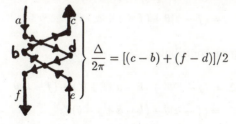

$$\frac{\Delta}{2\pi} = [(c-b) + (f-d)]/2$$

Figure 19

This explains the exponents in the cross-channel inversion statement for Theorem 15.1.

Proof of Theorem 15.1. By the Lemmas and discussion prior to this proof, it suffices to show that $\langle K \rangle$ is invariant under the move (IIIA) (Recall that IIA, IIB, IIIA together imply IIIB as in section 11^0.).

Now refer to Figure 20 and note that because of the standard angle form of the type III moves it will suffice to prove that the sum of the angle contributions are the same for the simplified diagram shown in Figure 20.

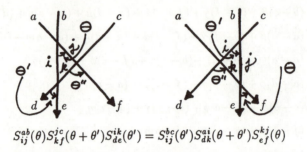

$$S_{ij}^{ab}(\theta)S_{kf}^{jc}(\theta + \theta')S_{de}^{ik}(\theta') = S_{ij}^{bc}(\theta')S_{dk}^{ai}(\theta + \theta')S_{ef}^{kj}(\theta)$$

Quantum Yang-Baxter Equation with Rapidity

Figure 20

In Figure 20 we see that for a given state σ with local spins $\{a, b, c, i, j, k, d, e, f\}$ the left hand diagram contribution to $2\pi \|\sigma\|$ by these spins is

$$\Delta = (j - a)\theta + (e - i)\theta' + (f - j)(\theta + \theta')$$
$$= (f - a)\theta + (f - j + e - i)\theta'.$$

The right hand diagram contributes

$$\Delta' = (j - b)\theta' + (k - a)(\theta + \theta') + (f - k)\theta$$
$$= (f - a)\theta + (k - a + j - b)\theta'.$$

Since $k + j = e + f$, we have $f - j + e - i = k - e + e - i = k - i$. Since $a + b = i + j$, we have $k - a + j - b = k + b - i - b = k - i$. Thus $\Delta = \Delta'$. Since $S_{cd}^{ab}(\theta) = R_{cd}^{ab}\lambda^{(d-a)\theta/2\pi}$, we see that this angle calculation, plus the fact that R satisfies the Yang-Baxter Equation implies that the model $\langle K \rangle$ is invariant under move IIIA. This completes the proof. //

Remark. In the course of this proof we have actually shown that if R_{cd}^{ab} satisfies the Yang-Baxter Equation, then $S_{cd}^{ab}(\theta) = R_{cd}^{ab}\lambda^{(d-a)\theta/2\pi}$ satisfies the Yang-Baxter Equation with rapidity parameter θ:

$$S_{ij}^{ab}(\theta)S_{kf}^{jc}(\theta + \theta')S_{de}^{ik}(\theta') = S_{ij}^{bc}(\theta')S_{dk}^{ai}(\theta + \theta')S_{ef}^{kj}(\theta).$$

In physical situations, from which this equation has been abstracted, the rapidity θ stands for the momentum difference between the particles involved in the interaction at this vertex.

A word of background about this rapidity parameter is in order. The underlying assumption is that the particles are interacting in the context of special relativity. This means that the momenta of two interacting particles can be shifted by a Lorentz transformation, and the resulting scattering amplitude must remain invariant. In the relativistic context (with light speed equal to unity) we have the fundamental relationship among energy (E), momentum (p) and mass (m):

$$E^2 - p^2 = m^2.$$

Letting mass be unity, we have

$$E^2 - p^2 = 1$$

and we can represent $E = \cosh(\theta)$, $p = \sinh(\theta)$. The Lorentz transformation shifts theta by a fixed amount. In light-cone coordinates (see section 9^0 of Part II) the Energy-Momentum has the form $[e^\theta, e^{-\theta}]$, and this is transformed via $[A, B] \mapsto [Ae^\phi, Be^{-\phi}]$ under the Lorentz transformation. Consequently, if two particles have respective **rapidities** (the θ parameter) θ_1 and θ_2, then the difference $\theta_1 - \theta_2$ is Lorentz invariant. The S-matrix must be a function of this difference of rapidities.

Now consider the diagram in Figure 20. Since momentum is conserved in the interactions, we see that for the oriented triangle interaction, the middle term has a rapidity difference that is the sum of the rapidity differences of the other two terms. Euclid tells us that (by the theorem on the exterior angle of a triangle) the angles between the corresponding lines satisfy just this relation.

Thus we identify rapidity differences with the angles between interaction lines, and obtain the correct angular relations from the geometry of the piecewise linear diagrams. This angular correspondence is remarkable, and it beckons for a deeper physical interpretation. This **is** a mystery of a sort because the Yang-Baxter Equation with rapidity parameter is the original form of the equation both in the context of field theory, and in the context of statistical mechanics. Some of the simplest solutions contain a nontrivial rapidity parameter. For example, let

$$S_{cd}^{ab} = \mathord{)}\mathord{(} + \theta \,\mathord{\times} = \delta_c^a \delta_d^b + \theta \delta_d^a \delta_c^b.$$

It is a nice exercise to check that $S_{cd}^{ab}(\theta)$ satisfies the QYBE with rapidity θ. Nevertheless, we do not know how to use this S to create a link invariant! The seemingly more complex form $S_{cd}^{ab}(\theta) = R_{cd}^{ab}\lambda^{(d-a)\theta/2\pi}$ is just suited to the knot theory.

Remark. To apply Theorem 15.1 one must adjust λ so that the condition for cross-channel unitarity is satisfied. We already know good examples of solutions to this problem from section 11^0 where

$$R_{cd}^{ab} = (q - q^{-1})\,\mathord{\downarrow}<\mathord{(} + q\,\mathord{)}=\mathord{(} + \,\mathord{\times}$$

and $\lambda = q$ for the index set $\mathcal{I}_n = \{-n, -n+2, \dots, n-2, n\}$. In this case it is easy to see that the state-evaluations match those of the model in section 11^0. There we regarded states σ as composed of loops of constant spin and each loop received (in $\|\sigma\|$) the product of the spin and rotation numbers. **The local angles in the pl model add up to give the same result.** To see this, first consider a 2-valent vertex with spin input a and output b: (We usually take $a = b$). Set the vertex weight to be $\lambda^{((b+a)/2)\theta/2\pi}$. This gives the usual result when $a = b$.

 $\longleftrightarrow \lambda^{((a+b/2))\theta/2\pi}$.

Now consider the angular contribution at a crossing:

 $\longleftrightarrow \lambda^{(d-a)\theta/2\pi}$ (angular contribution).

Here we assume that $a + b = c + d$. Thus if we break the crossing we find

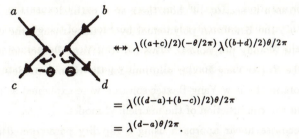

$$\longleftrightarrow \lambda^{((a+c)/2)(-\theta/2\pi)}\lambda^{((b+d)/2)\theta/2\pi}$$

$$= \lambda^{(((d-a)+(b-c))/2)\theta/2\pi}$$

$$= \lambda^{(d-a)\theta/2\pi}.$$

The angular contribution at a crossing is the same as the sum of the angular contributions from the corners obtained by splicing that crossing, and using the generalized vertex weights for the corners.

In the R-matrix of section 11^0 we have a correspondence between such splicings and the states as loops of constant spin. (Note that if $a = d$ and $b = c$ as in then $(d - a)\theta/2\pi = 0$.) Thus each loop receives the value ($a =$ spin value of the loop)

$$\|\text{loop}\| = \frac{a}{2\pi} \times (\text{the sum of the angles at the corners})$$

$$= \frac{a}{2\pi}(\theta_1 + \theta_2 + \ldots + \theta_n)$$

$$= a \cdot \text{rot(loop)}.$$

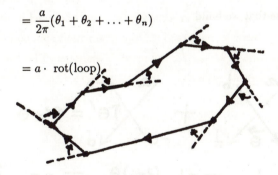

And this is exactly the evaluation we used in section 11^0. This completes the verification that the models of section 11^0 are special cases of this form of pl state model. It also shows how the pl state model can be construed as a vacuum-vacuum expectation, since we have already explained how the models of section 11^0 can be cast in this form.

A Remark on the Unoriented Model.

We have seen in section 14^0 how there arise naturally-state models for (sepcializations of) the Kauffman polynomial by "symmetrizing" the oriented state models for the Homfly polynomial. This led to a Yang-Baxter solution that was built from the $SL(n)$ Yang-Baxter solutions for the Homfly models. We saw how certain factors in the new Yang-Baxter solution were explained as rotation compensations in the construction of the unoriented model.

The piecewise linear approach using a rapidity parameter θ shows us the pattern of this tensor in a new light. Suppose we have

$$S^{ab}_{cd}(\theta) = R^{ab}_{cd}\lambda^{(d-a)\theta/2\pi}$$

corresponding to

and suppose that we build

$T^{ab}_{cd}(\theta) \leftrightarrow$

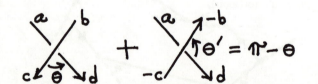

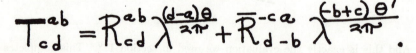

(We are using the convention that $\overset{a}{\rightarrow} \equiv \overset{-a}{\leftarrow}$.) Then

$$T_{cd}^{ab}(\theta) = R_{cd}^{ab}\lambda^{(d-a)\theta/2\pi} + \bar{R}_{d-b}^{-ca}\lambda^{(-b+c)(\pi-\theta)/2\pi}.$$

Here $a + b = c + d$ (i.e. we assume that R is spin-preserving). Hence $d - a = b - c$. Thus

$$T_{cd}^{ab}(\theta) = R_{cd}^{ab}\lambda^{(d-a)\theta/2\pi} + \bar{R}_{d-b}^{-ca}\lambda^{(d-a)(\theta-\pi)/2\pi}$$

$$= \left[R_{cd}^{ab} + \bar{R}_{d-b}^{-ca}\lambda^{(a-d)/2)}\right]\lambda^{(d-a)(\theta/2\pi)}.$$

Thus

$$T_{cd}^{ab}(\theta) = \widehat{R}_{cd}^{ab}\lambda^{(d-a)\theta/2\pi}$$

where

$$\widehat{R}_{cd}^{ab} = R_{cd}^{ab} + \bar{R}_{d-b}^{-ca}\lambda^{(a-d)/2}.$$

This tensor $\widehat{R}$ is exactly the tensor that we used in section 14^0 with $R = (q - q^{-1})\,\text{⌣}<\text{⌣} + q\,\text{⌣} = \text{⌣} + \text{✕}$, $\lambda = q$ and a special index set to produce our version of the Turaev models for the Kauffman polynomial. ($\widehat{R}$ is called M in section 14^0.)

It is a good research problem to determine general conditions on R that will guarantee that $\widehat{R}$ is a solution to the Yang-Baxter Equation. With this problem we close the lecture!

16^0. Three Manifold Invariants from the Jones Polynomial.

The purpose of this section is to explain how invariants of three dimensional manifolds can be constructed from the Jones polynomial. This approach, due to Reshetikhin and Turaev [RT2], instantiates invariants first proposed by Edward Witten [WIT2] in his landmark paper on quantum field theory and the Jones polynomial. (See section 17^0 for a description of Witten's point of view.)

The approach of Reshetikhin and Turaev uses quantum groups to construct the invariants. The invariants themselves are averages of link polynomials – adjusted so that the resulting summation is unchanged under the Kirby moves [KIRB]. We use the theorem of Dehn and Lickorish [LICK1] that represents three-dimensional manifolds by surgery on framed links in the three-sphere. Kirby [KIRB] gave a set of moves that can be performed on such links so that the classification of three dimensional manifolds is reduced to the classification of framed links up to ambient isotopy augmented by these moves.

In fact, I shall concentrate on a version of the Reshetikhin-Turaev invariant for the case of the classical Jones polynomial. In this case, Lickorish [LICK3] has given a completely elementary proof of the existence of the invariant based upon the bracket polynomial (section 3^0 and [LK4]) and certain properties of the Temperley-Lieb algebra in its diagrammatic form (as explained in section 7^0). Thus the present section only requires knowledge of the bracket polynomial for the construction of the three-manifold invariant.

I begin by describing the Kirby Calculus and the formula for the invariant. We then backtrack to fill in background on the Temperley-Lieb algebra.

Framed Links and Kirby Calculus.

Consider a single curve K embedded in three-dimensional space. We may equip this curve K with a normal vector field, and if the lengths of the normal vectors are small, then their tips will trace out a second embedded curve - moving near the first curve and winding about it some integral number of times. If K' is that second curve, endowed with an orientation parallel to that of K, then the winding number is equal to the linking number $\ell k(K, K')$. This number $n = \ell k(K, K')$ is called the **framing number of K**. Conversely, if we assign an integer n to K, then one can construct the normal vector field and curve K' so that $n = \ell k(K, K')$. The ambient isotopy class of K' in the complement of K is uniquely determined by this process.

By definition a **framed link** L is a link L together with an assignment of integers, one to each component of L. These integers can be used to describe normal framings of the components as indicated above. In working with link diagrams there is a natural framing that is commonly referred to as the **blackboard framing**: Assign to each component L_i of L the number $n_i = w(L_i)$ where $w(L_i)$ denotes the writhing number of the diagram L_i. That is, $w(L_i)$ is the sum of the signs of the crossings of L_i. (Note that this sum is independent of the orientation assigned to L_i, and hence makes sense for unoriented links.)

A normal vector field that produces the blackboard framing is obtained by using the plane on which the link is drawn. That is, except for the crossings, the vectors lie in the plane, and hence the push-off components L_i' are obtained by drawing essentially parallel copies of the L_i – as shown below:

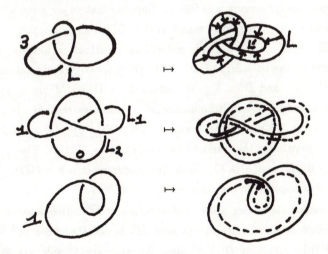

In each of these examples, I indicate the framing numbers on the link components, but note that via $w(L_i) = n_i$ these numbers are intrinsically determined by the diagram. In fact, any framing can be obtained as a blackboard framing since we can modify the writhe of a component by adding positive or negative curls. The

0-framed trefoil is the blackboard framing of the diagram

I denote a framing of the link L by the notation (L, f) where $f(L_i)$ is the integer assigned to the i-th component of L.

As we shall see, the study of three-dimensional manifolds is equivalent to a particular technical study of the properties of framed links. To be precise, Lickorish [LICK1] proved that every compact oriented three-dimensional manifold can be obtained via surgery on a framed link (L, f). Kirby [KIRB] gave a set of moves on framed links and proved that two three-manifolds are homeomorphic if and only if the corresponding links can be transformed into one another by these moves. Our aim here is to describe this surgery and the moves, and to show how the Jones polynomial can be used to produce invariants of three dimensional manifolds.

Surgery.

The basic idea of surgery is as follows. Suppose that we are given a 3-manifold M^3 bounding a four-dimensional manifold W^4. Given an embedding $\alpha : S^1 \times D^2 \to M$ we can use this embedding to **attach a handle**, $D^2 \times D^2$, to W^4. That is, the boundary of $D^2 \times D^2$ (D^2 is a two-dimensional disk) is the union of $S^1 \times D^2$ and $D^2 \times S^1$. We attach $S^1 \times D^2$ to M via α, and obtain a new four-manifold $\overline{W}^4$ with boundary $\overline{M}^3$. One says that $\overline{M}^3$ **is obtained via surgery on M^3 along the framed link** $\alpha(S^1 \times 0) \subset M$. The framing is given by the twisting of the standard longitude $\alpha(S^1 \times 1) \subset M$. That is, we have $K = \alpha(S^1 \times 0)$, $K' = \alpha(S^1 \times 1)$ and the framing number is $n = \ell k(K, K')$. (This linking number happens in the solid torus $\alpha(S^1 \times D^2)$.)

It is worthwhile having a direct description of $\overline{M}^3$. It is easy to see from our description of handle-attachment that $\overline{M}^3$ is obtained from $M^3 -$ Interior $(\alpha(S^1 \times D^2))$ by attaching $D^2 \times S^1$ along its boundary $S^1 \times S^1$ via the map α. This means that the twisting of the original **longitude** $S^1 \times 1 \subset S^1 \times D^2$ is now matched with a **meridian** $(S^1 \times 1) \subset D^2 \times S^1$ on the new attaching torus. In other words, **we cut-out the interior of $\alpha(S^1 \times D^2)$ and paste back a copy of $D^2 \times S^1$, matching the meridian of $D^2 \times S^1$ to the** (twisted by framing

number) longitude on the boundary torus in M.

Example 1. $K = \alpha(S^1 \times 0) \subset S^3$, unknotted and framing number is 3.

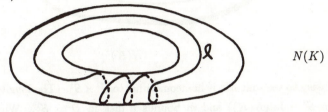

Let $N(K)$ denote $\alpha(S^1 \times D^2)$ and draw the longitude ℓ on the boundary of $N(K)$:
$\ell = \alpha(S^1 \times 1)$

$N(K)$

Let M^3 denote the manifold obtained by framed surgery along K with this framing. Then M is obtained from $S^3 - \text{Int}(N(K))$ by attaching $D^2 \times S^1$ so that the meridian $m = (D^2 \times 1)$ is attached to ℓ:

$(D^2 \times S^1)$

Since m bounds a disk in $(D^2 \times S^1)$, it is easy to see that M^3 has first homology group $\mathbf{Z}/3\mathbf{Z}$. M^3 is an example of a Lens space.

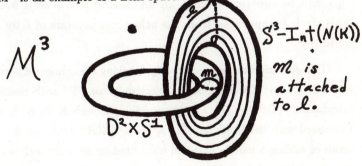

M^3

$D^2 \times S^1$

$S^3 - \text{Int}(N(K))$

m is attached to ℓ.

Example 2. Let this be the same as Example 1, but with framing number 0. Then ℓ is a simple longitude,

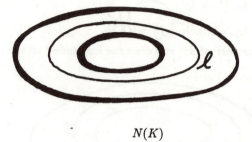

$$N(K)$$

and it is easy to see that M^3 is homeomorphic to $S^2 \times S^1$. (The longitude bounds a disk in $S^3 - \text{Int}(N(K))$ and m bounds a disk in $D^2 \times S^1$. With m and ℓ identified, we get a union of two disks forming a 2-sphere $S^2 \subset M^3$. The family of 2-spheres obtained by the family of longitudes $S^1 \times e^{i\theta}$ sweeps out M^3, creating a homeomorphism of M^3 and $S^2 \times S^1$.)

Kirby Moves.

We now describe two operations on framed links such that $\partial W_L \simeq \partial W_{L'}$ if and only if we can pass from L to L' by a sequence of these operations. (Notation: ∂W^4 denotes the boundary of the four manifold W^4. W_L^4 denotes the four manifold obtained from the 4-ball D^4 by doing handle attachments along the components of the framed link $L \subset S^3$.)

$\mathcal{O}_1$. **Add or subtract from L an unknotted circle with framing 1 or -1. This circle is separated from the other components of L by an embedded S^2 in S^3.**

This operation corresponds in W_L to taking the connected sum with (or splitting off) a copy of the complex projective space $\mathbf{CP}^2$ with positive or negative orientation. ($\mathbf{CP}^2 = D^4 \cup E$ where $\partial E = S^3$ and $E \to S^2$ is the D^2 bundle associated with the Hopf map $H : S^3 \to S^2$ [HOP]. One can show that E is the result of adding a handle along an unknotted circle in S^3 with $+1$ framing. This is a translation of the well-known property of the Hopf map that for $p \neq q$ in S^2, $H^{-1}\{p,q\}$ is the link ⬡ .)

$\mathcal{O}_2$. **Given two components L_i and L_j in L, we "add" L_i to L_j as follows:**

1. **Push L_i off itself (missing L) using the framing f_i on L_i to obtain L_i' with $\ell k(L_i, L_i') = f_i$.**

2. **Change L by replacing L_j with $\tilde{L}_j = L_i' \#_b L_j$ where b is any band missing the rest of L.**

The **band connected sum** $L_i' \#_b L_j$ is defined as follows: Let γ_0, γ_1 be two knots in S^3. Let $b : I \times I \to S^3$ be an embedding of $[0,1] \times [0,1]$ such that $b(I \times I) \cap \gamma_i = b(i \times I)$ for $i = 0, 1$. Then

$$\gamma_0 \#_b \gamma_1 = \gamma_0 \cup \gamma_1 - b(\partial I \times I) \cup b(I \times \partial I).$$

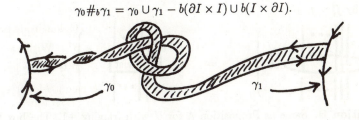

This second move corresponds in W_L to sliding the j-th handle over the i-th handle via the band b. In order to compute the framing $\tilde{f}_j$ of $\tilde{L}_j$, consider the intersection form on the second homology group $H_2(W_L; \mathbf{Z})$. If we orient each L_k then they determine a basis $\mathcal{B}$ for $H_2(W_L; \mathbf{Z})$ denoted by $\{[L_1], \ldots, [L_r]\}$. The matrix A_L of the intersection form on $H_2(W_L; \mathbf{Z})$ in this basis has the framing numbers f_k down the diagonal, and entries $a_{ij} = \ell k(L_i, L_j)$ for $i \neq j$. The move $\mathcal{O}_2$ corresponds to either adding $[L_i]$ or subtracting $[L_i]$ from $[L_j]$, depending on whether the orientations on L_i and L_j correspond under b. Thus

$$\tilde{f}_j = (\pm[L_i] + [L_j]) \cdot (\pm[L_i] + [L_j])$$
$$= [L_i] \cdot [L_i] + [L_j] \cdot [L_j] \pm 2[L_i] \cdot [L_j].$$

Hence

$$\boxed{\tilde{f}_j = f_i + f_j \pm 2a_{ij}}.$$

Call two framed links L and L' ∂-**equivalent** if we can obtain L' from L by a sequence of the operations $\mathcal{O}_1$ and $\mathcal{O}_2$ (plus ambient isotopy). We write $L \underset{\partial}{\sim} L'$.

Theorem 16.1. (Kirby-Craggs). Given two framed links L and L', then $L \underset{\partial}{\sim} L'$ if and only if ∂W_L is diffeomorphic to $\partial W_{L'}$ (preserving orientations). (Diffeomorphic and piecewise linearly homeomorphic are equivalent in this dimension.)

The next proposition gives a fundamental consequence of the Kirby moves: **A cable of lines encircled by an unknotted loop of framing ± 1 is ∂-equivalent to a twisted cable without this loop.**

Proposition A. Let L and L' be identical except for the parts shown in the figure below. Then $L \underset{\partial}{\sim} L'$. Here U is an unknot with framing -1 which disappears in L', and the box in L' denotes a full (2π) right handed twist. The framing on L'_i is given by $f'_i = f_i + 1$.

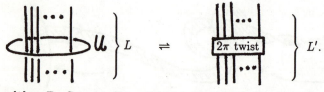

Proposition B. Same as Proposition A for U with framing $+1$. The box then denotes one full left hand twist and $f'_i = f_i - 1$.

Proof Sketch for A.

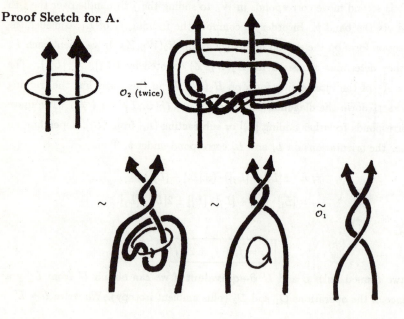

$//$

Since this proposition is so important, it is worthwhile discussing the proof sketch once again **in the language of blackboard framings and regular isotopy of diagrams.** In blackboard framing we have $f_i = w(L_i)$, the writhe of the i-th component. The equation for change of framing under operation $\mathcal{O}_2$ is (as remarked above) $\widetilde{f}_j = f_i + f_j \pm 2a_{ij}$. Hence we require

$$w(\widetilde{L}_j) = w(L_i) + w(L_j) \pm 2\ell k(L_i, L_j)$$

and this equation is **true whenever the band consists of parallel strands that contribute no extra writhe or linking!** Thus we can express the Kirby calculus in terms of the blackboard framing and regular isotopy of diagrams by simply stipulating that **band-sums are to be replaced by recombinations:**

Note that the parallel push-off from L_i creates a **copy of L_i** which then undergoes recombination with L_j to form $\widetilde{L}_j$. Processes of reproduction and recombination generate the underpinning of the Kirby calculus. [These formal similarities with processes of molecular biology deserve further study.] Now, in blackboard framing, let's return to the proof-sketch using, for illustration, a triple strand:

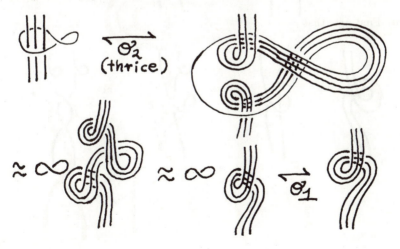

The generalization to n-strands is the move κ illustrated below.

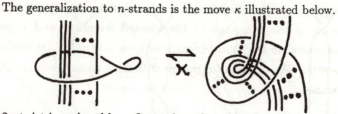

The 2π-twist is replaced by a flat curl on the cable of parallel strands.

We now move on to the beautiful theorem of Fenn and Rourke [FR]. They prove that **the move $\mathcal{O}_1$ together with the move κ of Proposition A/B generate the Kirby calculus.** This means that the move κ (above) (and its mirror image) together with blackboard $\mathcal{O}_1$ (∞ $\rightleftharpoons$ (nothing)) generate $\mathcal{O}_2$, and hence the entire Kirby calculus.

In order to see why we can generate move $\mathcal{O}_2$ using only $\mathcal{O}_1$ and the κ-move, we first observe that κ **lets us switch a crossing:**

Lemma 16.2.

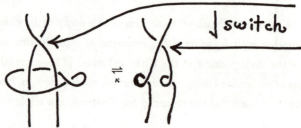

Proof. First note that

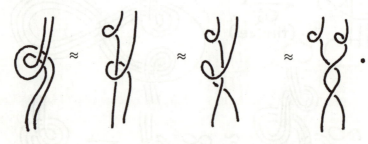

Thus

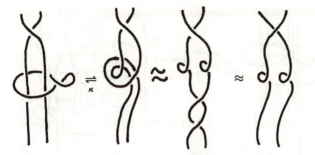

//

This lemma implies that we may take all the components in our link L to be **unknotted**. Furthermore, since

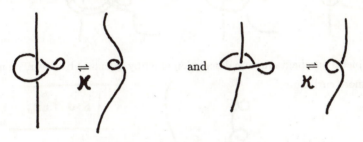

and

we can change the writhe of a single component until it is $+1$ or -1 or 0.

Thus it will suffice to use κ and $\mathcal{O}_1$ to obtain $\mathcal{O}_2$ for the situation.

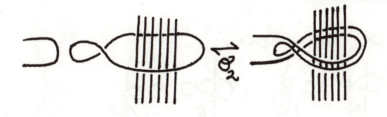

Here is a demonstration of this maneuver: (I replace the cable by a single strand.

The formalism for a full cable is identical in form.)

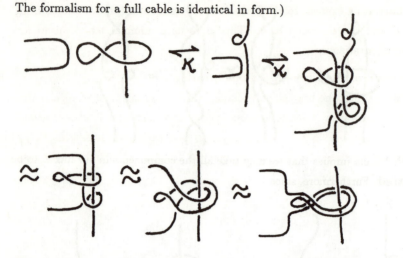

//

This completes our diagrammatic description of Kirby calculus. However, we must add one more curl-move:

$$\underset{c}{\rightleftharpoons}$$

With this move added, two unknotted curves are equivalent if and only if they have the same writhe (as desired). Thus our calculus is generated by

$\mathcal{O}_0$. regular isotopy

$\mathcal{O}_1$. $\infty \rightleftharpoons$ (blank) $\rightleftharpoons \infty$

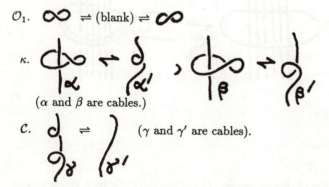

κ.

(α and β are cables.)

C. $\rightleftharpoons$ (γ and γ' are cables).

It is an interesting exercise to verify that the β-version of κ (above) follows from $\mathcal{O}_0$, $\mathcal{O}_1$, $\mathcal{C}$ and the α-version of κ. This gives us a minimal set of moves for investigating invariance.

Construction of the Three-Manifold Invariant.

We wish to use the bracket polynomial (which is an invariant of regular isotopy) and create from it a new regular isotopy invariant that is also invariant under the κ-move and the $\mathcal{C}$-move (a suitable normalization will take care of $\mathcal{O}_1$).

Since we must examine the behavior of the bracket on cables, it is best to take care of the $\mathcal{C}$-move first.

Proposition 16.3. Let ⌒$\overset{i}{\smile}$ denote an i-strand parallel cable. Let $\langle K \rangle$ denote the bracket polynomial of section 3^0. Then

$$\left\langle \overset{i}{} \right\rangle = \left\langle \overset{i}{} \right\rangle.$$

Hence the bracket polynomial itself satisfies move $\mathcal{C}$.

Remark. Recall that the bracket is determined by the equations

$$\left\langle \times \right\rangle = A \left\langle \asymp \right\rangle + A^{-1} \left\langle \supset\subset \right\rangle$$

$$\left\langle \bigcirc K \right\rangle = (-A^2 - A^{-2})\langle K \rangle$$

$$\left\langle \bigcirc \right\rangle = -A^2 - A^{-2} = \delta$$

and that it is an invariant of regular isotopy of unoriented link diagrams. It follows directly from these defining equations that (for single strands)

$$\left\langle \right\rangle = \left\langle \right\rangle = (-A^3) \left\langle \sim \right\rangle$$

and

$$\left\langle \right\rangle = \left\langle \right\rangle = (-A^{-3}) \left\langle \sim \right\rangle.$$

262

Proof. First recall that we have the regular isotopy

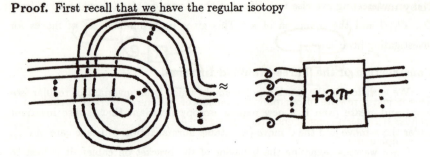

Thus we have the consequent regular isotopy

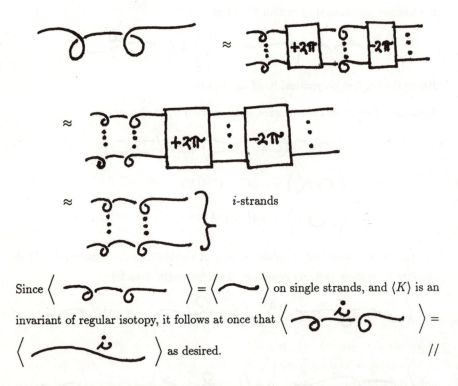

Since $\left\langle \text{—} \rangle = \langle \text{—} \right\rangle$ on single strands, and $\langle K \rangle$ is an invariant of regular isotopy, it follows at once that $\left\langle \overset{\iota}{\text{—}} \right\rangle = \left\langle \overset{\iota}{\text{—}} \right\rangle$ as desired.   //

As a result of Proposition 16.3, we can forget about move $\mathcal{C}$ and concentrate on making some combination of bracket evaluations invariant under the **basic**

Kirby move κ:

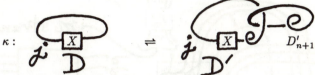

(This is just a rewrite of our previous version of κ. The j refers to a cable of j strands. If D has n components, then D' has $(n+1)$ components and D'_{n+1} is the new component.)

As it stands, the bracket will **not** be invariant under κ, but we shall consider the bracket values on cables

$$\widehat{T}_{i+j} = \underset{i+j}{\text{\raisebox{-2pt}{⊚}}} \approx \underset{i}{\text{⊚}}\;\underset{j}{\text{⊚}}$$

Let $T_{i+j} = \langle \widehat{T}_{i+j} \rangle$. Thus

$$T_1 = \left\langle \text{⊚} \right\rangle = -A^3$$

$$T_2 = \left\langle \text{⊚} \right\rangle = \left\langle \text{⊚} \right\rangle$$

$$= (-A^3)^2 \left\langle \text{⊚} \right\rangle$$

$$= +A^6(-A^4 - A^{-4})$$

$$T_2 = -A^{10} - A^2.$$

Of course, $T_0 = \langle \text{empty link} \rangle = 1$.

We are now in position to state the basic technique for obtaining invariance under the Kirby move κ. First consider two species of tangles that we shall denote by T_m^{in} and T_m^{out}. T_m^{in} consists in tangles with m inputs and m outputs so that the tangle is confined to the **interior** of a given tangle-box in the plane.

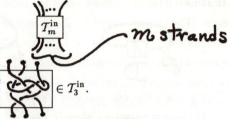

Thus

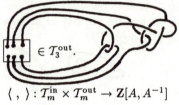

$\in T_3^{\text{in}}$.

Similarly, T_m^{out} consists of tangles restricted to the **outside** of this same box. For example,

$\in T_3^{\text{out}}$.

We have a pairing

$$\langle\ ,\ \rangle : T_m^{\text{in}} \times T_m^{\text{out}} \to \mathbf{Z}[A, A^{-1}]$$

defined by the formula

$$\langle x, y \rangle = \langle xy \rangle.$$

The second bracket denotes the regular isotopy invariant bracket polynomial, and the product xy denotes the link that results from attaching the free ends of $x \in T_m^{\text{in}}$ to the corresponding free ends of $y \in T_m^{\text{out}}$.

In this context it is convenient to consider formal linear combinations of elements in T_m^{in} and T_m^{out} (with coefficients in $\mathbf{Z}[A, A^{-1}]$). We can then write (by definition)

$$\langle X, aY + bZ \rangle = a\langle X, Y \rangle + b\langle X, Z \rangle$$

and

$$\langle aX + bW, Z \rangle = a\langle X, Z \rangle + b\langle W, Z \rangle$$

when $a, b \in \mathbf{Z}[A, A^{-1}]$, $X, W \in T_m^{\text{in}}$, $Y, Z \in T_m^{\text{out}}$. This makes $\langle\ -\ ,\ -\ \rangle$ a bilinear form on these modules of linear combinations. In the following discussion, **I shall let T_m^{in} and T_m^{out} denote this extension of tangles to linear combinations of tangles.**

With the help of this language, we shall now investigate when the following form of regular isotopy invariant, denoted $\ll K \gg$, is also invariant under the Kirby κ-move:

Let $\mathcal{I}_r$ be a fixed finite index set of the form $\mathcal{I}_r = \{0, 1, 2, \dots, r-2\}$ for $r \geq 3$ an integer. (The convenience of $r-2$ will appear later.) Let $\mathcal{C}(n, r)$ denote the set of all functions $c : \{1, 2, \dots, n\} \to \mathcal{I}_r$ where $n \geq 1$ is an integer. If $\{1, 2, \dots, n\}$ are the names of the components of a link L, then $c(1), c(2), \dots, c(n)$ assign labels to these components from the set $\mathcal{I}_r$.

Given an n-component link diagram L with components $L_1, L_2, \dots, L_n$ and given $c \in \mathcal{C}(n, r)$, **let $c * L$ denote the diagram obtained from L by replacing L_i by c_i parallel planar copies of L_i.** Thus

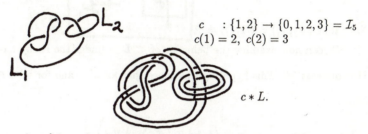

$$c \quad : \{1, 2\} \to \{0, 1, 2, 3\} = \mathcal{I}_5$$
$$c(1) = 2, \ c(2) = 3$$

$$c * L.$$

Let $\{\lambda_0, \lambda_1, \dots, \lambda_{r-2}\}$ be a fixed set of scalars. (Read scalar as an element of the complex numbers $\mathbf{C}$ - assuming we are evaluating knot polynomials at complex numbers.) Define $\ll L \gg$ via the formula

$$\ll L \gg = \ll L \gg (A, r, \lambda_0, \dots, \lambda_{r-2})$$
$$\ll L \gg = \sum_{c \in \mathcal{C}(n,r)} \lambda_{c(1)} \lambda_{c(2)} \cdots \lambda_{c(n)} \langle c * L \rangle.$$

Thus $\ll L \gg$ is a weighted average of bracket evaluations $\langle c * L \rangle$ for all the black-board cables that can be obtained from L by using $\mathcal{I}_r$. By using the blackboard framing, as described in this section, we can regard $\ll L \gg$ as a regular isotopy invariant of framed links.

We write the Kirby move in a form that can be expressed in products of tangles:

$$L \qquad\qquad L'$$

(L has n-components.)

Thus we have, in tangle language, $L = X\epsilon_j$, $L' = X\mu_{ij}$ where the tangles ϵ_j and μ_{ij} are defined by the diagrams below:

Thus $\epsilon_j \in T_j^{\text{out}}$ and $\mu_{ij} \in T_j^{\text{out}}$.

I add to this rogue's gallery of tangles the counterpart of ϵ_j in T_j^{in}, and call it 1_j:

Note that $\langle\, 1_j, \epsilon_j \,\rangle = \langle\, 1_j\epsilon_j \,\rangle = \delta^j$.

We can now examine the behavior of $\ll L \gg$ under the Kirby move:

Hypothesis $\mathcal{H}$. This hypothesis assumes that for all j and for all $x \in T_j^{\text{in}}$

$$\langle x, \epsilon_j \rangle = \Big\langle x, \sum_{i=0}^{r-2} \lambda_i \mu_{ij} \Big\rangle.$$

We can write hypothesis $\mathcal{H}$ symbolically by

Proposition 16.4. Under the assumption of hypothesis $\mathcal{H}$:

(i) For all j, $\delta^j = \sum_{i=0}^{r-2} \lambda_i T_{i+j}$ where $T_{i+j} = \Big\langle \text{image} \Big\rangle$.

(ii) $\ll L \gg$ is invariant under the Kirby move κ, and

$$\ll L \amalg \text{image} \gg\; = \tau \ll L \gg$$

(τ denotes $\sum_{i=0}^{r} \lambda_i \overline{T}_i$ where $\overline{T}_i = \Big\langle \text{image}\; i \Big\rangle$ and $\amalg$ denotes disjoint union.)

(iii) $\tau^{(\sigma+\nu-n)/2} \ll L \gg$ is an invariant of the 3-manifold obtained by framed surgery on L. Here σ and ν are the signature and nullity of the linking matrix associated with the n-component link L.

Proof. In this context, let $L' = x\mu_{1y}$ and $L = x\epsilon_j$ as explained above. Also, given $c : \{1,\ldots,n\} \to \mathcal{I}_r$, let $c * x$ denote the result of replacing the components of the tangle x by $c(i)$ parallels for the i-th component. Then if $x \in T_j^{in}$ we have $c * x \in T_{j'}^{in}$ where j' is the sum of $c(K)$ where K runs over the components of x that share input or output lines on the tangle. If we need to denote dependency on c, I shall write $j' = j(c)$. For $c \in \mathcal{C}(n,r)$ I shall abbreviate $\lambda_{\underset{c}{}}$ for $\lambda_{c(1)}\lambda_{c(2)}\ldots\lambda_{c(n)}$.

$$
\begin{aligned}
\ll L' \gg\; &= \;\ll x\mu_{1j} \gg \\
&= \sum_{c \in \mathcal{C}(n+1,r)} \lambda_{c(1)}\ldots\lambda_{c(n)}\lambda_{c(n+1)}\langle c * [x\mu_{1j}]\rangle \\
&= \sum_{c \in \mathcal{C}(n,r)} \lambda_{\underset{c}{}} \sum_{i=0}^{r-2} \lambda_i \langle (c * x)(\mu_{ij}(c))\rangle \\
&= \sum_{c \in \mathcal{C}(n,r)} \lambda_{\underset{c}{}} \left\langle c * x, \sum_{i=0}^{r-2} \lambda_i \mu_{ij(c)} \right\rangle \\
&= \sum_{c \in \mathcal{C}(n,r)} \lambda_{\underset{c}{}} \langle c * x, \; \epsilon_j(c)\rangle \qquad \text{(hypothesis } \mathcal{H}) \\
&= \sum_{c \in \mathcal{C}(n,r)} \lambda_{\underset{c}{}} \langle c * (x\epsilon_j)\rangle \\
&= \sum_{c \in \mathcal{C}(n,r)} \lambda_{\underset{c}{}} \langle c * L\rangle \\
&= \;\ll L \gg.
\end{aligned}
$$

This demonstrates the first part of (ii). To see the second part, let

$$L' = L\mathrm{II} \;\;\text{\reflectbox{$\circledcirc$}}$$

Then

$$\ll L' \gg = \sum_{c \in \mathcal{C}(n,r)} \lambda_{\frac{}{c}} \sum_{i=0}^{r-2} \lambda_i \langle c * L \, \text{II} \; \text{⬭} \quad i \rangle$$

$$= \sum_{c \in \mathcal{C}(n,r)} \lambda_{\frac{}{c}} \sum_{i=0}^{r-2} \lambda_i \langle c * L \rangle \langle \; \text{⬭} \quad i \rangle$$

$$= \left[\sum_{i=0}^{r-2} \lambda_i \langle \; \text{⬭} \quad i \rangle \right] \sum_{c \in \mathcal{C}(n,r)} \lambda_{\frac{}{c}} \langle c * L \rangle$$

$$\ll L' \gg = \tau \ll L \gg.$$

Now, turn to part (i): We have $T_{i+j} = \langle 1_j \mu_{ij} \rangle$. Therefore,

$$\delta^j = \langle 1_j \epsilon_j \rangle = \langle 1_j, \epsilon_j \rangle$$

$$= \left\langle 1_j, \sum_{i=0}^{r-2} \lambda_i \mu_{ij} \right\rangle \quad \text{(by } \mathcal{H})$$

$$\therefore \; \delta^j = \sum_{i=0}^{r-2} \lambda_i T_{i+j}.$$

Finally, to prove (iii) note that the $n \times n$ linking matrix of L has signature σ and nullity ν, so $\frac{1}{2}(n - \sigma - \nu)$ is the number of negative entries in a diagonalization of the matrix. That number is unchanged under the κ-move and increases by one when we add ⬭ . Hence (iii) follows directly from (ii). This completes the proof of the proposition. //

Now examine (i) of the above proposition. The set of equations

$$\delta^j = \sum_{i=0}^{r-2} \lambda_i T_{i+j} \qquad j = 0, 1, 2, \ldots$$

is an apparently overdetermined system. Nevertheless, as we shall see for the bracket, they can be solved when $A = e^{\pi i / 2r}$ and the result is a set of λ's satisfying hypothesis $\mathcal{H}$ and giving the desired three-manifold invariant. Before proceeding to the general theory behind this, we can do the example for $r = 3$ (and 4) explicitly. However, it is important to note that our tangle modules can be factored by relations corresponding to the generating formulas for the bracket:

$$\left\langle \times \right\rangle = A \left\langle \asymp \right\rangle + A^{-1} \left\langle \supset \subset \right\rangle$$

$$\left\langle \mathsf{LO} \right\rangle = \delta \langle L \rangle.$$

This means that we can replace the tangle modules T_m^{in} and T_m^{out} by their quotients under the equivalence on tangle diagrams generated by

(i) $\times = A \asymp + A^{-1} \supset \subset$

(ii) $OD = \delta D.$

We shall do this and **use the notation V_m^{in} and V_m^{out} for these quotient modules.**

With this convention, and assuming that A has been given a value in C, we see that V_m^{in} and V_m^{out} are finitely generated over C. For example, V_2^{in} has generators

$$\boxed{} \; 1 \quad , \quad \boxed{} \; e_1 \quad .$$

Note that V_m^{in} and V_m^{out} are abstractly isomorphic modules. I shall **use V_m for the corresponding abstract module** and **use pictures written horizontally to illustrate V_m.** Thus V_2 has generators $\rightrightarrows$ and $\supset\subset (1 \text{ and } e_1)$. We map $V_m \hookrightarrow V_{m+1}$ by adding a string on the top:

$$\boxed{X} \; \mapsto \; \boxed{X}$$

V_3 is generated by

$$\mathcal{B} = \left\{ \underset{1}{\equiv} \; , \; \underset{e_1}{\supset\subset} , \; \underset{e_2}{\supseteq\subseteq} , \; \underset{\alpha}{\supset\subseteq} , \; \underset{\beta}{\supseteq\subset} \right\}$$

That is, every element of V_3 is a linear combination of these elements. Under tangle multiplication V_m becomes an algebra over C. This is the **Temperley-Lieb algebra.** It first arose, in a different form, in the statistical mechanics of the Potts model (see [BA1]). This diagrammatic interpretation is due to the author ([LK4], [LK8], [LK10]) and is related to the way Jones constructed his V-polynomial [JO2] as we have explained in section 7^0. Multiplicatively, V_m is generated by the elements $\{1, e_1, e_2, \ldots, e_{m-1}\}$ where the e_i's are obtained by hooking together the i-th and $(i+1)$-th points on each end of the tangle:

It is easy to see that

$$
\left\{
\begin{array}{lll}
e_i^2 & = \delta e_i, & i = 1, \ldots, m-1 \\
e_i e_{i+1} e_i & = e_i, & i = 1, \ldots, m-2 \\
e_i e_{i-1} e_i & = e_i, & i = 0, 1, \ldots, m-1 \\
e_i e_j & = e_j e_i, & |i - j| > 1
\end{array}
\right\} \qquad (*)
$$

and one can take V_m as the algebra over $\mathbf{C}$ with these relations (δ has a chosen complex value for the purpose of discussion).

Remark. In Jones' algebra $e_i^2 = e_i$ while $e_i e_{i\pm 1} e_i = \delta^{-1} e_i$. This is obtained from the version given here by replacing e_i by $\delta^{-1} e_i$.

In V_3 we have

$$
e_1^2 = \;\overset{\cdot}{\rule{0pt}{0pt}}\;\diagram\; = \diagram = \delta e_1
$$

$$
e_1 e_2 e_1 = \diagram \approx \diagram = e_1.
$$

Note also that in V_3,

$$
e_1 e_2 = \diagram = \alpha
$$

$$
e_2 e_1 = \diagram = \beta.
$$

We have been concerned with the pairing $\langle\,,\,\rangle : \mathcal{T}_m^{\text{in}} \times \mathcal{T}_m^{\text{out}} \to \mathbf{C}$, $\langle x, y \rangle = \langle xy \rangle$. With respect to the Temperley-Lieb algebra this pairing can be written

$$
\langle\,,\,\rangle = V_m \times V_m \to \mathbf{C}
$$

$$
\langle x, y \rangle = \langle x\hat{y} \rangle
$$

where it is understood that x is identified with the corresponding element of T_m^{in} and $\hat{y}$, with the corresponding element of T_m^{out}. Thus

$$\langle \alpha, \beta \rangle = \left\langle \ \ , \ \ \right\rangle$$

$$= \left\langle \ \ \right\rangle = \delta^3.$$

In general,

$$\left\langle \ \ , \ \ \right\rangle = \left\langle \ \ \right\rangle$$

$$= \left\langle \ \ \right\rangle.$$

Thus, we can identify $x\hat{y}$ as the link obtained from the tangle product xy by closing with an identity tangle.

Example. The matrix for the pairing $V_2 \times V_2 \to \mathbf{C}$ is determined by

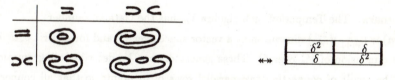

Note that this matrix is **singular** (for $\delta \neq 0$) when $\delta^4 - \delta^2 = 0 \Leftrightarrow \delta^2 - 1 = 0 \Leftrightarrow (-A^2 - A^{-2})^2 = 1 \Leftrightarrow A^4 + A^{-4} + 2 = 1$. Thus, we have $\langle \ , \ \rangle$ singular when $A = e^{i\pi/2r}$ for $r = 3$. For this same $r = 3$, we can consider the equations

$$\delta^j = \sum_{i=0}^{r-2} \lambda_i T_{i+j}$$

first for $j = 0, \ldots, r - 2$ and then for arbitrary j. For $j = 0, 1$, the system reads:

$$1 = \lambda_0 T_0 + \lambda_1 T_1$$

$$\delta = \lambda_0 T_1 + \lambda_1 T_2$$

and we know that

$$T_0 = 1$$
$$T_1 = -A^3$$
$$T_2 = -A^{10} - A^2.$$

Thus

$$\begin{bmatrix} \lambda_0 \\ \lambda_1 \end{bmatrix} = \frac{1}{T_2 - T_1^2} \begin{bmatrix} T_2 & -T_1 \\ -T_1 & 1 \end{bmatrix} \begin{bmatrix} 1 \\ \delta \end{bmatrix}.$$

If $A = e^{i\pi/6}$ then

$$T_2 - T_1^2 = -A^{10} - A^2 + A^6$$
$$= -A^4 A^6 - A^2 - 1$$
$$= A^4 - A^2 - 1$$
$$= e^{2\pi i/3} - e^{\pi i/3} - 1$$
$$\neq 0.$$

Hence we can solve for λ_0, λ_1. While it may not be obvious that these λ's provide a solution to the infinite system $\delta^j = \lambda_0 T_j + \lambda_1 T_{j+1}$ $j = 0, 1, 2, \ldots$, this is in fact the case! The singularity of the pairing $\langle \ , \ \rangle$ at $e^{\pi i/2r}$ is crucial, as we shall see.

Remark. The Temperley-Lieb algebra V_m has the Catalan number $d(m) = (\frac{1}{m+1})(\begin{smallmatrix} 2m \\ m \end{smallmatrix})$ dimension as a vector space over **C**; and for generic δ, $d(m)$ is the dimension of V_m over **C**. These generators can be described diagramatically as the result of connecting two parallel rows of m points so that all connecting arcs go between the rows, points on the same row can be connected to each other and no connecting arcs intersect one another. Thus for $m = 3$, we have

$$1 \qquad e_1 \qquad e_2 \qquad \alpha \qquad \beta$$

the familiar basis for V_3. For convenience, I shall refer to the non-identity generators of $V_m : \{e_1, e_2, \ldots, e_{m-1}, e_m, \ldots, e_{d(m)}\}$ where $e_m, \ldots, e_{d(m)}$ is a choice of labels for the remaining elements beyond the standard multiplicative generators $e_1, \ldots, e_{m-1}$. This notation will be used in the arguments to follow.

Theorem 16.5. Let $A = e^{\pi i/2r}$, $r = 3,4,5,\ldots$ Let V_m denote the Temperley-Lieb algebra for this value of $\delta = -A^2 - A^{-2}$. (Hence $\delta = -2\cos(\pi/r)$.) **Then there exists a unique solution to the system** $\delta^j = \sum\limits_{i=0}^{r-2} \lambda_i T_{i+j}$, $j = 0,1,2,\ldots$

In fact

(i) The matrix $M = (M_{ij})$ with $M_{ij} = T_{i+j-2}$, $1 \leq i,j \leq r$ is nonsingular for $A = e^{\pi i/2r}$.

and

(ii) The pairing $\langle\ ,\ \rangle : V_m \times V_m \to \mathbf{C}$ is **degenerate for** $A = e^{\pi i/2r}$ and $m \geq r-1$. In particular there exists a dependence so that $\langle - , m \rangle$ is a linear combination of pairings with the nonidentity elements e_i, $i = 1,\ldots,d(m)$.

Conditions (i) and (ii) imply the first statement of the theorem.

This theorem plus Proposition 16.4 implies an infinite set of 3-manifold invariants, one for each value of $r = 3,4,5,\ldots$ Before proving the theorem, I will show with two lemmas that conditions (i) and (ii) do imply that $\lambda_0,\ldots,\lambda_{r-2}$ solve the infinite system

$$\delta^j = \sum_{i=0}^{r-2} \lambda_i T_{i+j}.$$

This makes it possible for the reader to stop and directly verify conditions (i) and (ii) for small values of r ($r = 3,4,5,6$) if she so desires.

Assuming condition (i) we know that there exists a unique system $\lambda_0,\ldots,\lambda_{r-2}$ to the restricted system of equations

$$\delta^j = \sum_{i=0}^{r-2} \lambda_i T_{i+j}, \quad j = 0,1,\ldots,r-2.$$

Let $\phi_j = \sum\limits_{i=0}^{r-2} \lambda_i \mu_{ij} \in T_j^{\mathrm{out}}$ for any j. Thus $T_{i+j} = \langle 1_j \mu_{ij} \rangle = \langle 1_j, \mu_{ij} \rangle$, and therefore $\delta^j = \langle 1_j, \phi_j \rangle$ for $j = 0,1,\ldots,r-2$. We wish to show that this last equation holds for all j.

Lemma 16.6. $\langle x, \phi_j \rangle = \langle x, \epsilon_j \rangle$ for all $x \in V_j^{\mathrm{in}}$, $j = 0,1,\ldots,r-2$.

Proof. Since $\langle 1_j, \epsilon_j \rangle = \delta^j$, the lemma is true for $x = 1_j$ by condition (i). Let e_k be any other generating element of V_j^{in}. (V_j^{in} is T_j^{in} modulo the Temperley-Lieb relations.) e.g.

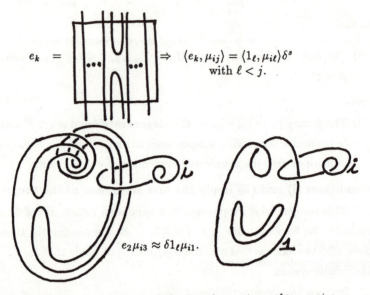

$e_k = \quad\Rightarrow\quad \langle e_k, \mu_{ij} \rangle = \langle 1_\ell, \mu_{i\ell} \rangle \delta^s$
with $\ell < j$.

$e_2 \mu_{i3} \approx \delta 1_\ell \mu_{i1}.$

By this basic annular isotopy, we conclude that $\langle e_k, \mu_{ij} \rangle = \delta^\ell \delta^s = \delta^j$ for each nonidentity Temperley-Lieb generator. The lemma follows at once from this observation. $\qquad$ //

Lemma 16.7. Assuming conditions (i) and (ii) as described in the statement of (16.5), $\langle x, \epsilon_j \rangle = \langle x, \phi_j \rangle$ for all $x \in V_j^{\text{in}}$ and for all $j = 0, 1, 2, \ldots$.

Proof. The proof is by induction on j. We assume that 16.7 is true for all integers less than j. By 16.6 we can assume that $j > r - 2$, since the hypothesis of induction is satisfied at least up to $r - 2$. By condition (ii) we can assume that there are coefficients α_k so that

$$\langle 1_j, - \rangle = \sum_{k=1}^{d(j)} \langle \alpha_k e_k, - \rangle$$

on V_j^{out}. Hence

$$
\begin{aligned}
\langle 1_j, \phi_j \rangle &= \sum_{k=1}^{d(j)} \alpha_k \langle e_k, \phi_j \rangle \\
&= \sum_{k=1}^{d(j)} \alpha_k \delta^{\ell(\kappa)} \langle 1_{j(\kappa)}, \phi_{j(\kappa)} \rangle \quad \text{(annular isotopy)} \\
&= \sum_{k=1}^{d(j)} \alpha_k \delta^{\ell(\kappa)} \langle 1_{j(\kappa)}, \epsilon_{j(\kappa)} \rangle \quad \text{(induction)} \\
&= \sum_{k=1}^{d(j)} \alpha_k \delta^{\ell(\kappa)} \delta^{j(\kappa)} \\
&= \sum_{k=1}^{d(j)} \alpha_k \langle e_k, \epsilon_j \rangle \\
\therefore \langle 1_j, \phi_j \rangle &= \langle 1_j, \epsilon_j \rangle.
\end{aligned}
$$

By using the annular isotopy for the other generators of V_j^{in} we complete this argument by induction. Hence $\langle x, \epsilon_j \rangle = \langle x, \phi_j \rangle$ for all j. //

Thus we have shown that conditions (i) and (ii) imply the existence of solutions $\lambda_0, \ldots, \lambda_{r-2}$ to the system $\delta^j = \sum_{i=0}^{r-2} \lambda_i T_{i+j}$ and hence have shown the existence of 3-manifold invariants for $A = e^{\pi i/2r}$. It remains to prove conditions (i) and (ii) of 16.5. This requires an excursion into the structure of the Temperley-Lieb algebra. This analysis of the Temperley-Lieb algebra will produce the following result:

Theorem 16.8. Let V_m denote the m-strand Temperley-Lieb algebra with multiplicative generators $1_m, e_1, e_2, \ldots, e_{m-1}$ as described in this section. Let δ denote the loop value for this algebra (assumed by convenience to be a complex number). Then there exist elements $f_n \in V_m$ for $0 \leq n \leq m-1$ such that

(i) $f_0 = 1_m$

(ii) $f_n = f_{n-1} - \mu_n f_{n-1} e_n f_{n-1}$ for $n \geq 1$

(iii) $e_i f_n = 0$ for $i \leq n$

(iv) $(e_{n+1}f_n)^2 = \mu_{n+1}^{-1} e_{n+1} f_n$ for $n \leq m - 2$

(v) $\mathrm{tr}(f_n) = \langle f_n \epsilon_n \rangle = \delta^{m-n-1} \Delta_{n+1}$.

Here μ_n is defined recursively by the formulas $\mu_1 = \delta^{-1}$, $\mu_n = (\delta - \mu_{n-1})^{-1}$, and
$\Delta_n = \prod_{k=1}^{n} (\delta - 2\cos(k\pi/(n+1)))$. //

We shall postpone discussion of the details of 16.8 to the end of this section. However, it is worth at least writing down f_1: According to 16.8(ii) we have $f_1 = 1_m - \delta^{-1} e_1$. Thus, in V_2 we have $f_1 =$ ⚎ $-\delta^{-1}$ ⊃⊄ and

$$f_1^2 = (1 - \delta^{-1} e_1)(1 - \delta^{-1} e_1)$$
$$= 1 - 2\delta^{-1} e_1 + \delta^{-2} e_1^2$$
$$= 1 - 2\delta^{-1} e_1 + \delta^{-2} e_1$$
$$= 1 - \delta^{-1} e_1$$
$$\therefore \; f_1^2 = f_1.$$

Also, $e_1 f_1 = e_1 - \delta^{-1} e_1^2 = 0$, and

$$\mathrm{tr}(f_1) = \left\langle \bigodot \right\rangle - \delta^{-1} \left\langle \otimes \right\rangle = \delta^2 - 1$$

$$= \delta^{2-1-1}(\delta^2 - 1). \qquad (m = 2, \; n = 1).$$

Note that by 16.8, $\Delta_2 = (\delta - 2\cos(\pi/3))(\delta - 2\cos(2\pi/3)) = (\delta - 2(1/2))(\delta - 2(-1/2)) = (\delta - 1)(\delta + 1) = \delta^2 - 1$.

Remark on the Trace. In the statement of 16.8 I have used the trace function $\mathrm{tr} : V_m \to \mathbf{C}$. It is defined by the formula, $\mathrm{tr}(x) = \langle x \epsilon_m \rangle$. We take the trace of an element of the Temperley-Lieb algebra by closing the ends of the corresponding tangle via the identity element, and then computing the bracket of the link so obtained.

The next result is a direct corollary of 16.8. Among other things, it provides the desired degeneracy of the form $V_m \times V_m \to \mathbf{C}$ for $A = e^{i\pi/2r}$, $r \geq 3$ and $m \geq r - 1$. In other words, 16.8 implies condition (ii) of 16.5.

Corollary 16.9. Let $A = e^{\pi i/r}$ for $r \geq 3$. Then

(i) If $m \leq r - 2$ then there exists an element $p(m) \in V_m$ such that $\langle -, p(m) \rangle = 1'_m$. [Here $1'_m$ is the element of V_m^* dual to $1_m \in V_m$. That is $1'_m(x)$ **equals the coefficient of 1_m in an expansion of x in the usual basis.**]

(ii) If $m \geq r - 1$, then the bilinear form $\langle \, , \, \rangle : V_m \times V_m \to \mathbf{C}$ is degenerate, and there exists an element $q(m) \in V_m$ such that $\langle -, q(m) \rangle = \langle -, 1_m \rangle$ and $q(m)$ is in the subalgebra generated by $\{e_1, e_2, \dots, e_{m-1}\}$.

Proof of 16.9. Recall from 16.8 that $\mathrm{tr}(f_n) = \delta^{m-n-1}\Delta_{n+1}$ for $f_n \in V_m$, and that $\Delta_n = \prod_{k=1}^{n} (\delta - 2\cos(k\pi/(n+1)))$. If $A^2 = e^{i\theta}$, then $\delta = -A^2 - A^{-2} = -2\cos(\theta)$. Thus $\Delta_n = \prod_{k=1}^{n} -2(\cos(\theta) + \cos(k\pi/(n+1)))$. The hypothesis of 16.9 is $\theta = \pi/r$ for $r \geq 3$. Hence $\Delta_1 \Delta_2 \dots \Delta_{r-2} \neq 0$, but $\Delta_{r-1} = 0$.

For $1 \leq m \leq r-2$ the element $f_{m-1} \in V_m$ is defined since $\Delta_1 \Delta_2 \dots \Delta_{m-1} \neq 0$. By 16.8, $e_i f_{m-1} = 0$ for $i = 1, 2, \dots, m-1$. Since the products of the e_i's generate a basis (other than 1_m) for V_m as a vector space over $\mathbf{C}$, we see that we can define $p(m) = \Delta_m^{-1} f_{m-1}$. This ensures that $p(m)$ projects everything but the coefficient of 1_m to zero, and

$$
\begin{aligned}
\langle 1_m, p(m) \rangle &= \langle 1_m, \Delta_m^{-1} f_{m-1} \rangle \\
&= \Delta_m^{-1} \mathrm{tr}(1_m f_{m-1}) \\
&= \Delta_m^{-1} \mathrm{tr}(f_{m-1}) \\
&= \Delta_m^{-1} \Delta_m \qquad \text{(by 16.8)} \\
&= 1.
\end{aligned}
$$

This completes the proof of 16.9 (i).

Now suppose that $m \geq r-1$. Then $\Delta_1 \Delta_2 \dots \Delta_{r-2} \neq 0$ so $f_{r-2} \in V_{r-1}$ is well-defined. Now f_{r-2} projects any basis element of V_{r-1} generated by $e_1, \dots, e_{r-2}$ to zero, **and** $\langle 1_{r-1}, f_{r-2} \rangle = \mathrm{tr}(f_{r-2}) = \Delta_{r-1} = 0$. Hence $\langle -, f_{r-2} \rangle : V_{r-1} \to \mathbf{C}$ is the zero map. For $m \geq r - 1$ we have the standard inclusion $V_{r-1} \hookrightarrow V_m$ obtained by adding $m - r + 1$ parallel arcs above the given element of V_{r-1} (in the convention of writing the diagrams of elements of V_m horizontally). It is then easy to see that $\langle -, f_{r-2} \rangle : V_m \to \mathbf{C}$ is the zero map for $m \geq r - 1$. Hence (since $f_{r-2} \neq 0$) the bilinear form is degenerate. Let $q(m) = 1_m - f_{r-2}$. This completes the proof of 16.9. //

Here is an example to illustrate 16.9 in the case $r = 4$. We have $f_1 = 1 - \delta^{-1}e_1$ and $e_1 f_1 = 0$.

$$f_2 = f_1 - \mu_2 f_1 e_2 f_1, \qquad \mu_2 = (\delta - \mu_1)^{-1}$$
$$= (1 - \delta^{-1}e_1)(1 - \mu_2 e_2(1 - \delta^{-1}e_1))$$
$$= (1 - \delta^{-1}e_1)(1 - \mu_2 e_2 + \mu_2 \delta^{-1}e_2 e_1)$$
$$= 1 - \mu_2 e_2 + \mu_2 \delta^{-1}e_2 e_1$$
$$\quad - \delta^{-1}e_1 + \delta^{-1}\mu_2 e_1 e_2 - \mu_2 \delta^{-2}e_1 e_2 e_1$$
$$= 1 - (\delta^{-1} + \mu_2 \delta^{-2})e_1 - \mu_2 e_2$$
$$\quad + \mu_2 \delta^{-1}e_2 e_1 + \delta^{-1}\mu_2 e_1 e_2$$
$$\therefore f_2 = 1 - \delta(\delta^2 - 1)^{-1}e_1 - \delta(\delta^2 - 1)^{-1}e_2$$
$$\quad + (\delta^2 - 1)^{-1}e_1 e_2 + (\delta^2 - 1)^{-1}e_2 e_1.$$

V_3 has basis $\{1, e_1, e_2, e_1 e_2, e_2 e_1\}$. Since $\delta = -2\cos(\pi/4) = -\sqrt{2}$, $f_2 \neq 0$. However, $\text{tr}(e_1) = \text{tr}(e_2) = \delta^2$, while $\text{tr}(e_1 e_2) = \text{tr}(e_2 e_1) = \delta$. Thus $\text{tr}(f_2) = (\delta/(\delta^2 - 1))(\delta^4 - 3\delta^2 + 2) = 0$, and $e_1 f_2 = e_2 f_2 = 0$. The element $q(3)$ has the formula

$$q(3) = \left(\frac{1}{\delta^2 - 1} \right)[\delta e_1 + \delta e_2 - e_1 e_2 - e_2 e_1]$$
$$q(3) = \sqrt{2}\, e_1 + \sqrt{2}e_2 - e_1 e_2 - e_2 e_1.$$

The beauty of 16.8 and 16.9 is that they provide specific constructions for these projections and degeneracies of the forms.

Since 16.9 implies condition (ii) of 16.5 (the degeneracy of the bilinear form) we have only to establish the uniqueness of solutions $\lambda_0, \ldots, \lambda_{r-2}$ of the equations

$$\delta^j = \sum_{i=0}^{r-2} \lambda_i T_{i+j}, \quad j = 0, 1, 2, \ldots, r - 2$$

in order to establish the existence of the three-manifold invariant. (Of course we will need to prove the algebraic Theorem 16.8.) That is, we must prove

Theorem 16.10. When $A = e^{\pi i/2r}$, $r \geq 3$, then the matrix $\{T_{i+j} : 0 \leq i, j \leq r-2\}$ is nonsingular.

Proof. Suppose that $\{T_{i+j}\}$ is singular. This means that there exist $\mu_i \in \mathbf{C}$, not all zero, such that

$$\sum_{i=0}^{r-2} \mu_i T_{i+j} = 0 \text{ for } j = 0, 1, \ldots, r-2.$$

Whence

$$\sum_{i=0}^{r-2} \mu_i \left\langle \vcenter{\hbox{[tangle diagram]}} \right\rangle = 0.$$

Now consider the functional on tangles given by the formula

$$\sum_{i=0}^{r-2} \mu_i \left\langle \vcenter{\hbox{[tangle diagram]}} \right\rangle.$$

Exactly the same "annulus trick" as we used in Lemma 16.6 then shows that this functional is identically zero on tangles. Hence

$$\sum_{i=0}^{r-2} \mu_i \left\langle \vcenter{\hbox{[tangle diagram]}} \right\rangle = 0.$$

Thus

$$\sum_{i=0}^{r-2} \mu_i \left\langle \vcenter{\hbox{[tangle diagram]}} \right\rangle = 0.$$

This means that the matrix

$$\left\{ \left\langle \vcenter{\hbox{[tangle diagram]}} \right\rangle : 0 \leq i, j \leq r-2 \right\}$$

is singular. Hence there exist constants $\nu_0, \nu_1, \ldots, \nu_{r-2}$ (not all zero) such that

$$\sum_{i=0}^{r-2} \nu_i \left\langle \vcenter{\hbox{[tangle diagram]}} \right\rangle = 0.$$

Now apply the annulus trick once more to conclude that

$$\sum_{i=0}^{r-2} \nu_i \left\langle \vcenter{\hbox{[tangle diagram]}} \right\rangle = 0,$$

280

and hence

$$\sum_{i=0}^{r-2} \nu_i \left\langle \begin{array}{c} \text{\raisebox{0pt}{\includegraphics}} \end{array} \right\rangle = 0$$

for all $j = 0, \ldots, r-2$ as a functional on tangles. In particular, we can put into this functional the projector element $p(j)$ of 16.9. Then

$$\left\langle \begin{array}{c} \boxed{p(j)} \end{array} \right\rangle$$

is equal to $1'_j(x^i_j)$, the coefficient of 1_j in the basis expansion for

$$x^i_j = \begin{array}{c} \text{\raisebox{0pt}{\includegraphics}} \end{array} \in V_j.$$

Note that since $\begin{array}{c}\text{\includegraphics}\end{array} = \begin{array}{c}\text{\includegraphics}\end{array}$ we actually have $x^i_j = (x_j)^i$, the i-th power of the element

$$x_j = \begin{array}{c} \text{\raisebox{0pt}{\includegraphics}} \end{array} \in V_j.$$

Therefore, to complete this analysis we need to compute $1'_j(x_j)$. (Note that $1'_j(x^i_j) = [1'_j(x_j)]^i$.) It is easy (bracket exercise!) to check by induction that if $A^2 = e^{i\theta}$, then $1'_j(x_j) = -2\cos(j+1)\theta$ for all j. Hence, in our case $e^{i\theta}$ is a primitive $2r$-th root of unity, whence $\{\cos(j+1)\theta \mid j = 0, 1, \ldots, r-2\}$ are all distinct. The matrix $\{(-2\cos(j+1)\theta)^i \mid 0 \le i, j \le r-2\}$ is a Vandermonde matrix, hence it has nonzero determinant. This contradicts the assumption that the original matrix was singular. //

The Temperley-Lieb Algebra.

It remains for us to prove Theorem 16.8. This is actually an elementary exercise in induction using the relations in the Temperley-Lieb algebra. Rather than do all the details, let's analyze the induction step to see why the relation $\mu_{n+1} = (\delta - \mu_n)^{-1}$ is needed, and see where the polynomials Δ_n come from. Thus we have $f_0 = 1$, $f_n = f_{n-1} - \mu_n f_{n-1} e_n f_{n-1}$. We assume inductively that $f_n^2 = f_n$ and that $(e_{n+1} f_n)^2 = \mu_{n+1}^{-1} e_{n+1} f_n$ ($n \leq m - 2$ for $f_n \in V_m$), and $e_i f_n = 0$ for $i \leq n$. Then

$$f_{n+1}^2 = (f_n - \mu_{n+1} f_n e_{n+1} f_n)^2$$

$$= f_n^2 + \mu_{n+1}^2 f_n e_{n+1} f_n^2 e_{n+1} f_n$$

$$- \mu_{n+1} f_n^2 e_{n+1} f_n - \mu_{n+1} f_n e_{n+1} f_n^2$$

$$= f_n + \mu_{n+1}^2 f_n (e_{n+1} f_n)^2 - 2\mu_{n+1} f_n e_{n+1} f_n$$

$$= f_n + \mu_{n+1} f_n e_{n+1} f_n - 2\mu_{n+1} f_n e_{n+1} f_n$$

$$= f_{n+1}.$$

Thus the crux of the matter is in the induction for $(e_{n+2} f_{n+1})^2$. We have

$$(e_{n+2}f_{n+1})^2 = (e_{n+2}(f_n - \mu_{n+1}f_n e_{n+1}f_n))^2$$

$$= (e_{n+2}f_n - \mu_{n+1}e_{n+2}f_n e_{n+1}f_n)^2$$

$$= e_{n+2}f_n e_{n+2}f_n$$

$$\quad + \mu_{n+1}^2 e_{n+2}f_n e_{n+1}f_n e_{n+2}f_n e_{n+1}f_n$$

$$\quad - \mu_{n+1}e_{n+2}f_n e_{n+2}f_n e_{n+1}f_n$$

$$\quad - \mu_{n+1}e_{n+2}f_n e_{n+1}f_n e_{n+2}f_n$$

$$= e_{n+2}^2 f_n^2$$

$$\quad + \mu_{n+1}^2 f_n e_{n+2}e_{n+1}e_{n+2}f_n^2 e_{n+1}f_n$$

$$\quad - \mu_{n+1}e_{n+2}^2 f_n^2 e_{n+1}f_n$$

$$\quad - \mu_{n+1}f_n e_{n+2}e_{n+1}e_{n+2}f_n^2$$

(Elements more than 1 unit apart in the $T - L$

algebra commute with one another.)

$$= \delta e_{n+2}f_n$$

$$\quad + \mu_{n+1}^2 f_n e_{n+2}f_n e_{n+1}f_n$$

$$\quad - \mu_{n+1}\delta e_{n+2}f_n e_{n+1}f_n$$

$$\quad - \mu_{n+1}f_n e_{n+2}f_n$$

$$= (\delta - \mu_{n+1})e_{n+2}f_n + (\mu_{n+1}^2 - \mu_{n+1}\delta)e_{n+2}f_n e_{n+1}f_n$$

$$= e_{n+2}(\delta - \mu_{n+1})(f_n - \mu_{n+1}f_n e_{n+1}f_n)$$

$$= (\delta - \mu_{n+1})e_{n+2}f_{n+1}.$$

Thus we require that

$$\mu_{n+2}^{-1} = (\delta - \mu_{n+1}), \text{ or}$$

$$\mu_1 = 1/\delta$$

$$\mu_{n+1} = 1/(\delta - \mu_n).$$

We have

$$\mu_1 = 1/\delta$$
$$\mu_2 = \delta/(\delta^2 - 1)$$
$$\mu_3 = (\delta^2 - 1)/(\delta^3 - \delta - 1)$$

and generally,

$$\mu_{n+1} = \Delta_n/\Delta_{n+1}$$

where

$$\Delta_0 = 1$$
$$\Delta_1 = \delta$$
$$\Delta_{n+2} = \delta\Delta_{n+1} - \Delta_n.$$

It is easy to see from this definition that

$$\Delta_n = \prod_{k=1}^{n}(\delta - 2\cos(k\pi/(n+1))).$$

(HINT: Let $\delta = x + x^{-1}$, then

$$\Delta_n = (x + x^{-1})\Delta_{n-1} - \Delta_{n-2}$$

$$\Rightarrow \Delta_n - x\Delta_{n+1} = x^{-1}(\Delta_{n-1} - x\Delta_{n-2})$$
$$\Rightarrow \Delta_n - x\Delta_{n-1} = x^{-n}$$
$$\Delta_n - x^{-1}\Delta_{n-1} = x^n$$
$$\Rightarrow (x - x^{-1})\Delta_n = x^{n+1} - x^{-(n+1)}.$$

I leave the calculation

$$\text{tr}(f_n) = \delta^{m-n-1}\Delta_{n+1}$$

as an exercise.

This completes the proof of Theorem 16.8, and hence the construction of the three-manifold invariant is now complete. //

Discussion. I recommend that the reader consult the papers [LICK3] of Lickorish, on which this discussion was based. The papers of Reshetikhin and Turaev ([RES], [RT1], [RT2]) are very useful for a more general viewpoint involving quantum groups. See also the paper by Kirby and Melvin [KM] for specific calculations for low values of r and a very good discussion of the quantum algebra. From the viewpoint of statistical mechanics, each multiple crossing

is a (twisted) version of a Potts model (see section 8^0 of Part II) interconnected with the rest of the knot diagram. This suggests that features of these invariants for $r \to \infty$ should be related to properties of the continuum limit of the Potts model. In particular the conjectural [AW4] relationships of the Potts model and the Virasoro algebra may be reflected in the behavior of these invariants.

17^0. Integral Heuristics and Witten's Invariants.

We have, so far, considered various ways to build link invariants as the combinatorial or algebraic analogues of partition functions or vacuum-vacuum expectations. These models tend to assume the form

$$\langle K \rangle = \sum_{\sigma} \langle K|\sigma \rangle \lambda^{\|\sigma\|}$$

where $\langle K|\sigma \rangle$ is a product of vertex weights, and σ runs over a (large) finite collection of states or configurations of the system associated with the link diagram. In this formulation, the summation $\sum_{\sigma} \lambda^{\|\sigma\|}$ is analogous to the bare partition function, and $\langle K \rangle$ is the analogue of an expectation value for the link diagram K. The link diagram becomes an observable for a system of states associated with its projection.

It is natural to wonder whether these finite summations can be turned into integrals. We shall see shortly that exactly this does happen in Witten's theory - using the Chern-Simons Lagrangian. The purpose of this section is to give a heuristic introduction to Witten's theory. In order to do this it is helpful to first give the **form** of Witten's definition and then play with this form in an elementary way. We then add more structure. Thus we shall begin the heuristics in **Level 0**; then we move to **Level 1** and beyond.

The Form of Witten's Definition.

Witten defines the link invariant for a link $K \subset M^3$, M^3 a compact oriented three-manifold, via a functional integral of the form

$$Z_K = \int dA e^{\mathcal{L}} \tau_K$$

where A runs over a collection of gauge fields on M^3. A gauge field (also called a gauge potential or gauge connection) is a Lie-algebra valued field on the three-manifold. $\mathcal{L}$ denotes the integral, over M^3, of a suitable multiple of the trace of a differential form (the Chern-Simons form). This differential form looks like $A \wedge dA + \frac{2}{3} A \wedge A \wedge A$. Finally, τ_K denotes the product of traces assigned to each component of K. If K has one component, then

$$\tau_K = \text{Trace}\left[\mathbf{P} \exp\left(\oint_K A\right)\right],$$

the trace of the path-ordered exponential of the integral of the field A around the closed circuit K. Now before going into the technicalities of these matters, we will look at the surface structure. For small loops K, the function τ_K measures curvature of the gauge field A [exactly how it does this will be the subject of the discussion at Level 1]. The Chern-Simons Lagrangian $\mathcal{L}$ also contains information about curvature. This information is encoded in the way $\mathcal{L}$ behaves when the gauge field is made to vary in the neighborhood of a point $x \in M^3$. [At Level 1, we shall see that the curvature tensor arises as the variation of the Chern-Simons Lagrangian with respect to the gauge field.] These two modes of curvature measurement interact in the integral Z_K to give rise to the link invariant.

We can begin to see this matter intuitively by looking at two issues from the knot theory. One issue is the behavior of Z_K on a "curl"  .
The other issue is the behavior of Z_K on a crossing switch

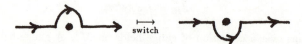

Since the curl involves a small loop, it is natural to expect that $\tau_{\curvearrowright}$ will differ from $\tau_{\rightarrow}$ by some factor involving curvature. These factors must average out to a constant multiple when the integration is performed.

The difference between $\times$ and $\times$ can be regarded as a small loop encircling one of the lines. Thus if we write one line perpendicular to the page (as a dot $\bullet$), then the crossing switch has the appearance

and the difference is a tiny loop around the line normal to the page:

Thus, we expect the curvature to be implicated in **both** the framing behavior **and** the change under crossing switch. [This approach to understanding the Witten integral is due to Lee Smolin [LS1], the author [LK29] and P. Cotta Ramusino,

Maurizio Martellini, E. Guadagnini and M. Mintchev [COTT].]

Level 0.

This level explores a formalism of implementing the ideas related to curvature that we have just discussed. Here the idea is to examine the simplest possible formalism that can hold the ideas. As we shall see, this can done very elegantly. But this is a case of form without content ("beautifully written but content free"). There is an advantage to pure form - it makes demands on the imagination, and maybe there is a - yet unknown - content that fits this form. We shall see.

To work. Let us try to obtain a formal model for the regular isotopy version of the homfly polynomial. That is, let's see what we can do to obtain a functional on knots and links that behaves according to the pattern:

$$\left\{ \begin{array}{rcl} \nabla\!\!\!\nearrow\!\!\!\nwarrow & - & \nabla\!\!\!\nwarrow\!\!\!\nearrow = z\nabla\!\!\!\rightrightarrows \\ \nabla\!\!\!\circlearrowright & = & a\nabla\!\!\!\rightarrow \\ \nabla\!\!\!\circlearrowleft & = & a^{-1}\nabla\!\!\!\rightarrow \end{array} \right\}$$

To this end we shall write

$$\nabla_K = \int dA e^{\mathcal{L}} \tau_K$$

and make the following assumptions:

(1) $\dfrac{\delta\mathcal{L}}{\delta A} = F$

(2) $\tau_{\nearrow\!\!\nwarrow} - \tau_{\nwarrow\!\!\nearrow} = -zF\tilde{\tau}_{\rightrightarrows}$

(3) $\tau_{\circlearrowright} = -aF\tilde{\tau}_{\rightarrow}$

$\tau_{\circlearrowleft} = -a^{-1}F\tilde{\tau}_{\rightarrow}$

(4) $\delta\tilde{\tau}/\delta A = \tau$

Let it be understood that F represents (the idea of) curvature - so that (1), (2) and (3) express the relationship of curvature to the Lagrangian $\mathcal{L}$, to the switching relation and to the presence of curls. In this Level 0 treatment we use the formalism of freshman calculus, differentiating as though A were a single variable.

Nevertheless, we do take into account the fact that a curve of finite size will not pinpoint the curvature of the field at its center. Thus the local skein relation

288

(2) involves a functional $\tilde{\tau}$ so that τ 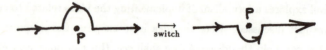 $= -zF\tilde{\tau}$. Strictly speaking, one should think of F as measuring curvature at the switch-point p:

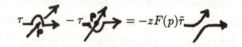

Thus

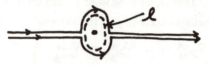

We have "rolled up" the extra geometry of the small finite loop ℓ

into the function $\tilde{\tau}$ and let the curvature evaluation happen at the center point p. Now in fact we can see approximately what this assumption (4) $(\delta\tilde{\tau}/\delta A = \tau)$ means, for

$$\tau\,\diagdown\!\!\diagup\; - \tau\,\diagdown\!\!\diagup\; = -zF(p)d\alpha\tau\,\longrightarrow$$

where $d\alpha$ is a function of the area enclosed by the loop ℓ. Thus, to first order of approximation, we are assuming a proportionality

$$\tau\delta A = d\alpha\delta\tau$$

where δA represents the variation of the gauge field in the neighborhood of the line, and $d\alpha$ represents a small area normal to the line. Thus the assumption $\delta\tilde{\tau}/\delta A = \tau$ is an **assumption of a correspondence between field change and geometry near the line.**

With these assumptions in place, and assuming that boundary terms in the integration by parts vanish, we can prove the global skein relation and curl identities:

Proposition 17.1. Under the assumptions (1), (2), (3) and (4) listed above, the formal integral $\nabla_K = \int dA e^{\mathcal{L}} \tau_K$ satisfies the following properties:

(i)
$$\nabla_{\nearrow\!\!\nwarrow} - \nabla_{\searrow\!\!\nwarrow} = z \nabla_{\rightrightarrows}$$

(ii)
$$\nabla_{\curvearrowright} = a \nabla$$

$$\nabla_{\curvearrowleft} = a^{-1} \nabla$$

Proof. To show (i), we take the difference and integrate by parts.

$$\nabla_{\nearrow\!\!\nwarrow} - \nabla_{\searrow\!\!\nwarrow} = \int dA e^{\mathcal{L}} \tau_{\nearrow\!\!\nwarrow} - \int dA e^{\mathcal{L}} \tau_{\searrow\!\!\nwarrow}$$

$$= \int dA e^{\mathcal{L}} [\tau_{\nearrow\!\!\nwarrow} - \tau_{\searrow\!\!\nwarrow}]$$

$$= \int dA e^{\mathcal{L}} (-z F \tilde{\tau}_{\rightrightarrows}) \tag{2}$$

$$= -z \int dA (e^{\mathcal{L}} F) \tilde{\tau}_{\rightrightarrows}$$

$$= -z \int dA \frac{\delta e^{\mathcal{L}}}{\delta A} \tilde{\tau}_{\rightrightarrows} \tag{1}$$

$$= z \int dA e^{\mathcal{L}} \frac{\delta \tilde{\tau}_{\rightrightarrows}}{\delta A} \qquad \text{(integration by parts)}$$

$$= z \int dA e^{\mathcal{L}} \tau_{\rightrightarrows} \tag{4}$$

$$= z \nabla_{\rightrightarrows} \,.$$

Part (ii) proceeds in a similar manner. We shall verify the formula for ∇⌒⃗ .

$$\nabla\text{⌒⃗} = \int dA\, e^{\mathcal{L}} \tau\text{⌒⃗}$$

$$= \int dA\, e^{\mathcal{L}}(-aF\tilde{\tau}\text{⌒⃗}) \tag{3}$$

$$= -a \int dA\, (e^{\mathcal{L}}F)\tilde{\tau}\text{⌒⃗}$$

$$= -a \int dA \left(\frac{\delta e^{\mathcal{L}}}{\delta A}\right)\tilde{\tau}\text{⌒⃗} \tag{1}$$

$$= a \int dA\, e^{\mathcal{L}}\, \frac{\delta\tilde{\tau}\text{⌒⃗}}{\delta A} \qquad \text{(integration by parts)}$$

$$= a \int dA\, e^{\mathcal{L}}\tau\text{⌒⃗} \tag{4}$$

$$= a\nabla\text{⌒⃗} .$$

This completes the proof. //

Discussion. Notice that in the course of this heuristic we have seen that a plausible condition on the relation of field behavior to geometry can lead the integral over such fields to average out all the local (curvature) behavior and to give global skein and framing identities. It would be interesting to follow up these ideas in more detail, as they suggest a particular physical interpretation of the meaning of the skein and framing identities. In fact, much of this elementary story does go over to the case of the full three-dimensional formalism of gauge fields. This we'll see in Level 1.

Remark. Note that in this picture of the behavior of $\nabla_K = \int dA\, e^{\mathcal{L}}\tau_K$, we see that ∇_K is **not** an ambient isotopy invariant. Its values depend crucially on the choice of framing, here indicated by the blackboard framing of link diagrams.

Remark. It appears that the best way to fulfill the promise of this heuristic is to use the full apparatus of gauge theory. Nevertheless, this does not preclude the possibility of some other solution to our conditions (1) → (4).

In any case, it is fun to play with the Level 0 heuristic. For example, we can expand

$$\tau_K = e^{\oint_K A} = \sum_{n=0}^{\infty}\left[\iint \cdots \int_{K^n} A^n\right]\Big/ n!.$$

Thus

$$\nabla_K = \sum_{n=0}^{\infty} \frac{1}{n!} \int dA e^{\mathcal{L}} \iint \cdots \int_{K^n} A^n$$

$$= \sum_{n=0}^{\infty} \frac{1}{n!} \iint \cdots \int_{K^n} \left[\int dA e^{\mathcal{L}} A^n \right].$$

Thus

$$\nabla_K = \sum_{n=0}^{\infty} \frac{1}{n!} \iint \cdots \int_{K^n} \langle A(x_1) \ldots A(x_n) \rangle$$

where $\langle A(x_1) \ldots A(x_n) \rangle$ denotes the correlation values of the product of these fields for n points on K. Thus, in this expansion the link invariant becomes an infinite sum of Feynman-diagram like evaluations of "self-interactions" of the link K. (See [BN], [GMM].)

Level 1.

Level 0 had the advantage that we could think in freshman calculus level. In order to do Level 1, a leap is required, and we must review some gauge theory. So let us begin by recalling $SU(2)$:

$SU(2)$ is the group of unitary, determinant 1 matrices over the complex numbers. Thus $U \in SU(2)$ has the form $U = \begin{pmatrix} a & b \\ c & d \end{pmatrix}$ with $ad - bc = 1$ an $\bar{d} = a, \bar{c} = -b$ where the bar denotes complex conjugation.

Let U^* denote the conjugate transpose of U, and note that

$$U = \begin{pmatrix} a & b \\ -\bar{b} & \bar{a} \end{pmatrix} \text{ with } a\bar{a} + b\bar{b} = 1$$

so that

$$UU^* = \begin{pmatrix} a & b \\ -\bar{b} & \bar{a} \end{pmatrix} \begin{pmatrix} \bar{a} & -b \\ \bar{b} & a \end{pmatrix} = \begin{pmatrix} 1 & 0 \\ 0 & 1 \end{pmatrix} = 1.$$

If we write $U = e^{i\eta}$ where η is a 2×2 matrix, then the conditions $UU^* = 1$ and $\text{Det}(U) = 1$ become $\eta = \eta^*$ and $\text{tr}(\eta) = 0$ (where tr denotes the usual matrix trace). A matrix of trace zero has the form

$$\eta = \begin{pmatrix} x & y \\ x & -x \end{pmatrix} = x \begin{pmatrix} 1 & 0 \\ 0 & -1 \end{pmatrix} + y \begin{pmatrix} 0 & 1 \\ 0 & 0 \end{pmatrix} + z \begin{pmatrix} 0 & 0 \\ 1 & 0 \end{pmatrix}$$

and the demand $\eta = \eta^*$ gives rise to the Pauli matrices

$$\tau_1 = \begin{pmatrix} 0 & 1 \\ 1 & 0 \end{pmatrix}, \tau_2 = \begin{pmatrix} 0 & -i \\ i & 0 \end{pmatrix}, \tau_3 = \begin{pmatrix} 1 & 0 \\ 0 & -1 \end{pmatrix}$$

whose linear combinations give rise to the matrices η.

It is customary to write η in the form $\eta = -\theta(\hat{n} \cdot \tau)/2$ where $\hat{n} = (n_1, n_2, n_3)$ is a real unit vector, and $\hat{n} \cdot \tau = n_1\tau_1 + n_2\tau_2 + n_3\tau_3$. Then

$$e^{i\eta} = e^{-i\theta(\hat{n}\cdot\tau)/2} = \lim_{N\to\infty}\left(1 - \frac{i}{2}\frac{\theta}{N}\hat{n}\cdot\tau\right)^N$$

and one regards $1 - \frac{i}{2}(d\theta)\hat{n} \cdot \tau$ as an infinitesimal unitary transformation. The Pauli matrices $\tau_a/2$ ($a = 1, 2, 3$) are called the "infinitesimal generators" of $SU(2)$. Note that they obey the commutation relations $[\frac{1}{2}\tau_a, \frac{1}{2}\tau_b] = i\epsilon_{abc}\frac{1}{2}\tau_c$ where ϵ_{abc} denotes the alternating symbol on three indices.

In general with a Lie group of dimension r one has an associated Lie algebra (the infinitesimal generators) with basis elements $t_1, t_2, \ldots, t_r$ and commutation relations

$$[t_a, t_b] = i f_{abc} t_c$$

where one sums over the repeated index c. The structure constants f_{abc} themselves assemble into a matrix representation of the Lie algebra called the **adjoint representation**. This is defined via $ad(A)(B) = [A, B]$ for A and B in the Lie algebra of G. Thus, let $T_a = ad(t_a)$, and note that

$$T_a(t_b) = [t_a, t_b] = i f_{abc} t_c.$$

Hence $(T_a)_{bc} = i f_{abc}$ with respect to the basis $\{t_1, t_2, \ldots, t_r\}$.

In the case of $SU(2)$, the adjoint representation gives the matrices

$$T_1 = \frac{1}{2}\begin{pmatrix} 0 & 0 & 0 \\ 0 & 0 & -i \\ 0 & i & 0 \end{pmatrix}, T_2 = \frac{1}{2}\begin{pmatrix} 0 & 0 & i \\ 0 & 0 & 0 \\ -i & 0 & 0 \end{pmatrix}, T_3 = \frac{1}{2}\begin{pmatrix} 0 & -i & 0 \\ i & 0 & 0 \\ 0 & 0 & 0 \end{pmatrix}.$$

These represent infinitesimal rotations about the three spatial axes in $\mathbf{R}^3$. These generators are normalized so that $\text{Trace}(T_a T_b) = \frac{1}{2}\delta_{ab}$ where δ_{ab} denotes the Kronecker delta.

With these remarks about $SU(2)$ in hand, let us turn to the beginnings of gauge theory: First, recall Maxwell's equation for the magnetic field $\vec{B}$: $\nabla \cdot \vec{B} = 0$. That the magnetic field is divergence-free leads us to write it as the curl of another field $\vec{A}$. Thus $\vec{B} = \nabla \times \vec{A}$ where $\vec{A}$ is called the vector potential. Since $\nabla \cdot (\nabla \times \vec{A}) = 0$ for any $\vec{A}$, this ensures that $\vec{B}$ is divergence-free. Since the curl of a gradient is zero, we can change $\vec{A}$ by an arbitrary gradient of a scalar field without affecting the value of the $\vec{B}$-field. Thus, if $\vec{A'} = \vec{A} + \nabla \Lambda$, then $\nabla \times (\vec{A'}) = \nabla \times \vec{A} = \vec{B}$. In the same way, the curl equation for the electric field is $\nabla \times \vec{E} = -\partial \vec{B}/\partial t$, whence $\nabla \times (\vec{E} + \partial \vec{A}/\partial t) = 0$. This suggests that the combination of the electric field and the time-derivative of the vector potential should be written as a gradient as in $\vec{E} + \partial \vec{A}/\partial t = \nabla V$. Here V is called the scalar potential. If $\vec{A'} = \vec{A} + \nabla \Lambda$, then we must set $V' = V - \partial \Lambda/\partial t$ in order to preserve the electric field.

In tensor form, one has the electromagnetic tensor

$$F^{\mu\nu} = \partial^\nu A^\mu - \partial^\mu A^\nu$$

constructed from the four-vector potential $A^\mu = (V; \vec{A})$. This tensor is unchanged by the **gauge transformation** $A^\mu \mapsto A^\mu - \partial^\mu \Lambda$ where Λ is any differentiable function of these coordinates.

This principle of gauge invariance provides a close tie between the formalisms of electromagnetic and quantum theory. Recall that a state in quantum mechanics is described by a complex-valued wave function $\psi(x)$ and that quantum mechanical observables involve integrations of the form $\langle E \rangle = \int \psi^* E \psi$. Such an average is invariant under a global phase change $\psi(x) \mapsto e^{i\theta}\psi(x)$. Thus it is natural to ask what happens to the quantum mechanics if we set $\psi'(x) = e^{i\theta(x)}\psi(x)$. We see at once that the derivatives of $\psi'(x)$ involve more than a phase change:

$$\partial_\mu \psi'(x) = e^{i\theta(x)}[\partial_\mu \psi(x) + i(\partial_\mu \theta(x))\psi(x)].$$

If we replace the derivative ∂_μ by the gauge-covariant derivative

$$\mathcal{D}_\mu = \partial_\mu + ieA_\mu$$

(Think of e as charge and A_μ as a four-vector potential for an electromagnetic field associated with $\psi(x)$.), and if we assume that A_μ transforms via $A_\mu \mapsto A'_\mu =$

$A_\mu - (1/e)\partial_\mu\theta(x)$ when $\psi \mapsto \psi' = e^{i\theta(x)}\psi$, then we have $\mathcal{D}_\mu\psi'(x) = e^{i\theta(x)}\mathcal{D}_\mu\psi(x)$. In this way, the principle of gauge invariance leads naturally to an interrelation of quantum mechanics and electromagnetism. This is the present form of a unification idea that originated with Herman Weyl [WEYL].

The Bohm-Aharonov Effect.

If the wave function $\psi^0(x,t)$ is a solution to the Schrödinger equation in the absence of a vector potential, then the solution in the presence of a vector potential will be $\psi(x,t) = \psi^0(x,t)e^{iS/\hbar}$ where $S = e\int dx \cdot \vec{A}$ (See [Q], p. 43.). At first it appears that since the new solution differs from the old by only a phase factor, the potential has no observable influence.

Imagine that a coherent beam of charged particles is split by an obstacle and forced to travel in two paths on opposite sides of a solenoid. After passing the solenoid the beams are recombined and the resulting interference pattern is the sum of the wave functions for each path. If current flows in the solenoid, a magnetic field $\vec{B}$ is created that is essentially confined to the solenoid. Since $\nabla \times \vec{A} = \vec{B}$ we see by Stoke's Theorem that the line integral of $\vec{A}$ around a closed loop containing the solenoid will be nonzero (and essentially constant - equal to the flux of $\vec{B}$). The interference in the two components of the wave function is then determined by the phase difference $(e/\hbar) \oint dx \cdot \vec{A}(x)$ consisting in this integral around a closed loop. The upshot is that the vector potential **does** have observable significance in the quantum mechanical context! Furthermore, the significant phase factor is the Wilson loop

$$\exp\left[(-ie/\hbar) \oint dx^\mu A_\mu\right].$$

This sets the stage for gauge theory and topology.

Parallel Transport and Wilson Loops.

The previous discussion has underlined the physical significance of the integral $R(C; A) = e^{i\int_{P_1}^{P_2} A \cdot dx}$ where A is a vector field. We find that when A undergoes a gauge transformation

$$A \mapsto A' = A - \nabla\Lambda$$

then

$$R(C; A') = e^{-i\Lambda(P_2)}R(C; A)e^{i\Lambda(P_1)}.$$

We can regard this integration as providing a parallel transport for the wave function ψ between the points p_1 and p_2. Specifically, suppose that we consider the displacement

$$p_1 = x, \ p_2 = x + dx.$$

Then

$$R(C; A) = 1 + iA \cdot dx.$$

The transport for the wave function $\psi(x)$ to the point $x + dx$ is given by $\psi^t(x) = (1 + iA \cdot dx)\psi(x)$. Thus, we define the **covariant derivative** $\mathbf{D}\psi$ via

$$\mathbf{D}\psi = \frac{\psi(x + dx) - \psi^t(x)}{dx}$$

Thus

$$\mathbf{D}\psi = \frac{\psi(x + dx) - \psi(x) + \psi(x) - \psi^t(x)}{dx} .$$

Hence

$$\mathbf{D}\psi = (\nabla - iA)\psi.$$

(This agrees (conceptually) with our calculation prior to the discussion of the Aharonov-Bohm effect.)

We see that if $\psi(x)$ transforms under gauge transformation as

$$\psi(x) \mapsto \psi'(x) = e^{-i\Lambda(x)}\psi(x),$$

then

$$\mathbf{D}'\psi'(x) = (\nabla - iA')e^{-i\Lambda(x)}\psi(x)$$
$$= (\nabla - iA + i\nabla\Lambda)e^{-i\Lambda(x)}\psi(x)$$
$$= e^{-i\Lambda(x)}(-i\nabla\Lambda + \nabla - iA + i\nabla\Lambda)\psi(x)$$
$$= e^{i\Lambda}(\nabla - iA)\psi$$
$$\mathbf{D}'\psi' = e^{-i\Lambda}\mathbf{D}\psi.$$

With these remarks, we are ready to generalize the entire situation to an arbitrary gauge group. (The electromagnetic case is that of the gauge group $U(1)$ corresponding to the phase $e^{i\theta}$.)

Thus we return now to a Lie algebra represented by Hermitian matrices $[T_a, T_b] = i f_{abc} T_c$. Group elements take the form $U = e^{-i \Lambda^a T_a}$ (sum on a) with Λ^a real numbers.

In this circumstance, the gauge field is a generalization of the vector potential. This gauge field depends on an **internal index** a (corresponding to a decomposition of the wave function into an ensemble of wave functions $\psi_a(x)$) and has values in the Lie algebra: $A_\mu(x) = T_a A_\mu^a(x)$. Thus $A(x) = (A_1(x), A_2(x), A_3(x))$ giving a Lie algebra valued field on three-dimensional space. The infinitesimal transport is given by $R(x + dx, x; A) = 1 - i dx^\mu A_\mu(x)$.

In order to develop this transport for a finite trajectory it is necessary to take into account the non-commutativity of the Lie algebra elements. Thus we must choose a partition of the path C from p_1 to p_2 and take an ordered product:

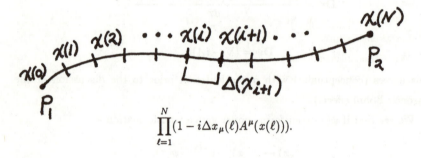

$$\prod_{\ell=1}^{N} (1 - i \Delta x_\mu(\ell) A^\mu(x(\ell))).$$

The limit (assuming it exists) of this ordered product, as $\Delta x(\ell) \to 0$ is by definition the **path-ordered exponential**

$$R(C; A) = \mathbf{P} \exp\left(-i \int_\alpha^\beta A(x) \cdot dx\right).$$

We want a notion of gauge transformation $A \mapsto A'$ so that

$$R(C; A') = U(\beta) R(C; A) U^{-1}(\alpha).$$

(This generalizes the abelian case.)

To determine the transformation law for A, consider the infinitesimal case.

We have

$$
\begin{aligned}
1 - i dx^\mu A'_\mu &= U(x + dx)(1 - i dx^\mu A_\mu)U^{-1}(x) \\
&= U(x + dx)U^{-1}(x) - i dx^\mu[U(x + dx)A_\mu U^{-1}(x)] \\
&= (U(x) + U'(x)dx)U^{-1}(x) - i dx^\mu[(U(x) + U'(x)dx)A_\mu U^{-1}(x)] \\
&= 1 + [(\partial_\mu U)dx^\mu]U^{-1} - i dx^\mu[(U + \partial_\mu U dx^\mu)A_\mu U^{-1}] \\
&\approx 1 + ((\partial_\mu U)U^{-1})dx^\mu - i dx^\mu[U A_\mu U^{-1}] \\
&= 1 + ((\partial_\mu U)U^{-1} - i U A_\mu U^{-1})dx^\mu \\
&= 1 - i dx^\mu[i(\partial_\mu U)U^{-1} + U A_\mu U^{-1}].
\end{aligned}
$$

Thus

$$
\boxed{A'_\mu = U A_\mu U^{-1} + i(\partial_\mu U)U^{-1}}.
$$

This equation defines the appropriate generalization of gauge transformation in the non-abelian context.

With this interpretation of covariant derivative and parallel transport via Wilson loops, we can recover the gauge theoretic analog of curvature via parallel transport around a small loop. That is, given the gauge potential

$$
A_\mu = A_\mu(x) = A^a_\mu(x)T_a
$$

let the **curvature tensor** $F_{\mu\nu}$ be defined by the formula

$$
F_{\mu\nu} = \partial_\mu A_\nu - \partial_\nu A_\mu - i[A_\mu, A_\nu]
$$

where $[A_\mu, A_\nu] = A_\mu A_\nu - A_\nu A_\mu$. Consider an infinitesimal circuit $\square$:

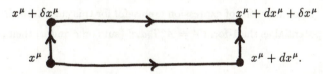

Let $R_\square = R(\square; A) = \mathbf{P}\exp(i \oint A dx)$ be the Wilson loop for this circuit. Then

Lemma 17.2. With $\square$ as above,

$$
R(\square; A) = \exp\{i F_{\mu\nu} dx^\mu \delta x^\nu\}.
$$

Proof.

$$R_\square = R(x, x+dx) \cdot R(x+dx, x+dx+\delta x) \cdot$$
$$= R(x+dx+\delta x, x+\delta x)R(x+\delta x, x).$$

Thus
$$R(x, x+dx)R(x+dx, x+dx+\delta x)$$
$$= \exp(iA_\mu(x)dx^\mu)\exp(iA_\nu(x+dx)\delta x^\nu)$$
$$= \exp(i[A_\mu dx^\mu + A_\nu \delta x^\nu + \partial_\mu A_\nu dx^\mu \delta x^\nu]$$
$$- \frac{1}{2}[A_\mu, A_\nu]dx^\mu \delta x^\nu).$$

Here we are using the (easily checked) identity
$e^{\lambda A}e^{\lambda B} = e^{\lambda(A+B)+(\lambda^2/2)[A,B]} + \mathcal{O}(\lambda^3)$ for matrices A and B and scalar λ).

Similarly, the return part of the loop yields

$$R(x+dx+\delta x, x+\delta x)R(x+\delta x, x)$$
$$= \exp(-iA_\mu(x+\delta x)dx^\mu)\exp(-iA_\nu(x)\delta x^\nu)$$
$$= \exp(-i(A_\mu dx^\mu + \partial_\nu A_\mu dx^\mu \delta x^\nu + A_\nu \delta x^\nu)$$
$$- \frac{1}{2}[A_\mu, A_\nu]dx^\mu \delta x^\nu).$$

Multiplying these two pieces, we obtain

$$R_\square = \exp(i[\partial_\mu A_\nu - \partial_\nu A_\mu - i[A_\mu, A_\nu]]dx^\mu \delta x^\nu)$$
$$= \exp(iF_{\mu\nu}dx^\mu x^\nu).$$

This completes the proof. //

Remark. Two directions of expression are useful for this curvature. If we express the gauge potential as the 1-form $A = A_\mu^a T_a dx^\mu$ (sum on a and μ) then $A = A_\mu dx^\mu$ and

$$dA - iA \wedge A = \frac{1}{2}(\partial_\mu A_\nu - \partial_\nu A_\mu - i[A_\mu, A_\nu])dx^\mu \wedge dx^\nu = \frac{1}{2}F_{\mu\nu}dx^\mu \wedge dx^\nu.$$

Thus the curvature can be expressed neatly in terms of differential forms. In this language it is customary to rewrite so that the i is not present, but we shall not do this since it is most natural to have the i as an expression of the transport $\psi(x + dx) = (1 + iA_\mu dx^\mu)\psi(x)$.

The other direction is to explicitly compute the terms of the curvature - using the Lie algebra: $[T_a, T_b] = if_{abc}T_c$ (sum on c). For this formalism, assume the f_{abc} is anti-symmetric in a, b, c. Then

$$
\begin{aligned}
F_{\mu\nu} &= \partial_\mu A_\nu - \partial_\nu A_\mu - i[A_\mu, A_\nu] \\
&= \partial_\mu A_\nu^a T_a - \partial_\nu A_\mu^b T_b - i[A_\mu^a T_a, A_\nu^b T_b] \\
&= (\partial_\mu A_\nu^a - \partial_\nu A_\mu^a)T_a - iA_\mu^a A_\nu^b[T_a, T_b] \\
&= \partial_\mu A_\nu^a - \partial_\nu A_\mu^a + A_\mu^a A_\nu^b f_{abc}T_c \\
&= \partial_\mu A_\nu^a - \partial_\nu A_\mu^a + A_\mu^a A_\nu^b f_{cab}T_c \\
&= \partial_\mu A_\nu^a - \partial_\nu A_\mu^a + A_\mu^b A_\nu^c f_{abc}T_a \\
F_{\mu\nu} &= (\partial_\mu A_\nu^a - \partial_\nu A_\mu^a + A_\mu^b A_\nu^c f_{abc})T_a.
\end{aligned}
$$

Therefore, we define $F_{\mu\nu}^a$ as the coefficient of T_a in the summation above: $F_{\mu\nu}^a = \partial_\mu A_\nu^a - \partial_\nu A_\mu^a + A_\mu^b A_\nu^c f_{abc}$.

The Chern-Simons Form.

We now turn to the Chern-Simons form, and its relationship with curvature. In the language of differential forms the Chern-Simons form is given by the formula

$$
CS = A \wedge dA - \frac{2i}{3} A \wedge A \wedge A
$$

where $A = A_\mu^a(x)T_a dx^\mu$ is a gauge potential on three-dimensional space. Here we assume that the T_a are Lie algebra generators, and that

(1) $\text{tr}(T_a T_b) = (\frac{1}{2})\delta_{ab}$ where δ_{ab} denotes the Kronecker delta, and tr is the usual matrix trace.

(2) $[T_a, T_b] = \sum_c i \, f_{abc}T_c$ with f_{abc} anti-symmetric in a, b, c.
   $([T_a, T_b] = T_a T_b - T_b T_a)$.

(Later in the chapter, we restrict to the special case of the Lie algebra of $SU(N)$ in the fundamental representation.)

The Chern-Simons form occurs in a number of contexts. An important fact, for our purposes, that I shall not verify is that, given a three-dimensional manifold

M^3 without boundary, and a gauge transformation $g : M^3 \to G$ (G denotes the Lie group in this context.) then the integral changes by a multiple of an integer. More precisely, if CS^g denotes the result of applying the gauge transformation to CS, then

$$\int_{M^3} \text{tr}(CS^g) = \int_{M^3} \text{tr}(CS) + 8\pi^2 n(g)$$

where $n(g)$ is the degree of the mapping $g : M^3 \to G$. See [JACI] for the proof of this fact.

Thus, it is sufficient to exponentiate the integral to obtain a gauge invariant quantity:

$$\exp\left(\frac{ik}{4\pi} \int_{M^3} \text{tr}(CS) \right).$$

We let $\mathcal{L}_M = \mathcal{L}_M(A) = \int_{M^3} \text{tr}(CS)$. Then $\exp((ik/4\pi)\mathcal{L}_M)$ is gauge invariant for all integers k.

Witten's definition for invariants of links and 3-manifolds is then given by the formula

$$W_K = \int dA e^{(ik/4\pi)\mathcal{L}_M} \prod_\lambda \text{tr}\left(\mathbf{P} \exp\left(i \oint_{K_\lambda} A \right) \right)$$

(The product is taken over link components K_λ.) where the integration is taken over all gauge fields on the three-manifold M^3, modulo gauge equivalence. The existence of an appropriate measure on this moduli space is still an open question (See [A2].). Nevertheless, it is possible to proceed under the assumption that such a measure does exist, and to investigate the formal properties of the integral.

One of the most remarkable formal properties of Witten's integral formula is the interplay between curvature as detected by the Wilson lines and curvature as seen in the variation of the Chern-Simons Lagrangian $\mathcal{L}_M(A)$. Specifically, we have the

Proposition 17.3. Let $CS = A \wedge A - i\frac{2}{3} A \wedge A \wedge A$ and

$$\mathcal{L}_{M^3}(A) = \int_{M^3} \text{tr}(CS)$$

where M^3 is a compact three-manifold and A is a gauge field as described above. Then the curvature of the gauge field at a point $X \in M^3$ is determined by the variation of $\mathcal{L}$ via the formula

$$\boxed{F_{\mu\nu}^a = \epsilon_{\mu\nu\lambda}\delta\mathcal{L}/\delta A_\lambda^a}.$$

The $\epsilon_{\mu\nu\lambda}$ is the "epsilon" on three indices with value $\epsilon_{123} = +1$, value ± 1 for $\mu\nu\lambda$ distinct according to the sign of the permutation, and value 0 otherwise. The derivative indicated in this formula is a functional derivative taken with respect to the gauge field varying in a neighborhood of the point X. The curvature tensor is given explicitly as

$$F_{\mu\nu}^a = \partial_\mu A_\nu^a - \partial_\nu A_\mu^a + A_\mu^b A_\nu^c f_{abc}.$$

Proof. In order to prove this proposition, we first obtain a local coordinate expression for the Chern-Simons form.

$$CS = A \wedge dA - (2i/3)A \wedge A \wedge A$$
$$A = A_k^a T_a dx^k$$
$$\Rightarrow dA = \partial_j A_k^a T_a dx^j \wedge dx^k$$
$$A \wedge dA = A_i^a T_a \partial_j A_k^b T_b dx^i \wedge dx^j \wedge dx^k$$
$$= A_i^a \partial_j A_k^b T_a T_b dx^i \wedge dx^j \wedge dx^k$$
$$A \wedge dA = \epsilon^{ijk} A_i^a \partial_j A_k^b T_a T_b dx^1 \wedge dx^2 \wedge dx^3.$$

Similarly $A \wedge A \wedge A = \epsilon^{ijk} A_i^a A_j^b A_k^c T_a T_b T_c dx^1 \wedge dx^2 \wedge dx^3$. Thus

$$CS = \epsilon^{jk\ell}\left[A_j^a \partial_k A_\ell^b T_a T_b - \frac{2i}{3}A_j^a A_k^b A_\ell^c T_a T_b T_c\right]dv$$

where $dv = dx^1 \wedge dx^2 \wedge dx^3$.

Now, we re-express $A \wedge A \wedge A$ using Lie algebra commutators:

$$A \wedge A \wedge A = \epsilon^{jk\ell} A_j^a A_k^b A_\ell^c T_a T_b T_c$$
$$= \sum_{\pi \in S_3} \text{sgn}(\pi) A_{\pi_1}^a A_{\pi_2}^b A_{\pi_3}^c T_a T_b T_c$$

where S_3 denotes the set of permutations of $\{1, 2, 3\}$ and $\text{sgn}(\pi)$ is the sign of the permutation π. Thus

$$A \wedge A \wedge A = \sum_{\pi \in \text{Perm}\{a,b,c\}} \text{sgn}(\pi) A_1^a A_2^b A_3^c T_{\pi(a)} T_{\pi(b)} T_{\pi(c)}.$$

Here we still sum over a, b, c and use the fact that A_μ^a is a commuting scalar. Thus

$$A \wedge A \wedge A = A_1^a A_2^b A_3^c \sum_{\pi \in \text{Perm}\{a,b,c\}} \text{sgn}(\pi) T_{\pi_a} T_{\pi_b} T_{\pi_c}$$

$$= A_1^a A_2^b A_3^c \{[T_a, T_b] T_c - [T_a T_c] T_b + [T_b, T_c] T_a\}.$$

Now use the fact that $\text{tr}(T_a T_b) = (\delta_{ab})/2$ to find that

$$\text{tr}([T_a, T_b] T_c) = \sum_k \text{tr}(i f_{abk} T_k T_c)$$

$$= \sum_k (i/2) f_{abk} \delta_{kc}$$

$$= \left(\frac{i}{2}\right) f_{abc}.$$

Thus

$$\text{tr}(A \wedge A \wedge A) = \left(\frac{i}{2}\right) A_1^a A_2^b A_3^c [f_{abc} - f_{acb} + f_{bca}]$$

$$= \left(\frac{i}{4}\right) A_1^a A_2^b A_3^c \left[\begin{array}{ccccc} f_{abc} & - & f_{acb} & + & f_{bca} \\ -f_{bac} & + & f_{cab} & - & f_{cba} \end{array}\right]$$

since f_{abc} is antisymmetric in a, b, c. Therefore, $\text{tr}(A \wedge A \wedge A) = \frac{i}{4} \epsilon^{jkl} A_j^a A_k^b A_l^c f_{abc}$.
Hence

$$\text{tr}(\mathcal{CS}) = \frac{\epsilon^{jkl}}{2} \left[\sum_a A_j^a \partial_k A_l^a + \frac{1}{3} A_j^a A_k^b A_l^c f_{abc} \right].$$

We are now ready to compute $\delta\mathcal{L}/\delta A_j^a$ where $\mathcal{L} = \int_{M^3} \text{tr}(\mathcal{CS})$. Note that

$$\frac{\delta}{\delta A_l^r} [\epsilon^{ijk} A_i^a A_j^b A_k^c f_{abc}]$$

$$= \epsilon^{ljk} A_j^b A_k^c f_{rbc} + \epsilon^{ilk} A_i^a A_k^c f_{arc} + \epsilon^{ijl} A_i^a A_j^b f_{abr}$$

$$= 3\epsilon^{ljk} A_j^b A_k^c f_{rbc}$$

(since both ϵ and f are antisymmetric), and

$$\frac{\delta}{\delta A_l^r} \left(\epsilon^{ijk} \sum_a A_i^a \frac{\partial A_k^a}{\partial x^j} \right)$$

$$= \epsilon^{ijk} \sum_a \left[\frac{\delta A_i^a}{\delta A_l^r} \partial_j A_k^a + A_i^a \frac{\partial_j [\delta A_k^a]}{\delta A_l^r} \right]$$

$$= \epsilon^{ljk} \partial_j A_k^r + \epsilon^{ijl} A_i^r \frac{\partial_j [\delta A_l^r]}{\delta A_l^r}.$$

In integrating over M^3, we assume that it is valid to integrate by parts and that boundary terms in that integration vanish. Thus

$$\int_{M^3} \frac{\delta}{\delta A_\ell^r} \left(\epsilon^{ijk} \sum_a A_i^a \partial_j A_k^a \right)$$
$$= \int_{M^3} \epsilon^{\ell jk} \partial_j A_k^r - \int_{M^3} \epsilon^{ij\ell} \partial_j A_i^r$$
$$= 2 \int_{M^3} \epsilon^{ijk} \partial_j A_k^r.$$

Therefore,

$$\frac{\delta \mathcal{L}}{\delta A_i^a} = \int_{M^3} \left[\epsilon^{ijk} \partial_j A_k^a + \frac{1}{2} \epsilon^{ijk} A_j^b A_k^c f_{abc} \right]$$
$$\epsilon_{rsi} \frac{\delta \mathcal{L}}{\delta A_i^a} = \int_{M^3} (\partial_r A_s^a - \partial_s A_r^a) + \frac{1}{2} [A_r^b, A_s^c] f_{abc}$$
$$= \int_{M^3} F_{rs}^a.$$

Note, however that it is understood in taking the functional derivative that we multiply by a delta function centered at the point $x \in M^3$. Thus

$$\epsilon_{\mu\nu\lambda} \frac{\delta \mathcal{L}}{\delta A_\lambda^a(x)} = F_{\mu\nu}^a(x). \qquad //$$

We are now ready to engage Level 1 of the integral heuristics. Let's summarize facts and notation. Let K be a knot, and let $\langle K|A \rangle$ denote the value of the Wilson loop for the gauge potential A taken around K. Thus

$$\langle K|A \rangle = \text{tr} \left(\mathbf{P} \exp \left(i \int_K A_\nu^a(x) T_a dx^\nu \right) \right).$$

We can write the Wilson loop symbolically as

$$\langle K|A \rangle = \prod_x (1 + i A_\nu^a(x) T_a dx^\nu)$$

where it is understood that this product stands for the limit

$$\lim_{n \to \infty} \prod_{k=1}^n (1 + i A_\nu^a(x_k) T_a dx_k^\nu)$$

where $\{x_1, x_2, \ldots, x_k\}$ is a partition of the curve K. With

$$\mathcal{L}(A) = \mathcal{L}_{M^3} = \int_{M^3} \mathcal{CS}(A)$$

($\mathcal{CS}$ denotes the Chern-Simon form.) we have the functional integral

$$W_K = \int dA e^{(ik/4\pi)\mathcal{L}} \langle K|A \rangle$$

and, if K has components $K_1, \ldots, K_n$ then

$$\langle K|A \rangle = \prod_{i=1}^{n} \langle K_i|A \rangle.$$

We shall determine a difference formula for $W_{\text{✗}} - W_{\text{✗}}$ that is valid in the limit of large k and infinitesimally close strings in the crossing exchange. Upon applying this difference formula to the gauge group $SU(N)$ in the fundamental representation, the familiar Homfly skein identity will emerge.

Before calculating, it will help to think carefully about the schematic picture of this model. First of all, the Wilson loop, written in the form

$$\langle K|A \rangle = \prod_{x \in K} (1 + iA_\nu(x)dx^\nu) = \prod_{x \in K} B(x)$$

is **already** a trace, since it is a circular product of matrices:

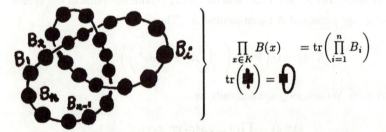

In this sense, the functional integral is similar to our previous state models in that its evaluation involves traced matrix products along the lines of the knot or link. But since the link is in three-space or in a three-manifold M^3, there is no preference for crossings. Instead, the matrices $B(x) = (1 + iA_\nu(x)dx^\nu)$ are arrayed all along the embedded curves of K. Each matrix $B(x)$ detects the local behavior of the gauge potential, and their traced product is the action of the link as an observable for this field.

It is definitely useful to think of these matrices as arrayed along the line and multiplied via our usual diagrammatic tensor conventions. In particular, if we differentiate the Wilson line with respect to the gauge, then the result is a **matrix insertion in the line:**

$$\frac{\delta\langle K|A\rangle}{\delta A_\nu^a(x)} = \frac{\delta}{\delta A_\nu^a(x)} \prod_{y\in K}(1 + iA_\nu^a(y)T_a dy^\nu)$$

$$= iT_a dx^\nu \langle K|A\rangle$$

$$\frac{\delta\langle K|A\rangle}{\delta A_\nu^a(x)} = idx^\nu T_a \langle K|A\rangle$$

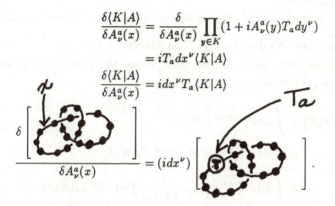

In this functional differentiation, the Lie algebra matrix T_a is inserted into the line at the point x. In writing $idx^\nu T_a\langle K|A\rangle$ it shall be **understood** that T_a is to be so inserted. Thus $dx^\nu T_a \mathcal{O}$ **means that T_a is inserted into the product $\mathcal{O}$ at the point x.** Such insertions can be indicated more explicitly, but at the cost of burdening the notation.

Curvature Insertion.

By Proposition 17.3 we know that $F_{\mu\nu}^a = \epsilon_{\mu\nu\lambda}\delta\mathcal{L}/\delta A_\lambda^a$, and by Lemma 17.2 we know that the curvature $F_{\mu\nu}^a$ can also be interpreted as the valuation of a small Wilson loop. More precisely, 17.2 tells us that

$$\langle\Box|A\rangle = \text{tr}R(\Box; A) = \text{tr}\exp(iF_{\mu\nu}dx^\mu dx^\nu)$$

$$= \text{tr}(1 + iF_{\mu\nu}dx^\mu dx^\nu) = \text{tr}(1 + iF_{\mu\nu}^a T_a dx^\mu dx^\nu).$$

It follows, that if we change the Wilson line by a small amount in the neighborhood of a point x, then the new line will differ from the old line by **an insertion of** $1 + iF_{\mu\nu}dx^\mu dx^\nu$ **at** x. Consequently, the difference between old and new lines, denoted by $\delta(K|A)$, is obtained by the **curvature insertion**

$$\delta(K|A) = idx^\mu dx^\nu F^a_{\mu\nu} T_a (K|A).$$

Once again, this is an insertion of the Lie algebra element T_a at x, the point of variation, multiplied by the curvature and by the infinitesimal area traced out by the moving curve.

We now work out the effect of this variation of the Wilson line on W_K.

Notation. Let $\langle K \rangle$ denote W_K. This follows our state-model conventions, but is at variance with statistical mechanics, where $\langle K \rangle$ is the quotient of W_K and $W_\phi = \int_M dA e^{(i/4\pi)\mathcal{L}_M}$. In this notation we have

$$\delta\langle K \rangle = \int dA e^{(ik/4\pi)\mathcal{L}} \delta(K|A)$$

$$= \int dA e^{(ik/4\pi)\mathcal{L}} idx^\mu dx^\nu F^a_{\mu\nu} T_a (K|A)$$

$$= \int dA e^{(ik/4\pi)\mathcal{L}} \left(\epsilon_{\mu\nu\lambda} \frac{\delta\mathcal{L}}{\delta A^a_\lambda(x)} \right) idx^\mu dx^\nu T_a (K|A)$$

$$= \frac{4\pi}{k} \int dA \frac{\delta\left(e^{(ik/4\pi)\mathcal{L}}\right)}{\delta A^a_\lambda(x)} \epsilon_{\mu\nu\lambda} dx^\mu dx^\nu T_a (K|A)$$

$$= -\frac{4\pi}{k} \int dA e^{(ik/4\pi)\mathcal{L}} \sum_a \epsilon_{\mu\nu\lambda} dx^\mu dx^\nu T_a \frac{\delta}{\delta A^a_\lambda(x)} (K|A)$$

$$= -\frac{4\pi i}{k} \int dA e^{(ik/4\pi)\mathcal{L}} [\epsilon_{\mu\nu\lambda} dx^\mu dx^\nu dx^\lambda] \left[\sum_a T_a T_a \right] (K|A).$$

Thus we have shown

Proposition 17.4. With $\langle K \rangle$ denoting the functional integral

$$\langle K \rangle = \int dA e^{(ik/4\pi)\mathcal{L}_M(A)} (K|A)$$

where $(K|A)$ denotes the product of Wilson loops for the link K, the variation in $\langle K \rangle$ corresponding to an infinitesimal deformation of the Wilson line K in the neighborhood of a point x is given by the formula,

$$\delta\langle K \rangle = \frac{-4\pi i}{k} \int dA e^{(ik/4\pi)\mathcal{L}} [\epsilon_{\mu\nu\lambda} dx^\mu dx^\nu dx^\lambda] \left[\sum_a T_a T_a \right] (K|A).$$

Discussion. In order to apply this variational formula, it is necessary to interpret it carefully. First of all, note that $\sum(x) = \epsilon_{\mu\nu\lambda}dx^\mu dx^\nu dx^\lambda$ is the volume element in three-space, with $dx^\mu dx^\nu$ corresponding to the **area** swept out by the deformed curve, and dx^λ to the **tangent direction** of the original curve. Thus a "flat" deformation (e.g. one that occurs entirely in a plane) will have zero variation. On the other hand $\delta\langle K \rangle$ will be nonzero for a twisting deformation such as

This is where the framing information appears in the functional integral. We see that for planar link diagrams **the functional integral is necessarily an invariant of regular isotopy.**

The term $\sum_a T_a T_a$ is the Casimir operator in the Lie algebra. We will use its special properties for $SU(N)$ shortly.

This entire discussion is only accurate for k large, and in this context it is convenient to normalize the volume form $[\epsilon_{\mu\nu\lambda}dx^\mu dx^\nu dx^\lambda]$ so that its values are $+1$, -1 or $\emptyset$. That is, I shall only use cases where the $\mu\nu\lambda$ frame is degenerate or orthogonal, and then multiply the integral by an (implicit) normalization constant. The normalized variation will be written:

$$\delta\langle K \rangle = \frac{-4\pi i}{k} \int dA e^{(ik/4\pi)\mathcal{L}} \sum(x)\left[\sum_a T_a T_a\right]\langle K|A\rangle$$

with $\sum(x) = \pm 1$ or $\emptyset$.

A Generalized Skein Relation.

We are now in a position to apply the variational equation of 17.4 to obtain a formula for the difference $\left\langle \vcenter{\hbox{$\nearrow\hspace{-6pt}\searrow$}} \right\rangle - \left\langle \vcenter{\hbox{$\nwarrow\hspace{-6pt}\nearrow$}} \right\rangle$. I call this a generalized skein relation for the invariant $\langle K \rangle$. The result is as follows.

Theorem 17.5. Let $\langle K \rangle$ denote the functional integral invariant of regular isotopy defined $\langle K \rangle = \int dA e^{(ik/4\pi)\mathcal{L}}\langle K|A\rangle$ as described above. Let $\{T_a\}$ denote the generators for the Lie algebra in the given representation of the gauge group G. Then for large k, the following switching identity is valid

$$\left\langle \vcenter{\hbox{$\nearrow\hspace{-6pt}\searrow$}} \right\rangle - \left\langle \vcenter{\hbox{$\nwarrow\hspace{-6pt}\nearrow$}} \right\rangle = z\left\langle \vcenter{\hbox{$\bigcirc$}} \right\rangle$$

where

(i) $z = e^{-(2\pi i/k)} - e^{(2\pi i/k)} \simeq -4\pi i/k$

(ii) $\mathcal{C} = \sum_a T_a T'_a$ denotes the insertion of T_a at x in the line ⟍ and a second T_a insertion in the line ⟋. (Each segment receives a separate insertion.)

Proof. Write $\langle \text{⤬} \rangle - \langle \text{⤬} \rangle = \Delta_+ - \Delta_-$ where $\Delta_+ = \langle \text{⤬} \rangle - \langle \text{⤬} \rangle$ and $\Delta_- = \langle \text{⤬} \rangle - \langle \text{⤬} \rangle$.

The notation ⤬ denotes the result of replacing the crossing by a graphical vertex. Let x denote the vertex:

In order to apply 17.4 to the case of a deformation from ⤬ to ⤬ (or to ⤬), we must reformulate the proof to include the conditions of self-crossing, since both parts of the Wilson loop going through x will participate in the calculation. We can assume that only one segment moves in the deformation. Therefore, the first part of the calculation, giving

$$\delta\langle K|A\rangle = idx^\mu dx^\nu F^a_{\mu\nu} T_a \langle K|A\rangle$$

refers to a T_a **insertion in the moving line.** After the integration by parts, the functional derivative $\delta\langle K|A\rangle/\delta A^a_\lambda(x)$ applies to **both** segments going through x. However, the resulting volume for $\sum(x)$ for the differentiation in the direction of the moving line is zero - since we assume that the line is deformed parallel to its tangential direction. Thus if we write $\langle K|A\rangle = \mathcal{O}\mathcal{O}'$ where $\mathcal{O}$ denotes that part of the Wilson loop (as a product of matrices $(1 + idx^\mu A^\mu)$) containing the moving segment and $\mathcal{O}'$ the stationary segment, then

$$\frac{\delta\langle K|A\rangle}{\delta A^a_\lambda(x)} = idx^\mu T_a \mathcal{O}\mathcal{O}' + idx'^\mu T_a \mathcal{O}\mathcal{O}'$$

where the two terms refer to insertions at x in $\mathcal{O}$ and $\mathcal{O}'$ respectively. Since the functional integration wipes out the first term, we conclude that

$$\delta\langle K\rangle = \frac{-4\pi i}{k} \int dA e^{(ik/4\pi)\mathcal{L}} \sum(x) \left[\sum_a T_a T'_a\right] \langle K|A\rangle$$

where T_a is inserted in the moving line, while $T'_a = T_a$ is inserted in the stationary line.

Now, in fact, we are considering two instances of $\delta\langle K \rangle$, namely Δ_+ and Δ_- as described at the beginning of this proof. Since the volume element $\sum(x)$ is positive for Δ_+ and negative for Δ_-, we find that $\Delta_- = -\Delta_+$ and that $\Delta_+ - \Delta_- = 2\Delta_+$. This difference is the significant one for the calculation. **By convention**, we take the full volume element from $\times$ to $\times$ to be $+1$ so that

$$\left\langle \times \right\rangle - \left\langle \times \right\rangle = \frac{-4\pi i}{k} \int dA e^{(ik/4\pi)\mathcal{L}} \left[\sum_a T_a T'_a \right] \langle K | A \rangle.$$

This formula is the conclusion of the theorem, written in integral form. This completes the proof. //

We now give two applications of Theorem 17.5. The first is the case of an abelian gauge (say $G = U(1)$). In this case the Lie algebra elements are commuting scalars. Hence the formula of 17.5 becomes

$$\left\langle \times \right\rangle - \left\langle \times \right\rangle = \frac{-4\pi i c}{k} \left\langle \times \right\rangle$$

where c is a constant, and there is no extra matrix insertion at the crossing. Now note that with $\Delta_+ = -\Delta_-$ (see the Proof of 17.5) we have $\Delta_+ + \Delta_- = 0$. Hence $0 = \left\langle \times \right\rangle + \left\langle \times \right\rangle - 2\left\langle \times \right\rangle$. Therefore, in abelian gauge,

$$\left\langle \times \right\rangle - \left\langle \times \right\rangle = \frac{-2\pi i c}{k} \left[\left\langle \times \right\rangle + \left\langle \times \right\rangle \right].$$

Hence

$$\left(1 + \frac{2\pi i c}{k}\right)\left\langle \times \right\rangle - \left(1 - \frac{2\pi i c}{k}\right)\left\langle \times \right\rangle = 0$$

or

$$\left\langle \times \right\rangle = e^{(4\pi i c/k)} \left\langle \times \right\rangle,$$

in this large k approximation.

Since switching a crossing just changes the functional integral by a phase factor, we see that linking and writhing numbers appear naturally in the abelian context: Letting $x = e^{-4\pi i c/k}$, we have that $\left\langle \diagdown\!\!\!\!\diagup \right\rangle = x\left\langle \diagup\!\!\!\!\diagdown \right\rangle$. By assumption in the heuristic, there is an $\alpha \in \mathbf{C}$ such that

$$\left\langle \partial\!\!\!\rightarrow \right\rangle = \alpha \left\langle \rightarrow \right\rangle$$

and

$$\left\langle \leftarrow\!\!\!\partial \right\rangle = \alpha^{-1} \left\langle \rightarrow \right\rangle .$$

Thus α is determined by the equation

$$\left\langle \partial\!\!\!\rightarrow \right\rangle = x \left\langle \leftarrow\!\!\!\partial \right\rangle$$

$$\Rightarrow \alpha = x\alpha^{-1}$$

$$\Rightarrow \alpha^2 = x$$

$$\Rightarrow \alpha = x^{1/2}.$$

We also have $\langle K\!\!\circledcirc \rangle = \delta \langle K \rangle$ for some δ, and can assume that $\left\langle \bigcirc \right\rangle = \delta$ as well. With this in mind, it is easy to see that for $K = K_1 \cup K_2 \cup \ldots \cup K_n$, a link of n components

$$\langle K \rangle = \left(x^{\sum_{i \neq j} \ell k(K_i, K_j) + \sum_i w(K_i)/2} \right) \delta^n$$

where ℓk denotes linking number and w denotes writhe.

All of these remarks can be extended to normalized versions for framed links in three-dimensional space (or an arbitrary three-manifold), but it is very interesting to see how simply the relationship of the abelian gauge with linking numbers appears in this heuristic.

The second application is the $SU(N)$ gauge group in the fundamental representation. Here we shall find the Homfly polynomial, and for $N = 2$ the original Jones polynomial. For $N = 2$, the fundamental representation of $SU(2)$ is given by $T_a = \sigma_a/2$ where σ_1, σ_2, σ_3 are the Pauli matrices.

$$\sigma_1 = \begin{pmatrix} 1 & 0 \\ 0 & -1 \end{pmatrix}, \sigma_2 = \begin{pmatrix} 0 & 1 \\ 1 & 0 \end{pmatrix}, \sigma_3 = \begin{pmatrix} 0 & -i \\ i & 0 \end{pmatrix}.$$

Note that $[\sigma_a, \sigma_b] = i\,\epsilon_{abc}\sigma_c$ and that $\mathrm{tr}(T_a T_b) = (\delta_{ab})/2$. In order to indicate the form of the fundamental representation for $SU(N)$, here it is for $SU(3)$:

$$\lambda_1 = \begin{pmatrix} 0 & 1 & 0 \\ 1 & 0 & 0 \\ 0 & 0 & 0 \end{pmatrix}, \lambda_2 = \begin{pmatrix} 0 & -i & 0 \\ i & 0 & 0 \\ 0 & 0 & 0 \end{pmatrix}$$

$$\lambda_3 = \begin{pmatrix} 1 & 0 & 0 \\ 0 & -1 & 0 \\ 0 & 0 & 0 \end{pmatrix}, \lambda_4 = \begin{pmatrix} 0 & 0 & 1 \\ 0 & 0 & 0 \\ 1 & 0 & 0 \end{pmatrix}$$

$$\lambda_5 = \begin{pmatrix} 0 & 0 & -i \\ 0 & 0 & 0 \\ i & 0 & 0 \end{pmatrix}, \lambda_6 = \begin{pmatrix} 0 & 0 & 0 \\ 0 & 0 & 1 \\ 0 & 1 & 0 \end{pmatrix}$$

$$\lambda_7 = \begin{pmatrix} 0 & 0 & 0 \\ 0 & 0 & -i \\ 0 & i & 0 \end{pmatrix}, \lambda_8 = \frac{1}{\sqrt{3}}\begin{pmatrix} 1 & 0 & 0 \\ 0 & 1 & 0 \\ 0 & 0 & -2 \end{pmatrix}.$$

Then $T_a = \lambda_a/2$.

The analog for $SU(N)$ should be clear. The nondiagonal matrices have two nonzero entries of i and $-i$ or 1 and 1. The diagonal matrices have a string of k 1's, followed by $-k$, followed by zeroes - and a suitable normalization so that $\mathrm{tr}(T_a T_b) = \frac{1}{2}\delta_{ab}$ as desired.

Now we use the **Fierz identity** for $SU(N)$ in the fundamental representation:

$$\sum_a (T_a)_{ij}(T_a)_{k\ell} = \frac{1}{2}\delta_{i\ell}\delta_{jk} - \frac{1}{2N}\delta_{ij}\delta_{k\ell}.$$

(**Exercise:** Check this identity.)

Applying this identity to 17.5, we find $\mathcal{C} = \sum_a T_a T_a'$

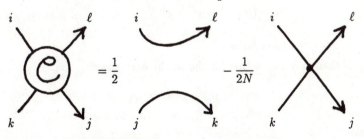

Hence, the generalized skein identity

$$\left\langle \nearrow\!\!\!\!\swarrow \right\rangle - \left\langle \searrow\!\!\!\!\nwarrow \right\rangle = \frac{-4\pi i}{k}\left\langle \mathcal{C} \right\rangle$$

becomes, for $SU(N)$,

$$\left\langle \vcenter{\hbox{⤢}} \right\rangle - \left\langle \vcenter{\hbox{⤡}} \right\rangle = \frac{-2\pi i}{k} \left\langle \vcenter{\hbox{)(}} \right\rangle - \frac{2\pi i}{Nk} \left\langle \vcenter{\hbox{)(}} \right\rangle.$$

Using $\left\langle \vcenter{\hbox{✕}} \right\rangle = \frac{1}{2}\left(\left\langle \vcenter{\hbox{⤢}} \right\rangle + \left\langle \vcenter{\hbox{⤡}} \right\rangle \right)$, as in the abelian case, we conclude

$$\left(1 - \frac{\pi i}{Nk}\right)\left\langle \vcenter{\hbox{⤢}} \right\rangle - \left(1 + \frac{\pi i}{Nk}\right)\left\langle \vcenter{\hbox{⤡}} \right\rangle = \frac{-2\pi i}{k}\left\langle \vcenter{\hbox{⇉}} \right\rangle.$$

Now we use this identity to determine α such that $\left\langle \vcenter{\hbox{↻}} \right\rangle = \alpha \left\langle \vcenter{\hbox{↱}} \right\rangle$: Let $\beta = 1 - \pi i/Nk$, $z = -2\pi i/k$. Then the identity above is equivalent to

$$\beta\left\langle \vcenter{\hbox{⤢}} \right\rangle - \beta^{-1}\left\langle \vcenter{\hbox{⤡}} \right\rangle = z\left\langle \vcenter{\hbox{⇉}} \right\rangle.$$

Hence

$$\beta\left\langle \vcenter{\hbox{↻}} \right\rangle - \beta^{-1}\left\langle \vcenter{\hbox{↺}} \right\rangle = z\left\langle \vcenter{\hbox{⇉}} \right\rangle.$$

The extra loop has value N since (compare with the Fierz identity) we are tracing an $N \times N$ identity matrix. Thus

$$\beta\alpha\left\langle \vcenter{\hbox{↝}} \right\rangle - \beta^{-1}\alpha^{-1}\left\langle \vcenter{\hbox{↝}} \right\rangle = zN\left\langle \vcenter{\hbox{↝}} \right\rangle$$

or

$$\beta\alpha - (\beta\alpha)^{-1} = \left(1 - \frac{\pi i N}{k}\right) - \left(1 + \frac{\pi i N}{k}\right)$$

$$\beta\alpha - (\beta\alpha)^{-1} = x^N - x^{-N}$$

with $z = x - x^{-1}$. Thus $\beta\alpha = x^N$.

It is then easy to see that for the writhe-normalized

$$P_K = \alpha^{-w(K)}\langle K \rangle$$

we have

$$(\beta\alpha)P_{\vcenter{\hbox{⤢}}} - (\beta\alpha)^{-1}P_{\vcenter{\hbox{⤡}}} = zP_{\vcenter{\hbox{⇉}}}.$$

Thus $x^N P_{\vcenter{\hbox{⤢}}} - x^{-N}P_{\vcenter{\hbox{⤡}}} = (x - x^{-1})P_{\vcenter{\hbox{⇉}}}$. This gives the specialization of the

Homfly polynomial. Note that they correspond formally to the types of special-ization available from the state models built via the Yang-Baxter Equation and the $SL(N)$ quantum groups in Chapter 11. A similar relationship holds for the Kauffman polynomial and the $SO(N)$ gauge in the fundamental representation. See [WIT3].

Beyond Integral Heuristics.

The next stage beyond these simple heuristics is to consider the large k limit for the three-manifold functional integrals

$$Z_{M^3} = \int dA e^{(ki/4\pi)\mathcal{L}_{M^3}}.$$

In Witten's paper [WIT2] he shows how this leads to the Ray-Singer torsion (ana-lytic Reidemeister torsion) for the three-manifold M^3, and to eta invariants related to the phase factor. See also [BN] and [GMM] for a discussion of the perturba-tion expansion of the functional integral. Quantization leads to relationships with conformal field theory: See Witten [WIT2] and also [MS], [CR1], [CR2]. In [MS] Witten explains his beautiful idea that can rewrite, via surgery on links in the three-manifold M^3, the functional integral as a sum over link invariants in the three-sphere. This expresses Z_{M^3} in a form that is essentially equivalent to our descriptions of three-manifold invariants in section 16^0.

It is worth dwelling on the idea of this surgery reformulation, as it leads directly to the patterns of topological quantum field theory and grand strategies for the elucidation of invariants. The surgery idea is this: Suppose that M' is obtained from M by surgery on an embedded curve $K \subset M$. Then we are looking at a way to change the functional integral $Z_M = \int dA e^{(ik/4\pi)\mathcal{L}_M}$ by changing the three-manifold via surgery. We already know other ways to change the functional integral relative to an embedded curve, namely by adding a Wilson loop along that curve. Therefore Witten suggested writing $Z_{M'}$ as a sum of integrals of the form

$$Z_M(K, G_\lambda) = \int dA e^{(ik/4\pi)\mathcal{L}_M} \langle K|A \rangle_\lambda$$

where $\langle K|A \rangle$ denotes the value of the Wilson loop in the representation G_λ of the gauge group G. Applied assiduously, this technique re-writes Z_M as a sum of link invariants of the surgery curves in S^3. Much remains to be understood about this point of view, but the specific constructions of Reshetikhin, Turaev and Lickorish's work with the bracket (see Chapter 16) show that this approach via surgery to the three-manifold invariants is highly non-trivial.

The idea of looking at the functional integral on a decomposition of the three-manifold is very significant. Consider a Heegard decomposition of the three-manifold M^3. That is, let $M^3 = M_1^3 \cup_\psi M_2^3$ where M_1^3 and M_2^3 are standard solid handlebodies with boundary a surface F and pasting map $\psi : F \to F$. The diffeomorphism ψ is used to glue M_1 and M_2 to form M. Then we can write (formally)

$$Z_M = \langle M_1 | \psi | M_2 \rangle$$

in the sense that these data are sufficient to do the functional integral and that $\langle M_1 |$ can be regarded as a **functional** on pairs (ψ, M_2) ready to compute an invariant by integrating over $M_1 \cup_\psi M_2$. In this sense, the functional integral leads to the notion of a Hilbert space $\mathcal{H}$ of functionals $\langle M_1 |$ and a dual space $\mathcal{H}^*$ with its ket $|M_2\rangle$. A handlebody M_1 gives rise to a "vacuum vector" $\langle M_1 | \in \mathcal{H}$ and the three-manifold invariant (determined up to a phase) is the inner product $\langle M_1 | \psi | M_2 \rangle$ essentially determined by the surface diffeomorphism $\psi : F \to F$.

Thus formally, we see that the Hilbert space $\mathcal{H}$ depends only on F and that there should be an intrinsic description of the three-manifold invariant via $\mathcal{H}(F)$, an inner product structure and the pasting data ψ. This goal has been accomplished via the use of conformal field theory on the surface. (See [CR1], [CR2], [CLM], [KO2], [SE], [MS].) The result is a definition of these invariants that does not directly depend upon functional integrals. Nevertheless, the functional integral approach is directly related to the conformal field theory, and it provides an organizing center for this entire range of techniques.

With these remarks, I reluctantly bring Part I to a close. The journey into deeper mysteries of conformal field theory and the functional integral must wait for the next time.

Remark. Figure 21 illustrates the form of the basic integration by parts maneuver that informs this section. This maneuver is shown in schematized form.

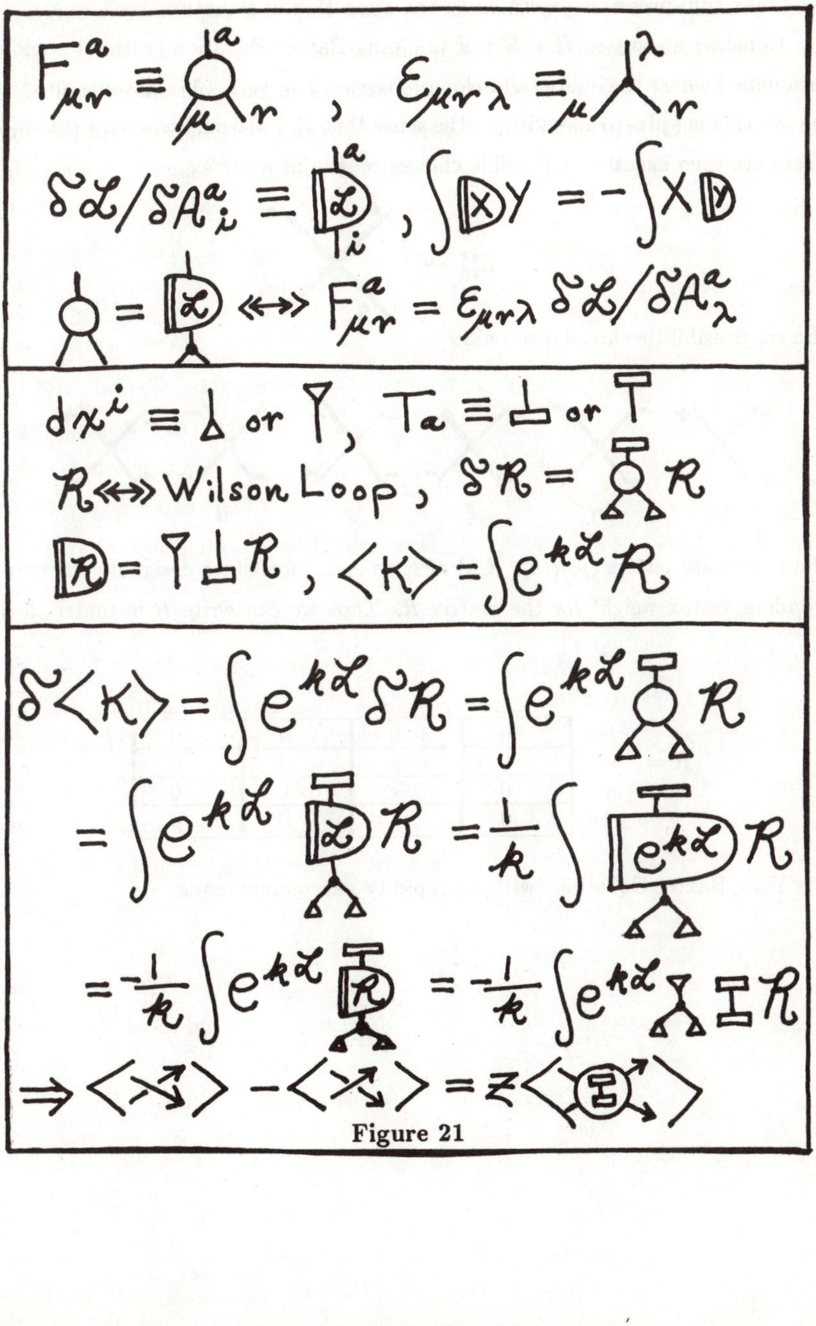

Figure 21

18^0. Appendix – Solutions to the Yang-Baxter Equation.

The purpose of this appendix is to outline the derivation of conditions for a two-index spin-preserving solution to the Yang-Baxter Equation.

Consider a solution $R = R_{cd}^{ab}$ of the Yang-Baxter Equation (without rapidity parameter) where the indices all belong to the set $\mathcal{I} = \{-1, +1\}$. Assume also that the solution is **spin-preserving** in the sense that $R_{cd}^{ab} \neq 0$ only when $a + b = c + d$. There are then exactly six possible choices of spin at a vertex

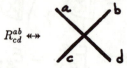

$$R_{cd}^{ab} \leftrightarrow$$

The six possibilities are shown below

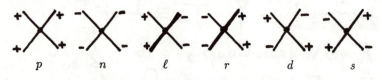

The lower-case letters (p, n, ℓ, r, d, s) next to each local state designate the corresponding vertex weight for the matrix R. Thus we can write R in matrix form as

$R =$		$--$	$-+$	$+-$	$++$
	$--$	n	0	0	0
	$-+$	0	r	d	0
	$+-$	0	s	ℓ	0
	$++$	0	0	0	p

The Yang-Baxter Equation (without rapidity parameter) reads

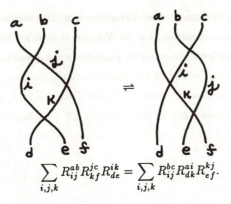

$$\sum_{i,j,k} R_{ij}^{ab} R_{kf}^{jc} R_{de}^{ik} = \sum_{i,j,k} R_{ij}^{bc} R_{dk}^{ai} R_{ef}^{kj}.$$

The main result of this appendix is the

Theorem. $R = \begin{bmatrix} n & 0 & 0 & 0 \\ 0 & r & d & 0 \\ 0 & s & \ell & 0 \\ 0 & 0 & 0 & p \end{bmatrix}$ is a solution to the Yang-Baxter Equation if and only if the following conditions are satisfied:

$$\left.\begin{cases} r\ell d = 0 \\ r\ell s = 0 \\ r\ell(\ell - r) = 0 \\ p^2 \ell = p\ell^2 + \ell ds \\ n^2 \ell = n\ell^2 + \ell ds \\ p^2 r = pr^2 + rds \\ n^2 r = nr^2 + rds \end{cases}\right\} \quad (*)$$

Proof (sketch). As we have noted above, the Yang-Baxter Equation has the form

$$\mathcal{L}_{def}^{abc} = \mathcal{R}_{def}^{abc}$$

where $\mathcal{L}$ and $\mathcal{R}$ denote the left and right halves of the equation and the indices $a, b, c, d, e, f \in \mathcal{I} = \{-1, +1\}$. Thus YBE in our case reduces to a set of 64 specific equations, one for each choice of a, b, c, d, e, f. These are further reduced by the requirement of spin-preservation. Thus, for any given choice of a, b, c, d, e, f we must, to construct $\mathcal{L}$, find i, j, k so that

$$\left.\begin{aligned} a + b &= i + j \\ j + c &= k + f \\ i + k &= d + e \end{aligned}\right\} \mathcal{L}$$

Let $(a, b, c/i, j, k/d, e, f)$ denote a specific solution to this condition. Since a configuration for $\mathcal{L}$ is also a configuration for $\mathcal{R}$ (turn it upside down!), it suffices to enumerate the 9-tuples for $\mathcal{L}$, in order to enumerate the equation. Here is the list of 9-tuples.

0. $(+++/+++/+++)$

1. $(++-/+++/++-)$
         $(++-/++-/+-+)$
         $(++-/++-/-++)$

2. $\left\{ \begin{array}{l} (+-+/+--/+-+) \\ (+-+/-++/+-+) \end{array} \right\}$
       $\left\{ \begin{array}{l} (+-+/-++/-++) \\ (+-+/+--/-++) \end{array} \right\}$
       $(+-+/+-+/++-)$

3. $\left\{ \begin{array}{l} (+--/+--/+--) \\ (+--/-++/+--) \end{array} \right\}$
       $\left\{ \begin{array}{l} (+--/-++/-+-) \\ (+--/+--/-+-) \end{array} \right\}$
       $(+--/-+-/--+)$

4. $\left\{ \begin{array}{l} (-++/-++/-++) \\ (-++/+--/-++) \end{array} \right\}$
       $\left\{ \begin{array}{l} (-++/+--/+-+) \\ (-++/-++/+-+) \end{array} \right\}$
       $(-++/+-+/++-)$

5. $\left\{ \begin{array}{l} (-+-/-++/-+-) \\ (-+-/+--/-+-) \end{array} \right\}$
       $\left\{ \begin{array}{l} (-+-/+--/+--) \\ (-+-/-++/+--) \end{array} \right\}$
       $(-+-/-+-/--+)$

6. $(--+/---/--+)$
         $(--+/--+/-+-)$
         $(--+/--+/+--)$

7. $(---/---/---)$

All the equations can be read from this list of admissible 9-tuples. For example, suppose that we want the equations for $(a, b, c) = (+, -, +)$ and $(d, e, f) = (+, +, -)$. Then the list tells us that the only $\mathcal{L}$ configuration available is $(+-+/+-+/++-)$ corresponding to

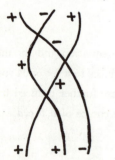

with product of
vertex weights ℓsp.

This look-up involves checking entries that begin with $(+ - +)$. Now for the

$\mathcal{R}$-configurations, we want

and this corresponds to the $\mathcal{L}$-configuration

We find $(-++/+--/+-+)$ and $(-++/-++/+-+)$ with configurations and vertex weights:

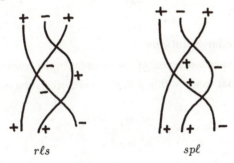

$$r\ell s \qquad\qquad sp\ell$$

Thus this specific equation $\mathcal{L} = \mathcal{R}$ is

$$\ell sp = r\ell s + sp\ell$$

or

$$0 = r\ell s.$$

Using this table of allowable configurations it is an easy task to enumerate the full list of equations. There are many repetitions, and the complete list of equations is exactly as given in the statement of the theorem. This completes the proof.   //

Remark. The six basic configurations for these solutions are in correspondence with the six basic local configurations of the six-vertex model [BA1] in statistical mechanics. In particular, they correspond to the local **arrow configurations** of the **ice-model**. In the ice-model the edges of the lattice (or more generally, a

4-valent graph) are given orientations so that **at each vertex two arrows go in and two arrows go out**. This yields six patterns

By choosing a direction (e.g. "up" as shown below) and labelling each line + or − as it goes with or against this direction, we obtain the original spin-preserving labellings.

Jones and Alexander Solutions.

We leave the full consequences of the theorem to the reader. But it should be noted that the special case $r = 0$, $\ell \neq 0$, $d = s = 1$ yields the equations

$$p^2 = p\ell + 1$$
$$n^2 = n\ell + 1$$

Thus assuming $p \neq 0 \neq n$, we have $p - p^{-1} = n - n^{-1}$. This has **two** solutions: $p = n$ and $pn = -1$. As we have previously noted (section 8^0) $p = n$ gives the YB solution for the $SL(2)$ quantum group and the Jones polynomial, while $pn = -1$ gives the YB solution derived from the Burau representation and yielding the Alexander-Conway polynomial (sections 11^0 and 12^0).

PART II

PART II. KNOTS AND PHYSICS – MISCELLANY.

This half of the book is devoted to all manner of speculation and rambling over the subjects of knots, physics, mathematics and philosophy. If all goes well, then many tales shall unfold in these pages.

1^0. Theory of Hitches.

This section is based on the article [BAY].

We give a mathematical analysis of the properties of **hitches**. A **hitch** is a mode of wrapping a rope around a post so that, with the help of a little friction, the rope holds to the post. And your horse does not get away.

First consider simple wrapping of the rope around the post in coil-form:

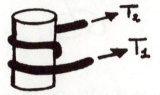

Assume that there are an integral number of windings. Let tensions T_1 and T_2 be applied at the ends of the rope. Depending upon the magnitudes (and relative magnitudes) of these tensions, the rope may slip against the post.

We assume that there is some friction between rope and post. It is worth experimenting with this aspect. Take a bit of cord and a wooden or plastic rod. Wind the cord one or two times around the rod. Observe how easily it slips, and how much tension is transmitted from one end of the rope to the other. Now wind the cord ten or more times and observe how little slippage is obtained – practically no counter-tension is required to keep the rope from slipping.

In general, there will be no slippage in the T_2-direction so long as

$$T_2 \leq \kappa T_1$$

for an appropriate constant κ. This constant κ will depend on the number of windings. The more windings, the larger the constant κ.

A good model is to take κ to be an exponential function of the angle (in radians) that the cord is wrapped around the rod, multiplied by the coefficient of friction between cord and rod. For simplicity, take the coefficient of friction to be unity so that

$$\kappa = e^{\theta/2\pi}$$

where θ is the total angle of rope-turn about the rod.

Thus, for a single revolution we need $T_2 \leq eT_1$ and for an integral number n of revolutions we need $T_2 \leq e^n T_1$ to avoid slippage.

A real hitch has "wrap-overs" as well as windings:

Clove Hitch

Here, for example, is the pattern of the clove hitch. In a wrap-over, under tension, the top part squeezes the bottom part against the rod.

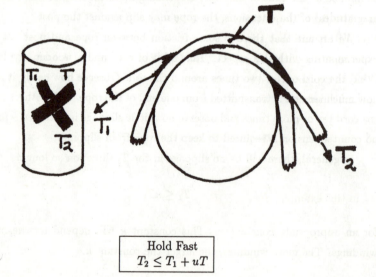

Hold Fast
$$T_2 \leq T_1 + uT$$

This squeezing produces extra protection against slippage. If, at such a wrap-over point, the tension in the overcrossing cord is T, then the undercrossing cord will hold-fast so long as $T_2 \leq T_1 + uT$ where u is a certain constant involving the friction of rope-to-rope, and T_2 and T_1 are the tensions on the ends of the undercrossing rope at its ends.

With these points in mind, we can write down a series of inequalities related to the crossings and loopings of a hitch. For example, in the case of the clove hitch we have

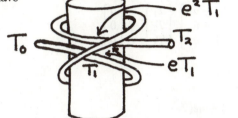

$$
\begin{array}{rcl}
T_1 & \leq & T_0 + ueT_1 \\
T_2 & \leq & e^2T_1 + ueT_1
\end{array}
$$

that the equations necessary to avoid slippage are:

$$T_1 \leq T_0 + ueT_1$$
$$T_2 \leq e^2T_1 + ueT_1.$$

Since **the first inequality holds whenever** $ue > 1$ or $u > 1/e$, we see that **the clove hitch will not slip** no matter how much tension occurs at T_2 just so long as the rope is sufficiently rough to allow $u > 1/e$.

This explains the efficacy of this hitch. Other hitches can be analyzed in a similar fashion.

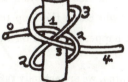

$$
\begin{array}{rcl}
T_1 & \leq & T_0 + ueT_3 \\
T_2 & \leq & T_1 + uT_3 \\
T_3 & \leq & eT_2 + ueT_3 \\
T_4 & \leq & eT_3 + uT_2
\end{array}
$$

In this example, if we can solve

$$T_1 \leq ueT_3$$
$$T_2 \leq T_1 + uT_3$$
$$T_3 \leq eT_2 + ueT_3$$

then the hitch will hold.

	T_1	T_2	T_3
	1	0	$-ue$
	-1	1	$-u$
	0	$-e$	$1 - ue$

In matrix form, we have

$$\begin{bmatrix} 1 & 0 & -ue \\ -1 & 1 & -u \\ 0 & -e & 1-ue \end{bmatrix} \begin{bmatrix} T_1 \\ T_2 \\ T_3 \end{bmatrix} \leq \begin{bmatrix} 0 \\ 0 \\ 0 \end{bmatrix}.$$

The determinant of this matrix is

$$1 - ue(2 - e).$$

Thus the critical case is $u = 1/e(2 + e)$. For $u > 1/e(2 + e)$, the hitch will hold.

Remark. Let's go back to the even simpler "hitch":

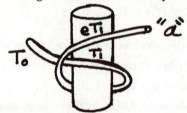

$$T_1 \leq T_0 + ueT_1$$

Our abstract analysis would suggest that this will hold if $ue > 1$. However, there is no stability here. A pull at "a" will cause the loop to rotate and then the "u-factor" disappears, and slippage happens. A pull on the clove hitch actually tightens the joint.

This shows that in analyzing a hitch, we are actually taking into account some properties of an already-determined-stable mechanical mechanism that happens to be made of rope. [See also Sci. Amer., Amateur Sci., Aug. 1983.]

There is obviously much to be done in understanding the frictional properties of knots and links. These properties go far beyond the hitch to the ways that ropes interplay with one another. The simplest and most fascinating examples are the **square knot** and the **granny knot**. The square knot pulls in under tension, each loop constricting itself and the other - providing good grip:

Square Knot

Construct this knot and watch how it grips itself.

The **granny** should probably be called the **devil**, it just won't hold under tension:

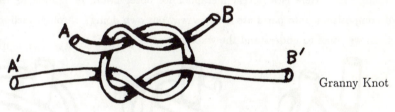

Granny Knot

Try it! Ends *A* and *B*, are twisted perpendicular to ends *A'* and *B'* and the rope will feed through this tangle if you supply a sufficient amount of tension.

The fact of the matter is that splices and hitches are fantastic sorts of mechanical devices. Unlike the classical machine that is composed of well-defined parts that interact according to well-understood rules (gears and cogs), the sliding interaction of two ropes under tension is extraordinary and interactive, with tension, topology and the system providing the form that finally results.

Clearly, here is an arena where a deeper understanding of **topology instantiated in mechanism** is to be desired. But here we are indeed upon untrodden ground. The topology that we know has been obtained at the price of initial abstraction from these physical grounds. Nevertheless, it is the intent of these notes to explore this connection.

We do well to ponder the knot as whole system, decomposed into parts only via projection, or by an observer attempting to ferret out properties of the interacting rope. Here is a potent metaphor for observation, reminding us that the decompositions into parts are in every way our own doing - through such explications we come to understand the whole.

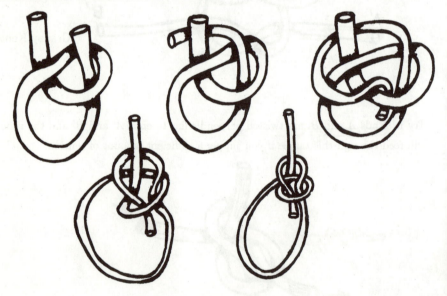

Tying the Bowline

2^0. The Rubber Band and Twisted Tube.

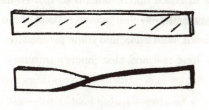

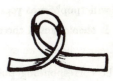

(one ordinary supercoil)

"all rolled up"

"the first "knurls" "

all "knurled up"

If you keep twisting the band it will "knurl", a term for the way the band gets in its own way after the stage of being all rolled up. A knurl is a tight super-coil and if you relax the tension (◄━━━━►) on the ends of the band, knurls will "pop" in as you feel the twisted band relax into some potential energy wells. It then takes a correspondingly long pull and more energy to remove the knurls.

The corresponding phenomena on a twisted – spring-loaded tube are even easier to see:

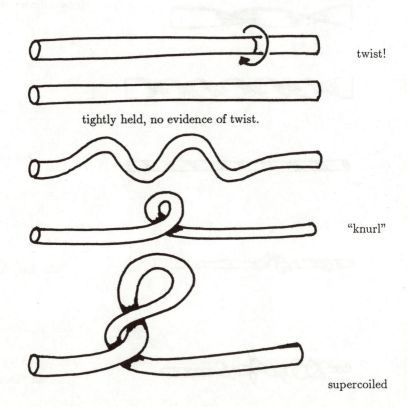

twist!

tightly held, no evidence of twist.

"knurl"

supercoiled

In all these cases, it is best to do the experiment. The interesting phenomenon is that once the first knurl has formed, it takes a lot of force to undo it due to the interaction of the tube against itself.

The energetics of the situation demand much experimentation. I am indebted to Mr. Jack Armel for the following experiment.

The Armel Effect. Take a flat rubber band. Crosscut it to form a rubber strip. Paint one side black. Tape one end to the edge of a table. Take the other end in your hand and twist gently counterclockwise until the knurls hide one color. Repeat the experiment, but turn the band clockwise. Repeat the entire experiment using a strip of paper.

3^0. On a Crossing.

This section is a musing about the structure of a diagrammatic crossing.

This (locally) consists in three arcs: The overcrossing arc, and two arcs for the undercrossing line. These two are drawn near the overcrossing line, and on opposite sides of it – indicating a line that continuously goes under the overcrossing line.

What happens if we vary the drawing? How much lee-way is there in this visual convention?

Obviously not much lee-way in regard to the meeting of the arcs for opposite sides. And the gap??

Here there seems to be a greater latitude.

Too much ambiguity could get you into trouble.

A typical arc in the diagram begins at one undercross and travels to the next. Along the way it is the overcrossing line for any other arcs. Could we generalize knot theory to include "one-sided originations"?

What are the Reidemeister moves for the likes of these diagrams?

Will you allow:

What about

Surely

must be forbidden!

Would you like some rules?? Ok. How about

1. ⟳ left alone (we'll do regular isotropies)

2.

3.

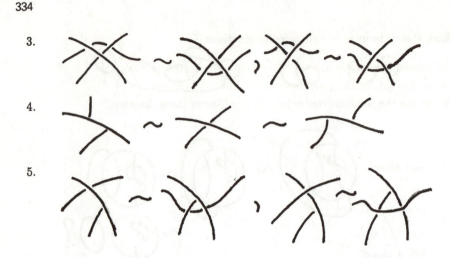

4.

5.

Call this **Imaginary Knot Theory**. Invariants anyone? (Compare with [K] and also with [TU4].)

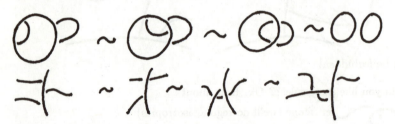

This idea of loosening the restrictions on a crossing has been independently invented in braid form by M. Khovanov [K]. In his system one allows

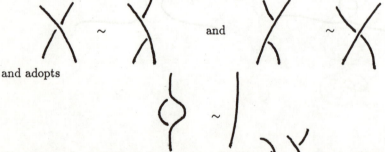

and adopts

This makes the individual three-vertex elements $\wedge$, $\curlyvee$ into analogs of three-vertices in recoupling theory (compare sections 12^0 and 13^0 of Part II). Outside of braids, it seems worthwhile to include the equivalence 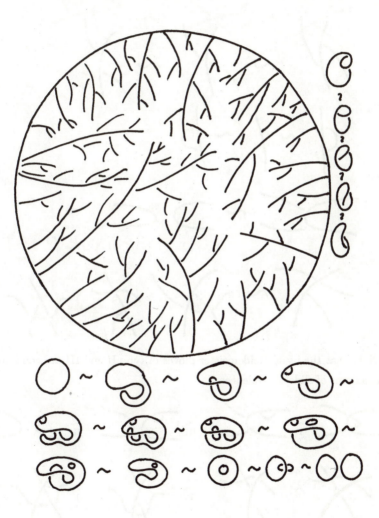 as well. We formalize this as **slide equivalence** in the next section.

4^0. **Slide Equivalence.**

Let's formalize **slide equivalence** for unoriented diagrams as described in the last section. Let slide equivalence (denoted $\sim$) be generated by:

I'.

II'.

III'.

IV'.

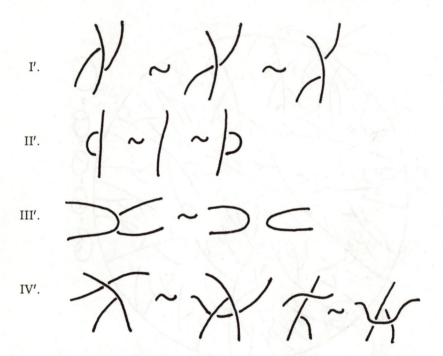

Note that the usual Reidemeister moves of type II and III are consequences of the axioms of slide equivalence:

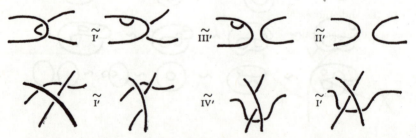

The oriented versions of the moves for slide equivalence should be clear. Moves I' and II' hold for any choice if orientations. Moves III' and IV' require that paired

arcs have compatible orientations. Thus

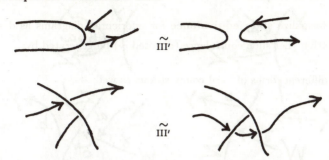

and

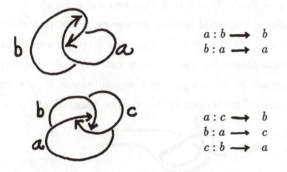

Definition. A category [M] C is said to be a **mirror category** if every object in C is also a morphism in C **and** every morphism in C is a composition of such objects.

We see that any oriented slide diagram can be regarded as a diagram of objects and morphisms for a mirror category. Thus

$$a : b \longrightarrow b$$
$$b : a \longrightarrow a$$

$$a : c \longrightarrow b$$
$$b : a \longrightarrow c$$
$$c : b \longrightarrow a$$

Of course the diagrams, being in the plane, have extra structure. Nevertheless, each of the slide-moves (I' to IV') expresses a transformation of diagrams that makes sense categorically. Thus I' asserts that

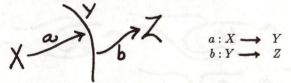

$$a : X \longrightarrow Y$$
$$b : Y \longrightarrow Z$$

morphisms can be composed without diagrammatic obstruction. II' allows the insertion or removal of self-maps that are not themselves target objects. III' is another form of composition.

IV' gives different forms of local compositions as well:

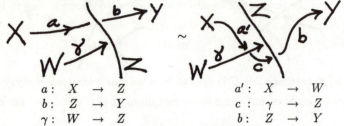

$$a: \quad X \rightarrow Z \qquad\qquad a': \quad X \rightarrow W$$
$$b: \quad Z \rightarrow Y \qquad\qquad c: \quad \gamma \rightarrow Z$$
$$\gamma: \quad W \rightarrow Z \qquad\qquad b: \quad Z \rightarrow Y$$

Thus the abstract version of IV' in mirror category language is that given morphisms $a: X \rightarrow Z$, $b: Z \rightarrow Y$ and an **object-morphism** $\gamma: W \rightarrow Z$ then there exist morphisms $c: \gamma \rightarrow Z$, $a': X \rightarrow \gamma$ such that $b \circ c \circ a' = b \circ a$.

This suggests defining a **special mirror category** as one that satisfies the various versions of IV' and the analog of II'. The slide diagrams are depictions of the generating object-morphisms for special mirror categories of planar type.

To return to diagrams in the category of side-equivalence, it is a very nice problem to find invariants. This is open territory. (Use orientation!)

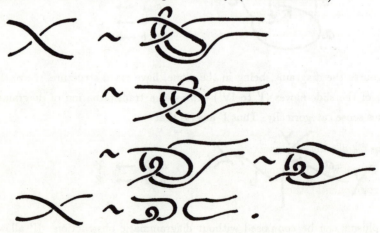

5⁰. Unoriented Diagrams and Linking Numbers.

Let's note the following: **The linking number of a two-component link in absolute value depends only on the unoriented link.**

Yet we normally go through all the complexities of putting down orientations to determine the linking number. In this section I will show how to compute the absolute value of the linking number from an unoriented diagram. Let K be any unoriented link diagram. Let $K^{\mathcal{O}}$ be the result of orienting K. Thus $\mathcal{O}$ belongs to one of the $2^{|K|}$ orientations of K where $|K|$ denotes the number of components of K.

Define $[K]$ by the formula

$$[K] = \sum_{\mathcal{O}} \alpha^{w(K^{\mathcal{O}})}.$$

Thus $[K]$ is the sum over all possible orientations of K of a variable α raised to the writhe of K with that orientation.

Remark. It is useful for display purposes to use a bracket in many different contexts and to denote different polynomials. Let us adopt the convention that a given bracket is specified by its last definition, unless otherwise specified.

Certainly, $[K] = \sum_{\mathcal{O}} \alpha^{w(K^{\mathcal{O}})}$ is a regular isotopy invariant.

If K_0 is a specific orientation of K then $\Lambda(K_0) = \alpha^{-w(K_0)}[K_0]$ will be an ambient isotopy invariant of the oriented link K_0. Now

$$\Lambda(K_0) = \sum_{\mathcal{O}} \alpha^{w(K^{\mathcal{O}})-w(K_0)}$$

$$= 1 + \sum_{\mathcal{O} \neq \mathcal{O}_0} \alpha^{w(K^{\mathcal{O}})-w(K_0)}$$

where $\mathcal{O}_0$ denotes the orientation of K_0.

To see what this means, note the following: if $K_0 = L_1 \cup L_2 \cup \ldots \cup L_n$ (its link components), then

$$w(K_0) = \sum_{i=1}^{n} w(L_i) + \sum_{i<j} 2\ell k(L_i, L_j).$$

Hence if $K^{\mathcal{O}}$ results in the changing of orientations on some subset of the L_i, we can denote this by

$$K^{\mathcal{O}} = L_1^{\epsilon_1} \cup L_2^{\epsilon_2} \cup \ldots \cup L_n^{\epsilon_n}$$

where $\epsilon_i = \pm 1$ (1 for same, -1 for changed). But $w(L_i^+) = w(L_i^-)$, hence

$$w(K^{\mathcal{O}}) - w(K_0) = \sum_{i<j} 2(\ell k(L_i^{\epsilon_i}, L_j^{\epsilon_j}) - \ell k(L_i, L_j))$$

and

i) $\left\{ \begin{array}{c} \epsilon_i = -1, \ \epsilon_j = +1 \\ \text{or} \quad \epsilon_i = +1, \ \epsilon_j = -1 \end{array} \right\} \Rightarrow \ell k(L_i^{\epsilon_i}, L_j^{\epsilon_j}) = -\ell k(L_i, L_j)$. Thus

$$w(K^{\mathcal{O}}) - w(K_0) = -4 \sum_{\substack{i<j \\ \epsilon_i \neq \epsilon_j}} \ell k(L_i^{\epsilon_i}, L_j^{\epsilon_j}).$$

Thus

$$\Lambda(K_0) = 1 - 4 \sum_{L, L' \subset K^{\sigma}} \ell k(L, L')$$

where L, L' are a pair of different components of $K^{\mathcal{O}}$ such that **exactly one** of them has orientation reversed from the reference K_0.

This is an exact description of the associated ambient isotopy invariant to $[K]$.

The simplest case is a link of two components $K = K_1 \cup K_2$. Then there are four possible orientations.

$$\begin{array}{rclcl} K_0 & = & K_1^+ \cup K_2^+ & & \text{Say } w(K_0) = w_0 \\ K_1 & = & K_1^- \cup K_2^- & \Rightarrow & w(K_1) = w(K_0) \\ K_2 & = & K_1^- \cup K_2^+ & & w(K_2) = w(K_3) \\ K_3 & = & K_1^+ \cup K_2^- & & \end{array}$$

Thus

$$[K] = 2(\alpha^{w(K_1^+ \cup K_2^+)} + \alpha^{w(K_1^+ \cup K_2^-)})$$

$$\Rightarrow [K] = 2(\alpha^{d_1} + \alpha^{d_2})$$

where

$$|d_1 - d_2| = 4|\ell k(K_1^+, K_2^+)|.$$

Thus, **for a two-component link, [K] calculates** (via the difference of two exponents) **the absolute linking number (≥ 0) of the two curves.**

What I want to do now is show how $[K] = \sum_{O} \alpha^{w(K^{\sigma})}$ can be calculated recursively from an unoriented link diagram **without assigning any orientations.**

In order to accomplish this feat, first extend the writhe to link diagrams containing a (locally) 4-valent graphical vertex in the oriented forms

Do this by summing over ± 1 contributions **from the crossings.** Thus

$$w\left(\text{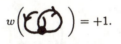}\right) = +1.$$

Definition 5.1. $\left[\times\right]_{+} = \sum_{O \in O_{+}(\times)} \alpha^{w(K^{O})}$

$$\left[\times\right]_{-} = \sum_{O \in O_{-}(\times)} \alpha^{w(K^{O})}$$

where $O_{+}(\times)$ denotes all orientations of the link that give a locally $+$ orientation of the two lines at the site indicated. Thus

$$O_{+}(\times) \leftrightarrow \times \cup \times$$
$$O_{-}(\times) \leftrightarrow \times \cup \times$$

where this diagram indicates all orientations of K with this local configuration at the crossing under surveillance.

Proposition 5.2. $\left[\times\right]_{+} = \frac{\alpha}{2}\left(\left[\times\right] + \left[\asymp\right] - \left[)(\right]\right)$

$$\left[\times\right]_{-} = \frac{\alpha^{-1}}{2}\left(\left[\times\right] - \left[\asymp\right] + \left[)(\right]\right)$$

Proof. Consider the possibilities.

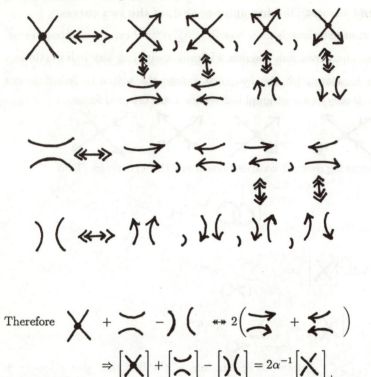

Therefore $\times + \asymp -)($ $\leftrightarrow 2\left(\overrightarrow{} + \overleftarrow{}\right)$

$$\Rightarrow \left[\times\right] + \left[\asymp\right] - \left[)(\right] = 2\alpha^{-1}\left[\times\right]_+$$

and similarly

$$\left[\times\right] - \left[\asymp\right] + \left[)(\right] = 2\alpha^{+1}\left[\times\right]_- \qquad //$$

Corollary 5.3.

(a) $\left[\times\right] = \left(\dfrac{\alpha - \alpha^{-1}}{2}\right)\left[\asymp\right] + \left(\dfrac{\alpha^{-1} - \alpha}{2}\right)\left[)(\right] + \left(\dfrac{\alpha + \alpha^{-1}}{2}\right)\left[\times\right].$

(b) If G is a planar graph with 4-valent vertices then

$$[G] = 2^{|G|}$$

where $|G|$ denotes the number of knot-theoretic circuits in G (i.e. the number of components in the link that would be obtained by creating a crossing at each vertex).

Proof. (a) follows from the previous proposition, coupled with the fact that $[\!\propto\!] = [\!\propto\!]_{+} + [\!\propto\!]_{-}$. (b) follows from the fact that a link with r components has 2^r orientations. //

Remark. Note that $[O] = 2$. The formulas of this corollary give a recursive procedure for calculating $[K]$ for any unoriented K. Hence we have produced the desired mode of computing absolute linking number from an unoriented diagram.

$$\Big[\textbf{X}\Big] = \Big(\tfrac{\alpha - \alpha^{-1}}{2}\Big)\Big[\textbf{⌣⌢}\Big] + \Big(\tfrac{\alpha^{-1} - \alpha}{2}\Big)\Big[)(\Big] + \Big(\tfrac{\alpha + \alpha^{-1}}{2}\Big)\Big[\textbf{X}\Big]$$

$$\Big[\textbf{X}\Big] = \Big(\tfrac{\alpha^{-1} - \alpha}{2}\Big)\Big[\textbf{⌣⌢}\Big] + \Big(\tfrac{\alpha - \alpha^{-1}}{2}\Big)\Big[)(\Big] + \Big(\tfrac{\alpha + \alpha^{-1}}{2}\Big)\Big[\textbf{X}\Big].$$

It follows from these equations that

$$\Big[\textbf{X}\Big] - \Big[\textbf{X}\Big] = \Big(\tfrac{\alpha - \alpha^{-1}}{2}\Big)\Big(\Big[\textbf{⌣⌢}\Big] - \Big[)(\Big]\Big)$$

$$\Big[\textbf{ʊ}\Big] = \alpha\Big[\textbf{∼}\Big], \quad \Big[\textbf{ʊ}\Big] = \alpha^{-1}\Big[\textbf{∼}\Big].$$

Thus $[K]$ of this section is a special case of the Dubrovnik polynomial discussed in section 14^0 of Part I.

We can also view the expansion formulas for $[K]$ in terms of a **state summation** of the form

$$[K] = \sum_{G}[K|G]2^{|G|}$$

where each **state** is obtained from the diagram for K as follows:

1. Replace each crossing $\textbf{X}$ of K by one of the three local configurations

$$\textbf{X} \quad C \qquad \textbf{⌣⌢} \quad A \qquad)(\quad B$$

labelled as indicated. Note that the label A corresponds to the A-split

$$\textbf{X} \;\mapsto\; \textbf{⌣⌢}$$

2. The resulting labelled state G is a planar graph (with 4-valent vertices).
   Let $[K|G]$ denote the product of the labels for G.

Let $|G|$ denote the number of **crossing circuits** in G, where a crossing circuit is obtained by walking along G and **crossing at each crossing**, e.g.

$$\text{e.g.} \quad \left|\; \vcenter{\hbox{}}\;\right| = 3.$$

It then follows from our discussion that if we set

$$C = \frac{1}{2}(\alpha + \alpha^{-1}) \quad \Longleftrightarrow \quad$$
$$B = \frac{1}{2}(\alpha^{-1} - \alpha) \quad \Longleftrightarrow \quad$$
$$A = \frac{1}{2}(\alpha - \alpha^{-1}) \quad \Longleftrightarrow \quad$$

then

$$[\times] = A[\asymp] + B[)(] + C[\times],$$

$$[K] = \sum_G [K|G] 2^{|G|}.$$

I will leave the details of verification as an exercise. However, here is an example:

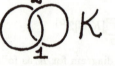

Rows are labelled with the state choice for vertex 1, columns for vertex 2.

Thus

$$\sum_G [K|G]2^{|G|} = A^2 \cdot 2^2 + AB \cdot 2^1 + AC \cdot 2^1$$
$$+ BA \cdot 2^1 + B^2 \cdot 2^2 + BC \cdot 2^1$$
$$CA \cdot 2^1 + CB \cdot 2^1 + C^2 \cdot 2^2$$
$$= 4(A^2 + B^2 + C^2) + 4(AB + AC + BC)$$
$$= (\alpha - \alpha^{-1})^2 + (\alpha^{-1} - \alpha)^2 + (\alpha + \alpha^{-1})^2$$
$$+ (\alpha^{-1} - \alpha)(\alpha - \alpha^{-1}) + (\alpha - \alpha^{-1})(\alpha + \alpha^{-1})$$
$$+ (\alpha^{-1} - \alpha)(\alpha + \alpha^{-1})$$
$$= (\alpha - \alpha^{-1})^2 + (\alpha + \alpha^{-1})^2$$
$$= 2(\alpha^2 + \alpha^{-2}) = [K] \text{ as expected!}$$

$$\left\{ \begin{array}{l} \bigcirc\!\!\!\bigcirc K \ , \ [K] = 2\left(\alpha^{d_1} + \alpha^{d_2}\right) \\[2mm] |\mathscr{lk}(K)| = \frac{1}{4}|d_1 - d_2| \end{array} \right\}$$

346

6^0. The Penrose Chromatic Recursion.

A diagram recursion very closely related to the formalism of 5^0 occurs in the number of edge 3-colorings for planar trivalent graphs. We wish to color the edges with three colors - say red (R), blue (B) and purple (P) so that **three distinct colors occur at each vertex.**

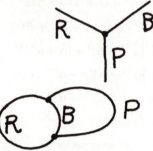

For example,

is such a coloring. In [PEN1] Penrose gives the following recursion formula:

$$\left[\,\big\lgroup\!\big\rgroup\,\right] = \left[\,)(\,\right] - \left[\,\times\,\right]$$

where $[G] = 3^{|G|}$ for a 4-valent planar graph G, and $|G|$ denotes the number of crossing circuits of G. For example,

$$\left[\,\Phi\,\right] = \left[\,00\,\right] - \left[\,\infty\,\right] = 3^2 - 3 = 6.$$

The proof of this recursion formula is actually quite simple. I will give a state summation that models it.

In the plane, we may distinguish two types of oriented vertex.

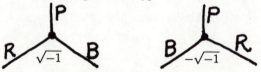

These two types correspond to the cyclic order RBP or BRP (counterclockwise) at the vertex. I shall **label** a vertex of type RBP by $\sqrt{-1}$ and a vertex of type BRP by $+\sqrt{-1}$.

Thus,

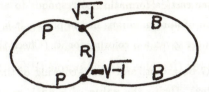

We shall need to distinguish between maps with and without singularities. Thus

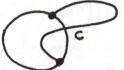

has a singularity (the fourfold (valent) crossing at c).

Definition 6.1. A **cast** is any (possibly singular) embedding of a graph (with trivalent vertices) into the plane. All cast singularities are ordinary double points as in

A **planar cast** or **map** is a cast that is free of singularities.

Now there is a nice way to think about an edge coloring of a cast by the three colors R, B, P. This is called a **formation**. (The terminology and is due to G. Spencer-Brown [SB1].) In a formation we have **red** (R) circuits and **blue** (B) circuits. Two circuits of opposite color can share an edge and

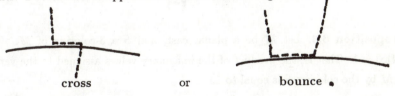

cross or **bounce** .

Here I let ──────── denote **red** (R) and I let ─ ─ ─ ─ ─ ─ denote **blue** (B). I also regard the superpositions of red and blue as purple. Thus

─ ─ ─ ─ ─ ─ ─ ─ ─

is **purple** (P). For planar casts it is a very easy matter to draw formations. Just draw a disjoint collection of red circles and interlace them with blue circles (crossing and bouncing).

Thus for a **map** (planar cast) a formation corresponds to an edge coloring and it is a certain collection of Jordan curves in the plane. It is easy to see how to formate any cast if you are given a 3-coloring for it. (Check this!)

Definition 6.2. Let G be a cast, and S a 3-coloring of the edges of G (three distinct colors per vertex). Define the **value of S with respect to G**, denoted $[G|S]$, by the formula

$$[G|S] = \prod_{j=1}^{n} \epsilon(j)$$

where $\{1, 2, \ldots, n\}$ denotes the vertices of G, and $\epsilon(j) = \pm\sqrt{-1}$ is the label assigned to this vertex by the coloring S.

Examples.

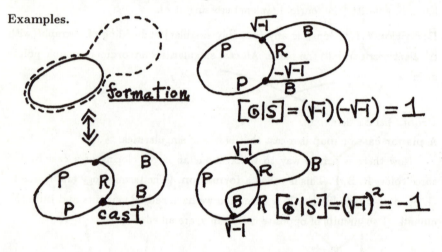

Proposition 6.3. Let M be a planar cast, and S a 3-coloring of M. Then $[M|S] = 1$. That is, the product of the imaginary values assigned to the vertices of M by the coloring S is equal to 1.

Proof.

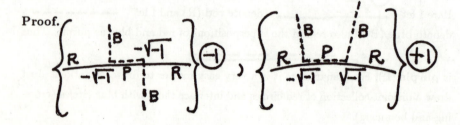

349

Take the formation $\mathcal{F}$ associated to S. Note that bounces contribute $(\sqrt{-1})(-\sqrt{-1}) = +1$ while crossings contribute $(\pm\sqrt{-1})(\pm\sqrt{-1}) = -1$. Thus $[G|S] = (-1)^c$ where c is the number of crossings between red curves and blue curves in the formation. But c is even by the Jordan curve theorem.   //

Definition 6.4. Let $[G] = \sum_S [G|S]$ where G is any cast, and S runs over all 3-colorings of G.

Proposition 6.5. If G is a planar cast, then $[G]$ equals the number of distinct 3-colorings of the edges of G with three colors per vertex.

Proof. By the previous proposition, when G is planar, then $[G|S] = +1$ for each S. This completes the proof.   //

Proposition 6.6. Let G be any cast. Let 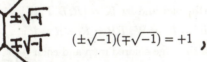 denote a specific edge of G and let $)($ and $\times$ denote the two casts obtained by replacing this edge as indicated. Then

$$\left[\times\right] = \left[)(\right] - \left[\times\right].$$

Proof. Let S_+ denote the states (colorings) S contributing to the edge $\times$ in the form

$(\pm\sqrt{-1})(\mp\sqrt{-1}) = +1$,

and let S_- denote the states S contributing to the edge in the form

$\times \begin{array}{c}\pm\sqrt{-1}\\\pm\sqrt{-1}\end{array}$.

Then

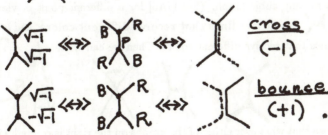

Thus the sum over all states becomes the sum over crossing types plus bounces.

$$\therefore \left[\; \chi \;\right] = - \left[\; \times \;\right] + \left[\;)(\;\right] ,$$

since $\mathcal{S}_+$ is in 1-1 correspondence with states of $)($ and $\mathcal{S}_-$ is in 1-1 correspondence with states of $\times$.   //

Examples.

$$\left[\; \Phi \;\right] = \left[\; 00 \;\right] - \left[\; \infty \;\right] = 3^2 - 3 = 6$$

$$\left[\; \mathcal{C} \;\right] = \left[\; \mathcal{G} \;\right] - \left[\; \Theta \;\right] = 3 - 3^2 = -6.$$

Note that $[G]$ does not necessarily count the number of colorings for a non-planar cast.

Theorem 6.7. (Kempe). Coloring a planar cast with 3 colors on the edges is equivalent to coloring the associated map with 4 colors so that regions sharing a boundary receive different colors.

Proof. Let W be a fourth color and let $\mathbf{K} = \{R, B, P, W\}$ with group structure given by: $W = 1$, $R^2 = B^2 = P^2 = W$, $RB = P$, $BR = P$.

Choose any region of the edge-colored map and color it W. Now color an adjacent region X so that $WY = X$ where Y is the color of their common boundary. Continue in this fashion until done.   //

A Chromatic Expansion.

There is a version of this formalism that yields the number of edge-three-colorings for any cubic graph. (See [JA6] for a different point of view.) To see this, let $)\!-\!($ denote **two lines that receive different colors**, and let $\times$ also denote lines that receive different colors. Then we can write

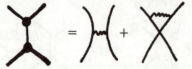

in the sense that whenever either of the graphs on the right is colored with distinct local colors, then we can amalgamate to form a coloration of the graph on the left.

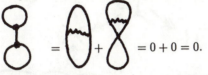

$$= 3 \cdot 2 + 0 = 6.$$

Since loops can not be colored differently from themselves, many states in this expansion contribute zero. Let us call this state summation the **chromatic expansion** of a cubic graph. Note how the dumbbell behaves:

In fact, it is easy to see that the 4-color theorem is equivalent to the following statement about Jordan curves in the plane: **Let J be a disjoint collection of Jordan curves in the plane. Assume that J is decorated with prohibition markers)⊸(, (locally as shown, and never between a curve and itself). Then either the loops of J can be colored with three colors in the restrictions indicated by the prohibition markers or some subset of markers can be switched (**)⊸(↦ₛwᵢₜ𝒸ₕ ✗ **) to create a colorable configuration.**

For example, let J be as shown below

Then J has no coloring, but J', obtained by switching all the markers, can be colored.

From the chromatic expansion we see that if ✗ represents an edge in any **minimal** uncolorable graph then both)(and ✗ must receive the **same** colors in any coloring of these smaller graphs. The simplest known (non-planar) example is the Petersen graph shown below:

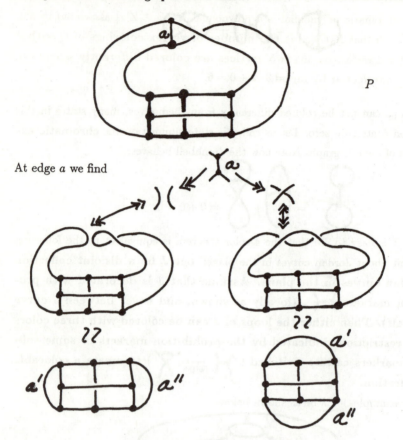

P

At edge a we find

and in each case a'' and a' are forced to receive the same color.

Remark. The Petersen is usually drawn in the pentagonal form [SA]:

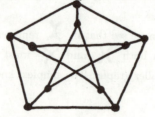

7⁰. The Chromatic Polynomial.

The chromatic polynomial is a polynomial $K(G) \in \mathbf{Z}[x]$ associated to any graph G such that $K(G)(x)$ **is the number of vertex colorings of** G **with** x **colors.** (In a vertex coloring **two vertices are colored differently whenever they are connected by an edge.**)

Examples.

$$K\left(\, \bullet \, \right) = x$$

$$K\left(\bullet\!\!-\!\!\bullet \right) = x(x-1) = x^2 - x$$

$$K\left(\, \mathcal{Q} \, \right) = 0$$

$$K\left(\triangle \right) = x(x-1)(x-2) = x^3 - 3x^2 + 2x.$$

Note that, by convention, we take $K(G) = 0$ if G has any self-loops.

To begin this tale, we introduce the well-known [WH2] recursive formula for $K(G)$:

$$K\left(\;\rightarrowtail\!\!-\!\!\leftarrowtail\; \right) = K\left(\;\rightarrowtail\;\;\leftarrowtail\; \right) - K\left(\;\maltese\; \right)$$
$$K(G) = K(G-a) - K(G/a).$$

This formula states that the number of colorings of a graph G with a specified edge a is the difference of the number of colorings of $(G-a)$ (G with a deleted) and (G/a) (G with a collapsed to a point). This is the logical identity:

$$\text{Different} = \text{All - Same.}$$

The formula may be used to recursively compute the chromatic polynomial.

354

For example:

$$K\left(\triangle\right) = K\left(\text{path}\right) - K\left(\triangle'\right)$$

$$= K\left(\cdot\cdot\right) - K\left(\cdot\cdot\cdot\right) - \left(K\left(\cdot\cdot\right) - K\left(\triangle\right)\right)$$

$$= \left\{ \begin{array}{ll} K\left(\cdot\cdot\cdot\right) & -K\left(\cdot\cdot\cdot\right) \\ -K\left(\cdot\cdot\right) & +K\left(\cdot\cdot\right) \\ -K\left(\cdot\cdot\right) & +K\left(\cdot\cdot\right) \\ -K\left(\cdot\cdot\right) & +K\left(\triangle\right) \end{array} \right\}$$

$$= \left\{ \begin{array}{lll} x^3 & - & x^2 \\ -x^2 & + & x \\ -x^2 & + & x \\ -x & + & x \end{array} \right\} = x^3 - 3x^2 + 2x$$

$$\therefore K\left(\triangle\right) = x(x-1)(x-2).$$

In this computation we have used the **edge-bond** •——•— to indicate a collapsed edge. Thus

$$K\left(\succ\!\!-\!\!\prec \right) = K\left(\succ\ \prec \right) - K\left(\succ\!\!-\!\!\prec \right)$$

As the middle section of the computation indicates, the recursion results in a sum of values of K on the **bond-graphs** obtained from G by making a choice for every edge to either delete it or to bond it. Clearly, the contribution of such a graph is $(-1)^{\#(\text{bonds})} x^{\#(\text{components})}$.

Thus, for any graph G we may define a **bond state** B of G to be the result of choosing, for each edge to delete it or to bond it. Let $\|B\|$ denote the number of components of B and $i(B)$ the number of bonds.

Thus B •—•—• is a bond state of ▭ , and $\|B\| = 2$, $i(B) = 2$. We have shown that the chromatic polynomial is given by the formula

$$K(G) = \sum_{B} (-1)^{i(B)} x^{\|B\|}.$$

The first topic we shall discuss is a reformulation of this chromatic formula in terms of states of a universe (planar 4-valent graph) associated with the graph G. In fact, **there is a one-to-one correspondence between universes and planar graphs.** The correspondence is obtained as follows (see Figure 7.1).

(i) Given a universe U, we obtain an associated graph G by first shading U in checkerboard fashion so that the unbounded region is colored white. (Call the colors white and black.) For each black region (henceforth called shaded) place a vertex in its interior. Join two such vertices by an edge whenever they touch at a crossing of U. Thus

A graph G and its universe U.
Figure 7.1

Compare this description with Figure 7.1.

(ii) Conversely, given a graph, we obtain a universe by first placing a crossing on each edge in the form

356

and then connecting these up by following the pattern indicated below for a single vertex:

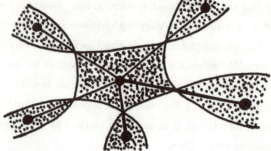

Exercise. Formulate this correspondence precisely, taking care of examples such as  .

(In the literature, the universe associated with a planar graph is sometimes referred to as its **medial graph**.) By replacing the graph G by its associated universe U, we obtain a reformulation of the chromatic polynomial. Instead of deleting and bonding edges of the graph, we split the crossings of U in the two possible ways:

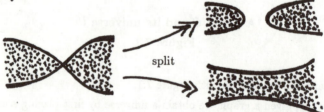

split

Since coloring the vertices of the graph G corresponds to **coloring the shaded regions of U so that any two regions meeting at a crossing receive different colors**, we see that the chromatic identity becomes:

$$K\left(\rlap{—}\bowtie\right) = K\left(\rlap{—}\bowtie\right) - K\left(\rlap{—}\bowtie\right)$$

By splitting the crossings so that the crossing structure is still visible we can classify **exterior and interior vertices of a chromatic state.** A chromatic state of a shaded universe U is a choice of splitting at each vertex (crossing). A split vertex is **interior** if it locally creates a connected shading, and it is exterior if there is a local disconnection. See Figure 7.2.

exterior   interior

Figure 7.2

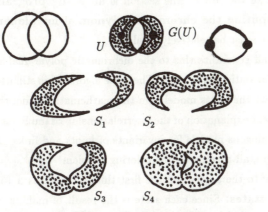

Chromatic States of U

$$\|S_1\| = 2, \|S_2\| = 1, \|S_3\| = 1, \|S_4\| = 1$$

$$i(S_1) = 0, i(S_2) = 1, i(S_3) = 1, i(S_4) = 2$$

$$K(U) = \sum_S (-1)^{i(S)} x^{\|S\|}$$

$$= (-1)^0 x^2 + (-1)^1 x + (-1)^1 (x) + (-1)^2 x$$

$$\therefore\ K(U) = x^2 - x$$

Figure 7.3

In Figure 7.3 we have listed all the chromatic states for a small universe U. Since interior vertices correspond to bonded edges in the associated graph we see that the coloring formula for the universe is:

$$K(U) = \sum_S (-1)^{i(S)} x^{\|S\|} \qquad (*)$$

where $i(S)$ is the number of interior vertices in the chromatic state S, and $\|S\|$ is the number of shaded components in S.

The subject of the rest of this section is **an easily programmable algorithm for computing the chromatic polynomial for planar graphs.** (Of course the graphs must be relatively small.)

Later we shall generalize this to the dichromatic polynomial and relate it to the Potts model in statistical physics. There the algorithm is still useful and can be used to investigate the Potts model. On the mathematical side, this formulation yields a transparent explanation of the interrelationship of the Potts model, certain algebras, and analogous models for invariants of knots and links. There is a very rich structure here, all turning on the coloring problem.

Turning now to the algorithm, the first thing we need is **a method to list the chromatic states.** Since each state is the result of making a binary choice (of splitting) at each vertex of U, it will be sufficient to use any convenient method for listing the binary numbers from zero (0) to $2^n - 1$ ($n =$ the number of vertices in U). In particular, it is convenient to use the **Gray code.** The Gray code lists binary numbers, changing only one digit from step to step. Figure 7.4 shows the first terms in the Gray code.

$$
\left.
\begin{array}{ccccc}
 & & & & 0 \\
 & & & & 1 \\
 & & & 1 & 1 \\
 & & & 1 & 0 \\
 & & 1 & 1 & 0 \\
 & & 1 & 1 & 1 \\
 & & 1 & 0 & 1 \\
 & & 1 & 0 & 0 \\
 & 1 & 1 & 0 & 0 \\
 & 1 & 1 & 0 & 1 \\
 & 1 & 1 & 1 & 1 \\
 & 1 & 1 & 1 & 0 \\
 & 1 & 0 & 1 & 0 \\
 & 1 & 0 & 1 & 1 \\
 & 1 & 0 & 0 & 1 \\
 & 1 & 0 & 0 & 0 \\
1 & 1 & 0 & 0 & 0 \\
 & & \vdots & &
\end{array}
\right\} \quad \text{Gray Code}
$$

Figure 7.4

The code alternates between changing the right-most digit and changing the digit just to the **left** of the **right-most string** of the forms 1 0 0 ... 0.

Since we are changing one site (crossing) at a time, the signs in this expansion will alternate $+ - + - + \ldots$ (These are the signs $(-1)^{i(S)}$ in the formula (∗).) The crucial item of information for each state is $\|S\|$, the number of shaded components. Thus **we need to determine how $\|S\|$ and $\|S'\|$ differ when S' is obtained from S by switching the split at one vertex.**

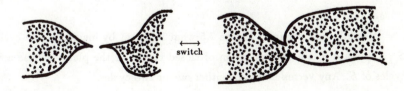

We have

360

Lemma 7.1. Let S and S' be chromatic states of a universe U. Suppose that S' is obtained from S by switching the split at one vertex v.

Then

1) $\|S'\| = \|S\| - 1$ when v is an exterior vertex incident to disjoint cycles of S.
2) $\|S'\| = \|S\| + 1$ when v is an interior vertex incident to a single cycle of S.
3) $\|S'\| = \|S\|$ in any other case.

(Refer to Figure 7.5.)

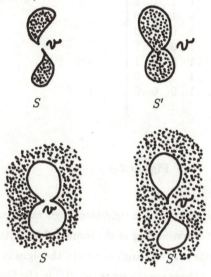

$$S \qquad\qquad S' \qquad\qquad \|S'\| = \|S\| - 1$$

$$S \qquad\qquad S' \qquad\qquad \|S\| = \|S'\|$$

Figure 7.5

It should be clear from Figure 7.5 what is meant by incidence (to) cycles of S. Any state S is a disjoint union of closed curves in the plane. These are the cycles of S. Any vertex is met by either one or two cycles.

We see from this lemma that the construction of an algorithm requires that we know how many cycles there are in a given state. In fact, it is a simple matter to **code a state by a list of its cycles** (plus a list of vertex types). Just list in clockwise order the vertices of each cycle. For example, (view Figure 7.6),

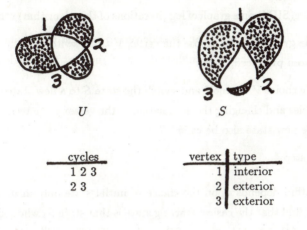

cycles	vertex	type
1 2 3	1	interior
2 3	2	exterior
	3	exterior

Figure 7.6

In Figure 7.6 we have illustrated a single trefoil state S, the list of cycles and the list of vertex types. From cycle and vertex information alone, we can deduce that switching vertex 3 reduces the number of components by 1 (It is an exterior vertex incident to **two** cycles.).

Thus we have given the outline of a computer-algorithm for computing the chromatic polynomial for a planar graph. The intermediate stages of this algorithm turn out to be as interesting as its end result! In order to discuss this I give below an outline form of the algorithm adapted to find the chromatic number of the graph:

1. Input the number of colors *(NUM)* (presumably four).

2. Input state information for **one** state S: number of vertices and their types, list of cycles, number of shaded regions.

3. Initialize the SUM to 0. (SUM will be the partial value of the state summation.)

4. SUM $=$ SUM $+(-1)^{i(S)}(\text{NUM}^{\|S\|})$. (Here $i(S)$ denotes the number of interior vertices in S, $\|S\|$ is the number of shaded regions.)

5. Print log(SUM) on a graph of log (iterations of this algorithm) versus log(SUM).

6. Use the **gray code** to choose the vertex V to be switched next. (If all states have been produced, stop!)

7. Use the choice V of step 6 and switch the state S to a new state by re-splicing the cycles and changing the designation of the vertex v. (interior $\rightleftarrows$ exterior). Let the new state also be called S.

8. Go to step 4.

This algorithm depends upon the choice of initial state only in its intermediate behavior. I find that the easiest starting state is that state S_0 whose cycles are the boundaries of the shaded regions of the universe U. Then all vertices are external and $\|S_0\| = (\text{NUM}^M)$ where M equals the number of shaded regions. This is the maximal value for S_0.

By analyzing the gray code, one can see that under this beginning the intermediate values of SUM are always **non-negative. That they are always positive is a conjecture generalizing the four color theorem** (for NUM $= 4$).

In fact, we find empirically that the graph of the points of the $\log - \log$ plot for 2^K iterations $K = 1, 2, \ldots$ is approximately linear (these values correspond to chromatic numbers of a set of subgraphs of the dual graph $\Gamma(U)$)! Clearly much remains to be investigated in this field.

Figure 7.7 illustrates a typical plot produced by this algorithm. (See [LK10].)

$\log_2(SUM)$

$\log_2(\#iterations)$

Figure 7.7

The **minima** in this plot correspond to chromatic numbers of subgraphs (i.e. they occur at 2^K iterations). The peaks correspond to those states where the gray code has disconnected most of the regions (as it does periodically in the switching process, just before switching a new bond.).

It is amusing and very instructive to use the algorithm, and a least squares approximation to the minima, as a "chromatic number predictor."

364

8⁰. The Potts Model and the Dichromatic Polynomial.

In this section we show how the partition function for the Potts model (See [BA2].) can be translated into a bracket polynomial state model. In particular we show how an operator algebra (first constructed by Temperley and Lieb from the physics of the Potts model) arises quite naturally in computing the dichromatic polynomial for universes (four-valent planar graphs).

Before beginning a review of the Potts model, here is a review of the operator algebra:

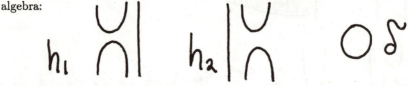

The diagrams h_1 and h_2 represent the two generators of the 3-strand **diagram algebra** D_3. The diagrams are multiplied by attachment:

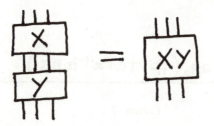

Elsewhere, we have used e_i for h_i and called this algebra (directly) the Temperley-Lieb algebra. Strands may be moved topologically, but they are not allowed to move above or below horizontal lines through the **fixed** end-points of the strands. Multiplication by δ is accomplished by disjoint union.

In D_n the following relations hold:

$$\left\{ \begin{array}{ll} h_i^2 = \delta h_i & i = 1, 2, \ldots, n-1 \\ h_i h_{i+1} h_i = h_i & \\ h_{i+1} h_i h_{i+1} = h_{i+1} & \\ h_i h_j = h_j h_i, & |i - j| > 1 \end{array} \right\}$$

See Figure 8.1 for an illustration.

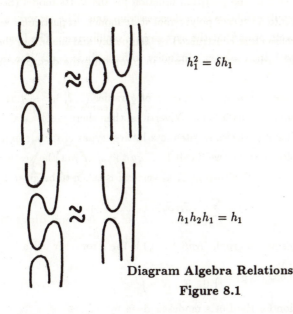

$$h_1^2 = \delta h_1$$

$$h_1 h_2 h_1 = h_1$$

Diagram Algebra Relations

Figure 8.1

The Potts Model.

In this model we are concerned with calculating a partition function associated with a planar graph G. In general for a graph G with vertices $i, j, \ldots$ and edges $\langle i, j \rangle$ a partition function has the form $Z_G = \sum_S e^{-E(S)/kT}$ where S runs over states of G, $E(S)$ is the energy of the state S, k is Boltzmann's constant and T is the temperature.

This presumes that there is a physical system associated to the graph and a collection of combinatorially described states S, each having a well-defined energy. The partition function, if sufficiently well-known, contains much information about the physical system. For example, the probability for the system to be in the state S is given by the formula $p(S) = e^{-E(S)/kT}/Z$.

This exponential model for the probability distribution of energy states is based on a homogeneity assumption about the energy: If $\hat{p}(E)$ denotes the probability of occurrence of a state of energy E then $\hat{p}(E + K)/\hat{p}(E' + K) = \hat{p}(E)/\hat{p}(E')$

for a pair of energy levels E and E'. Furthermore, states with the same energy level are equally probable.

The Boltzmann distribution follows from these assumptions. The second part of the assumption is an assumption that the system is in equilibrium. The homogeneity really has to do with the observer's inability to fix a "base-point" for the energy.

In the Potts model, the states involve choices of q values $(1, 2, 3, \ldots, q$ say) for each vertex of the graph G. Thus if G has N vertices, then there are q^N states. The q values could be spins of particles at sites in a lattice, types of metals in an alloy, and so on. The choice of spins (we'll call $1, \ldots, q$ the spins) at the vertices is free, and the energy of a state S with spin S_i at vertex i is taken to be given by

$$E(S) = \sum_{\langle i,j \rangle} \delta(S_i, S_j).$$

Here we sum over all edges in the graph. And $\delta(x, y)$ is the Kronecker delta:

$$\delta = (x,y) = \left\{ \begin{matrix} 1 & \text{if } x = y \\ 0 & \text{if } x \neq y \end{matrix} \right\}.$$

Thus, the partition function for the Potts model is given by

$$Z_G = \sum_S e^{K \sum_{\langle i,j \rangle} \delta(S_i, S_j)} \qquad (K = -1/kT)$$

$$= \sum_S \prod_{\langle i,j \rangle} e^{K\delta(S_i, S_j)}$$

$$\therefore Z_G = \sum_S \prod_{\langle i,j \rangle} (1 + v\delta(S_i, S_j))$$

where $v = e^K - 1$

(i.e. $e^{K\delta(x,y)} = 1 + v\delta(x,y)$).

This last form of the partition function is particularly convenient because it implies the structure:

$$Z(\text{⟩⟩}\!\!-\!\!\text{⟨⟨}) = Z(\text{⟩⟩} \; \text{⟨⟨}) + vZ(\text{⟩⟩⟨})$$

$$Z(\bullet \sqcup H) = qZ(H).$$

Here ⇢• ◄ denotes the graph $G($ ⇢•—◄ $)$ after deletion of one edge, while ⇢◄ means the graph obtained by collapsing this edge.

Hence, $Z(G)$ is a **dichromatic polynomial of G in variables v, q.**

Remark. In some versions of the Potts model the value of v is taken to vary according to certain edge types. Here we have chosen the simplest version.

Remark. One reason to calculate $Z(G)$ is to determine critical behavior corresponding to phenomena such as phase transitions (liquids to gas, ice to liquid, ...). Here one needs the behavior for large graphs. The only thoroughly analyzed cases are for low q (e.g. $q = 2$ in Potts is the Ising model) and planar graphs.

Remark. For $v = -1$, $Z(G)$ is the chromatic polynomial (variable q). Physically, $v = -1$ means $-1 = e^{-1/kT} - 1 \Rightarrow e^{-1/kT} = 0$. Hence $T = 0$.

Dichromatic Polynomials for Planar Graphs.

Just as we did for the chromatic polynomial, we can replace a planar graph with its associated universe (the medial graph) and then the dichromatic polynomial can be computed recursively on the shaded universe via the equations:

(i) $Z\left(\ \right) = Z\left(\ \right) + vZ\left(\ \right)$

(ii) $Z\left(\ \sqcup\ \right) = qZ\left(\ \right)$. Here ● stands for any connected shaded region, and $\sqcup$ denotes disjoint union.

Defining, as before, a **chromatic state S** to be any splitting of the universe U so that every vertex has been split, then **with the shading of U** we have the numbers

$$\|S\| = \text{number of shaded regions of } S,$$

and

$$i(S) = \text{number of interior vertices of } S.$$

Then, as a state summation,

$$Z(U) = \sum_{S} q^{\|S\|} v^{i(S)}.$$

368

We will now translate this version of the dichromatic polynomial so that the shaded universe is replaced by an **alternating link diagram** and the count $\|S\|$ **of shaded regions is replaced by a component (circuit) count** for S.

Proposition 8.1. Let U be a universe (the shadow of a knot or link diagram). Let U be shaded in checkerboard fashion. Let N denote the number of shaded regions in U (These are in one-to-one correspondence with the vertices of graph $\Gamma(U)$. Hence we say N is the number of **dual** vertices.). Let S be a chromatic state of U with $\|S\|$ connected shaded regions, and $i(S)$ internal vertices. Let $|S|$ denote the number of circuits in S (i.e. the number of boundary cycles for the shaded regions in S). Then

$$\|S\| = \tfrac{1}{2}(N - i(S) + |S|).$$

Proof. Associate to each state S a graph $\Gamma(S)$: The vertices of $\Gamma(S)$ are the vertices of $\Gamma(U)$, one for each shaded region of U. Two vertices of $\Gamma(S)$ are connected by an edge exactly when this edge can be drawn in the gap of an interior vertex of S. Thus

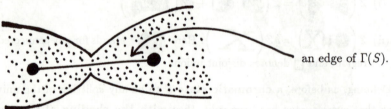

an edge of $\Gamma(S)$.

Thus the edges of $\Gamma(S)$ are in one-to-one correspondence with the interior vertices of S. $\Gamma(S)$ is a planar graph, and we assert that **the number of faces of $\Gamma(S)$** is $|S| - \|S\| + 1$.

To see this last assertion, note that each cycle in $\Gamma(S)$ surrounds a white island interior to one of the black components of S. Thus we may count faces of $\Gamma(S)$ by counting circuits of S, but those circuits forming outer boundaries to shaded parts of S must be discarded. This accounts for the subtraction of $\|S\|$. We add 1 to count the unbounded face.

Clearly, $\Gamma(S)$ has $\|S\|$ components. Therefore we apply the Euler formula for planar graphs:

$$(\text{Vertices}) - (\text{Edges}) + (\text{Faces}) = (\text{Components}) + 1$$

$$N - i(S) + (|S| - \|S\| + 1) = \|S\| + 1.$$

Hence $N - i(S) + |S| = 2\|S\|$. This completes the proof. //

Example.

$$N = 4$$

U

S

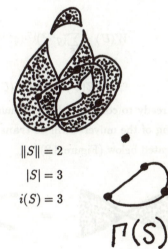

$$\|S\| = 2$$
$$|S| = 3$$
$$i(S) = 3$$

$\Gamma(S)$

Thus

$$\frac{N - i(S) + |S|}{2} = \frac{4 - 3 + 3}{2} = 2\,,$$

$$\|S\| = 2.$$

Note that in the dichromatic polynomial, S contributes the term $q^{\|S\|}v^{i(S)} = q^2v^3$ here. Since the number of v-factors corresponds to the positive energy contributions in the Potts model, we see that these contributions correspond to the clumps of vertices in $\Gamma(S)$ that are connected by edges. These vertices all mutually agree on choice of spin state.

Now the upshot of Proposition 8.1 is that we can replace $\|S\|$ by $|S|$ in our formulas! Let's do so!

$$
\begin{aligned}
Z(U) &= \sum_S q^{\|S\|}v^{i(S)} \\
&= \sum_S q^{1/2(N-i(S)+|S|)}v^{i(S)} \\
\therefore Z(U) &= q^{N/2}\sum_S (q^{1/2})^{|S|}(q^{-(1/2)}v)^{i(S)}
\end{aligned}
$$

Let

$$
W(U) = \sum_S (q^{1/2})^{|S|}(q^{-1/2}v)^{i(S)}.
$$

Thus

$$
Z(U) = q^{N/2}W(U).
$$

We are now ready to complete the reformulation of the dichromatic polynomial. The **shading** of the universe will be **translated into crossings of a link diagram** as indicated below (Figure 8.2):

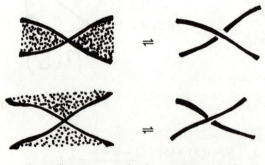

Crossing Transcriptions of Local Shading

Figure 8.2

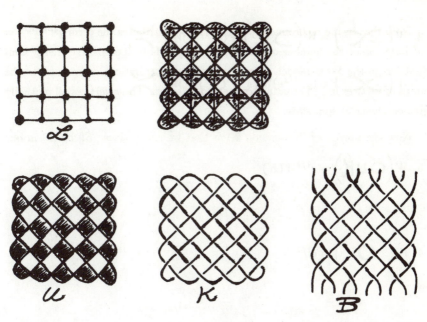

Figure 8.3

Since W satisfies

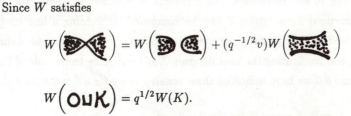

$$W\left(\text{OuK}\right) = q^{1/2}W(K).$$

(Note the extra component is not shaded!) We can write the

W-Axioms (The Potts Bracket).

1. Let K be any knot or link diagram. Then $W(K) \in \mathbf{Z}[q^{1/2}, q^{-1/2}, v]$ is a well-defined function of q and v.

2. $W\left(\asymp \right) = W\left(\supset \subset \right) + (q^{-1/2}v)W\left(\mathopen{\rotatebox{90}{$\asymp$}} \right)$

$W\left(\rotatebox{90}{$\asymp$} \right) = (q^{-1/2}v)W\left(\supset \subset \right) + W\left(\mathopen{\rotatebox{90}{$\asymp$}} \right)$

3. $W\left(\bigcirc \sqcup K \right) = q^{1/2}W(K)$

4. $W\left(\bigcirc \right) = q^{1/2}.$

and the Z-Definition: $Z(K) = q^{N/2}W(K)$ where N is the number of shaded regions in K. (Compare with [LK10].)

According to our discussion, the dichromatic polynomial (hence the Potts partition function) for a lattice $\mathcal{L}$ can be computed by forming a link diagram $K(\mathcal{L})$ obtained by first forming the medial graph of $\mathcal{L}$, regarding it as a shaded universe U, and configuring the knot diagram $K(\mathcal{L})$ according to the rule of Figure 8.2. In Figure 8.3 we have indicated these constructions for a 5×5 lattice $\mathcal{L}$.

Here is a worked example for the 2×2 lattice:

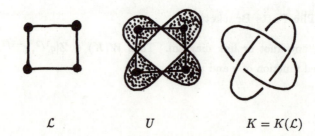

$$\mathcal{L} \qquad\qquad U \qquad\qquad K = K(\mathcal{L})$$

(a) Direct computation of $Z(\mathcal{L}) = Z_1$.

$$Z_1\left(\square\right) = Z_1\left(\sqcup\right) + vZ_1\left(\triangle\right)$$
$$= Z_1\left(\vdots\sqcup\right) + vZ_1\left(\dashv\right)$$
$$+ vZ_1\left(\diagup\right) + v^2 Z_1\left(\bigcirc\right)$$
$$= (q+v)Z_1\left(\dashv\right) + vZ_1\left(\diagup\right) + v^2 Z_1\left(\bigcirc\right)$$
$$= (q+v)(q+v)^2 q + v(q+v)^2 q + v^2(q+v)q$$
$$= (q+v)^3 q + (q+v)^2 vq + (q+v)v^2 q$$
$$= (q^3 + 3q^2 v + 3qv^2 + v^3)q$$
$$+ (q^2 + 2qv + v^2)vq + q^2 v^2 + v^3 q$$
$$= q^4 + 3q^3 v + 3q^2 v^2 + qv^3$$
$$+ q^3 v + 2q^2 v^2 + qv^3 + q^2 v^2 + qv^3$$
$$\therefore Z\left(\square\right) = q^4 + 4q^3 v + 6q^2 v^2 + 3qv^3$$

(b) Computing $W(K)$ first.

In order to compute W, we use the rules

$$W\left(\asymp\right) = W\left(\supset\subset\right) + xW\left(\curlyvee\right)$$
$$W\left(\bigcirc \sqcup L\right) = yW(L), \quad W(\mathbf{0}) = y$$

with $x = q^{-1/2}v$, $y = q^{1/2}$.

Before, doing the 2×2 lattice, let's get some useful formulas:

$$
\left\{
\begin{aligned}
W\left(\text{⌀} \right) &= W\left(\text{ひ} \right) + xW\left(\text{ᴐ} \right) \\
&= W\left(\sim \right) + xyW\left(\sim \right) \\
\therefore W\left(\text{ᴐ⁻} \right) &= (1+v)W\left(\sim \right) = (1+xy)W\left(\sim \right). \\
W\left(\text{⌀} \right) &= W\left(\text{ᴐ} \right) + xW\left(\text{ひ} \right) \\
&= yW\left(\sim \right) + xW\left(\sim \right) \\
\therefore W\left(\text{⌀} \right) &= (q^{1/2} + q^{-1/2}v)W\left(\sim \right) = (x+y)W\left(\sim \right).
\end{aligned}
\right.
$$

Returning to our lattice/knot diagram K, we have

$$
\begin{aligned}
W\left(\text{⊗} \right) &= W\left(\text{⊗} \right) + xW\left(\text{⊗} \right) \\
&= (x+y)^3 W\left(\bigcirc \right) + x\left[W\left(\text{⊗} \right) + xW\left(\text{⊗} \right) \right] \\
&= (x+y)^3 y + x(x+y)^2 y + x^2 W\left(\text{⊗} \right) \\
&= (x+y)^3 y + x(x+y)^2 y + \\
&\quad + x^2 \left(W\left(\text{⊗} \right) + xW\left(\text{⊗} \right) \right) \\
&= (x+y)^3 y + x(x+y)^2 y + x^2((x+y)y + x(x+y)y) \\
\therefore W\left(\text{⊗} \right) &= (x+y)^3 y + (x+y)^2 xy + (x+y)(x^2 y + xy).
\end{aligned}
$$

We leave as an exercise for the reader to show that $Z(\mathcal{L}) = q^{N/2} W(K(\mathcal{L}))$ for this case!

Comment. Obviously, this translation of the dichromatic polynomial into computations using knot diagrams does not directly make calculations easier. It does however reveal the geometry of the Temperley-Lieb algebra in relation to the dichromatic polynomial (via the Potts model).

I will show how the W-formulation of the dichromatic polynomial naturally gives rise to a mapping of braids into the Temperley-Lieb algebra.

The idea is as follows: If we expanded W on a braid diagram out into the corresponding states, we would find ourselves looking at an element of the Temperley-Lieb algebra. For example, if we were calculating $W(\beta)$ for

$$\beta = \quad$$

then one possible local state configuration is

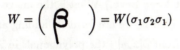

and this is $h_1 h_2 h_1$ in D_3. In general, if we expand (for example)

$$W = \left(\; \beta \; \right) = W(\sigma_1 \sigma_2 \sigma_1)$$

we can obtain the expansion algebraically by replacing

$$\sigma_i \quad \text{by} \quad \mathcal{I} + x h_i$$
$$\sigma_i^{-1} \quad \text{by} \quad x\mathcal{I} + h_i$$

and formally multiplying these replacements.

For example, in the above case we have

$$(\mathcal{I} + x h_1)(\mathcal{I} + x h_2)(\mathcal{I} + x h_1)$$
$$= (\mathcal{I} + x h_1 + x h_2 + x^2 h_1 h_2)(\mathcal{I} + x h_1)$$
$$= \mathcal{I} + x h_1 + x h_2 + x^2 h_1 h_2$$
$$\quad + x h_1 + x^2 h_1^2 + x h_2 h_1 + x^3 h_1 h_2 h_1$$
$$= \mathcal{I} + 2(x h_1) + (x h_2) + (x^2 h^2 + 1) + (x^2 h_1 h_2)$$
$$\quad + (x h_2 h_1) + (x^3 h_1 h_2 h_1).$$

$\leftrightarrow h_1$ $\leftrightarrow h_2$ $\leftrightarrow h_1^2$

$\leftrightarrow h_1 h_2$ $\leftrightarrow h_2 h_1$ $\leftrightarrow h_1 h_2 h_1$

Now it is **not** the case that W is a topological invariant of braids (when we put a braid in W we mean it to indicate a larger diagram containing this pattern), however **this correspondence does respect the relations in the Temperley-Lieb algebra.**

Thus

$$W(h_1 h_2 h_1) = W\left[\begin{array}{c}\text{⋃⋃}\\\text{⋂⋂}\end{array}\right] = W\left[\begin{array}{c}\text{⋃}\\\text{⋂}\end{array}\middle|\right] = W(h_1)$$

and

$$W(h_1^2) = W\left[\begin{array}{c}\text{⋃}\\\text{○}\\\text{⋂}\end{array}\middle|\right] = yW\left[\begin{array}{c}\text{⋃}\\\text{⋂}\end{array}\middle|\right] = yW(h_1)$$

($h_1^2 = \delta h_1$, let $W(\delta X) = yW(X)$ be the algebraic version of $W\left(\text{O⋃X}\right) = yW(X)$. **Therefore let $\delta = y$ here.**) What is the image algebra here? Formally, it is $\mathcal{U}_n = \mathbf{Z}[x,y](D_n)$. That is, we are formally adding and multiplying elements of D_n with coefficients from $\mathbf{Z}[x,y]$ - the polynomials in x and y.

In creating this formulation we have actually reformulated the calculation of the rectangular lattice Potts model **entirely** into an algebra problem about $\mathcal{U}_n$. To see this, think of the link diagram $K(\mathcal{L})$ of the lattice $\mathcal{L}$ as a braid that has been closed to form $K(\mathcal{L})$ by adding maxima and minima:

$$P(\beta) = K(\mathcal{L})$$

This way of closing a braid β to form $P(\beta)$ is called a **plat.**

Now note. Let R denote the diagram

$$\begin{array}{ccccc}\text{⋃} & \text{⋃} & \cdots & \text{⋃}\\\text{⋂} & \text{⋂} & \cdots & \text{⋂}\end{array} \approx h_1 h_3 h_5 \ldots h_{2n-1} \in D_{2n}$$

Then, **if β is any element of B_{2n} we have $R\beta R = P(\beta)R$.**

Proof.

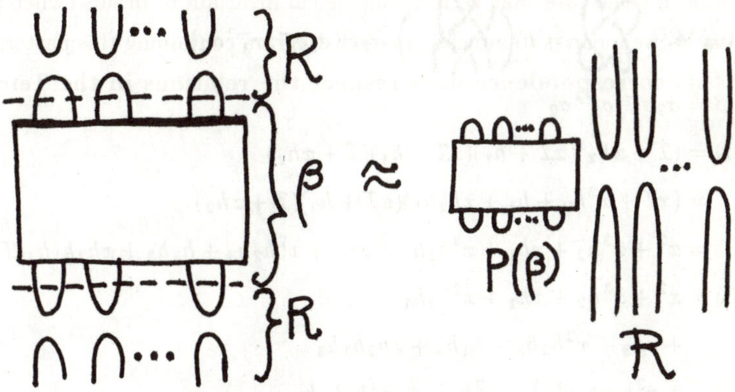

$$R\beta R = P(\beta)R. \qquad //$$

The upshot of this formula is that all of the needed component counting will happen **automatically** if we substitute for β the product expression in $\mathcal{U}_n$. Let $X(\beta)$ denote the corresponding element of $\mathcal{U}_n$:

$$X : \widehat{B}_n \to \mathcal{U}_n$$

$$\left\{\begin{array}{lcl} X(\sigma_i) & = & \mathcal{I} + xh_i \\ X(\sigma_i^{-1}) & = & x\mathcal{I} + h_i \\ X(\mathcal{I}) & = & \mathcal{I} \end{array}\right\}$$

$$\{X(ab) = X(a)X(b)\}.$$

Here $\widehat{B}_n$ is the collection of braid diagrams, described by braid words. On the diagrammatic side we are not including ambient isotopy. On the algebraic side $\widehat{B}_n$ is the **free group** on $\sigma_1, \ldots, \sigma_{n-1}, \sigma_1^{-1}, \ldots, \sigma_{n-1}^{1}, \mathcal{I}$ (modulo $\sigma_i\sigma_j = \sigma_j\sigma_i$ for $|i-j| > 1$). Then if $R = h_1h_3 \ldots h_{2n-1}$ and $\beta \in \widehat{B}_{2n}$ we have

$$RX(\beta)R = W(P(\beta))R,$$

a completely algebraic calculation of W using the operator algebra $\mathcal{U}_n$.

Example. $= P\left(\begin{array}{c}\end{array}\right) = P(\sigma_2\sigma_1^{-1}\sigma_3^{-1}\sigma_2)$

$\beta = \sigma_2\sigma_1^{-1}\sigma_3^{-1}\sigma_2$

$$X(\beta) = (I + xh_2)(xI + h_1)(xI + h_3)(I + xh_2)$$
$$= (xI + x^2h_2 + h_1 + xh_2h_1)(xI + h_3)(I + xh_2)$$
$$= x^2 + x^3h_2 + xh_1 + x^2h_2h_1 + xh_3 + x^2h_2h_3 + h_1h_3 + xh_2h_1h_3(I + xh_2)$$

$$X(\beta) = x^2 + x^3h_2 + xh_1 + x^2h_2h_1$$
$$+ xh_3 + x^2h_2h_3 + h_1h_3 + xh_2h_1h_3$$
$$+ x^3h_2 + x^4h_2^2 + x^2h_1h_2 + x^2h_2h_1h_2$$
$$+ x^2h_3h_2 + x^3h_2h_3h_2 + xh_1h_3h_2$$
$$+ x^2h_2h_1h_3h_2.$$

These correspond to the sixteen states of K. Let's take one term and look at it algebraically and geometrically. For this problem, $\beta \in B_4$ and

$$R = h_1h_3 \longleftrightarrow$$

Take $RX(\beta)R$ and look at the term

$$R[x^2h_2h_3h_2]R$$
$$= x^2h_1h_3h_2h_3h_2h_1h_3$$
$$= x^2h_1h_3h_2h_1h_3$$
$$= x^2h_1h_3h_2h_3h_1$$
$$= x^2h_1h_3h_1$$
$$= x^2h_1^2h_3$$
$$= (x^2y)h_1h_3$$
$$= (x^2y)R.$$

Geometrically:

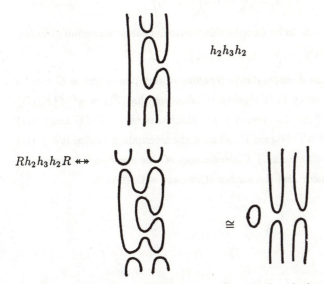

$$h_2 h_3 h_2$$

$$Rh_2 h_3 h_2 R \longleftrightarrow$$

and the extra component gives rise to the y in the algebra.

To summarize, we have shown that for any braid diagram $\beta \in \widehat{B}_{2n}$, the W-polynomial is given algebraically via

$$W(P(\beta))R = RX(\beta)R.$$

This yields an operator-algebraic formulation of the Potts model that is formally isomorphic to that given by Temperley and Lieb (see [BA1]). Our method constructs the operator algebra purely geometrically. In fact, this formulation makes some of the statistical mechanical features of the model quite clear. For example, in the anti-ferromagnetic case of a large rectangular lattice one expects the critical point to occur when there is a symmetry between the partition function on the lattice and the dual lattice. In the bracket reformulation of the Potts model this corresponds to having $W(\diagdown\!\!\!\diagup) = W(\diagup\!\!\!\diagdown)$ and this occurs when $q^{-(1/2)}v = 1$, Hence the critical temperature occurs (conjecturally) at

$$e^{(1/kT)} - 1 = \sqrt{q}$$

or

$$T = \frac{1}{k \ln(1 + \sqrt{q})} \; .$$

Obviously, more work needs to be done in this particular interconnection of knots, combinatorics and physics.

Exercise. Show that the dichromatic polynomial of any planar graph G can be expressed via the Temperley-Lieb algebra by showing that $Z_G = q^{N/2}\{\tilde{K}(G)\}$ where $\{L\} = W_L$ is the Potts bracket, N is the number of vertices of G and $\tilde{K}(G)$ is a link in plat form. (HINT: Obtain $\tilde{K}(G)$ from the alternating medial link $K(G)$ by arranging maxima and minima.) Consequences of this exercise will appear in joint work of Hubert Saleur and the author. Compare with [SAL2].

9^0. Preliminaries for Quantum Mechanics, Spin Networks and Angular Momentum.

This section is preparatory to an exposition of the Penrose theory of spin-networks. These networks involve an evaluation process very similar to our state summations and chromatic evaluations. And by a fundamental result of Roger Penrose, there is a close relation to three-dimensional geometry in the case of certain large nets. This is remarkable: a three-dimensional space becomes the appropriate **context** for the interactive behaviors in a large, purely combinatorially defined network of spins.

We begin with a review of spin in quantum mechanics.

Spin Review. (See [FE].)

Recall that in the quantum mechanical framework, to every **process** there corresponds an **amplitude**, a complex number ψ. ψ is usually a function of space and time: $\psi(x,t)$. With proper normalization, the probability of a process is equal to the absolute square of this amplitude:

$$\text{prob}(\psi) = \psi^*\psi$$

where ψ^* denotes the complex conjugate of ψ. I am using the term process loosely, to describe a condition that includes a physical process **and** a mode of observation. Thus if we observe a beam of polarized light, there is an amplitude for observing polarization in a given direction θ (This could be detected by an appropriately oriented filter.).

Consider a beam of light polarized in a given direction. Let the axis of an analyzer (polaroid filter or appropriate prism) be placed successively in two perpendicular directions x, y to measure the number of photons passing a corresponding lined-up polarizer. [x and y are orthogonal to the direction of the beam.] Let a_x

and a_y be the amplitudes for these directions. Then, if the analyzer is rotated by θ degrees, with respect to the x-direction, the amplitude becomes the superposition

$$a(\theta) = \cos(\theta)a_x + \sin(\theta)a_y$$

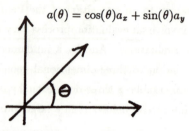

This result follows from the two principles:

1) Nature has no preferred axis. [Hence the results must be invariant under rotation.]

2) Quantum amplitudes for a mixed state undergo linear superposition.

The polarizer experiment already contains the essence of quantum mechanics. Note, for example, that it follows from this description that the amplitude for the process

x-filter followed by y-filter

is zero, while the process

x-filter / 45°-filter / y-filter

has a non-zero amplitude in general. Two polarizing filters at 90° to one another block the light. Insert a third filter between them at an angle, and light is transmitted.

The two principles of **non-preferential axis** and **linear superposition** are the primary guides to the mathematics of the situation.

In the more general situation of a beam of particles there may be a vector of amplitudes

$$\vec{a} = (a_1, a_2, \dots, a_n).$$

Rotation of the analyzer will produce a new amplitude

$$\vec{a}' = D(R)\vec{a}.$$

where R denotes the rotation in three-dimensional space of the analyzer.

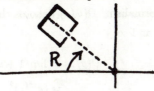

If S and R are two rotations, let SR denote their composition (first R, then S). We require that $D(SR) = D(S)D(R)$ **and** that $D(R)$ and $D(S)$ are linear transformations.

Linearity follows from the superposition principle, and composition follows essentially from independence of axis. That is, suppose $\vec{a}$ is the amplitude for a standard direction and that we rotate by R to get $\vec{a}' = D(R)\vec{a}$ and **then** rotate by S to get $\vec{a}'' = D(S)D(T)\vec{a}$. This is the same experiment as rotating by SR and so $\vec{a}'' = D(SR)\vec{a}$ also, whence $D(SR) = D(S)D(R)$. (This is true up to a phase factor.)

In order to discuss the possibilities for these representations, recall that over the complex numbers a rotation of the plane can be represented by

$$e^{i\theta} = \lim_{n\to\infty} (1 + i\theta/n)^n.$$

Note that this is a formal generalization of the real limit $e^x = \lim_{n\to\infty} (1 + x/n)^n$, and note that

$$(1 + i\epsilon)z = z + i\epsilon z$$

Multiplication of a complex number z by $(1 + i\epsilon)$ has the effect of rotating it by approximately ϵ radians in a counter-clockwise direction in the complex plane. Thus we can regard $(1 + i\epsilon)$ as an "infinitesimal rotation".

Following this lead, we shall try

$$D(\epsilon/x) = 1 + i\epsilon M_x$$
$$D(\epsilon/y) = 1 + i\epsilon M_y$$
$$D(\epsilon/z) = 1 + i\epsilon M_z$$

where these represent small angular rotations about the x, y and z axes respectively. For this formalism, $D(\theta/z)$ denotes the matrix for a rotation by angle θ about the z-axis, and so

$$D(\theta/z) = \lim_{\epsilon \to 0}(1 + i\epsilon M_z)^{\theta/\epsilon}$$
$$= (e^{iM_z})^{\theta}$$
$$D(\theta/z) = e^{i\theta M_z}.$$

For a given vector direction $\vec{v}$, we have $D(\theta/\vec{v}) = e^{i\theta(\vec{v}\cdot\vec{M})}$ where

$$\vec{M} = (M_x, M_y, M_z)$$
$$\vec{v}\cdot\vec{M} = v_x M_x + v_y M_y + v_z M_z.$$

Proposition 9.1. $D(R)$ is a unitary matrix. That is, $D(R)^{-1} = D(R)^*$ where $*$ denotes conjugate transpose.

Proof. It is required that $\psi\psi^*$ be invariant under rotation. But if $\psi' = D\psi$, then

$$\psi'\psi'^* = D\psi(D\psi)^* = \sum_i (D\psi)_i (\psi^* D^*)_i$$

$$= \sum_i \left(\sum_j D_{ji}\psi_j\right)\left(\sum_k \psi_k^* D_{ki}^*\right)$$

$$= \sum_{jk}\sum_i D_{ji}\overline{D}_{ik}\psi_j\overline{\psi}_k$$

$$= \sum_{j,k}(DD^*)_{jk}\psi_j\overline{\psi}_k.$$

In order for this last formula to equal $\psi\psi^*$ for all possible, ψ we need $DD^* = I$.

$$//$$

Since $D = e^{i\theta M}$ is unitary, we conclude that

$$1 = DD^* = e^{i\theta(M-M^*)}.$$

Thus $M = M^*$ and so M is **Hermitian**.

Thus the matrices M are candidates for **physical observables** in the quantum theory. [Since they have real eigenvalues, and the quantum mechanical model for observation is $M\psi = \lambda\psi$ for some amplitude ψ.]

Consequently, we want to know more about M_x, M_y and M_z. In particular, we want their eigenvalues and how they commute.

To obtain commutation relations for M_x and M_y, consider a **rotation by** ϵ **about** x**-axis** followed by **rotation by** ϵ **about** y**-axis** followed by **rotation by** $-\epsilon$ **about** x**-axis** followed by **rotation by** $-\eta$ **about** y**-axis.**

1. Up to 1st order the z-axis is left fixed.

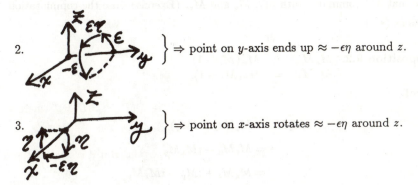

2. $\quad\Rightarrow$ point on y-axis ends up $\approx -\epsilon\eta$ around z.

3. $\quad\Rightarrow$ point on x-axis rotates $\approx -\epsilon\eta$ around z.

Thus

$$(1 - i\epsilon\eta M_z) \approx D(-\eta/y)D(-\epsilon/x)D(\eta/y)D(\epsilon/x)$$

$$= (1 - i\eta M_y)(1 - i\epsilon M_x)(1 + i\eta M_y)(1 + i\epsilon M_x)$$

$$\Rightarrow -iM_z = -M_y M_x - M_y M_x + M_y M_x + M_x M_y.$$

So

$$M_x M_y - M_y M_x = -iM_z.$$

Because of this minus sign it is convenient to rewrite to the conjugate representation. Then we have:

$$
\begin{array}{rcl}
M_x M_y - M_y M_x &=& iM_z \\
M_y M_z - M_z M_y &=& iM_x \\
M_z M_x - M_x M_z &=& iM_y
\end{array}
\qquad (*)
$$

Eigenstructure.

Let

$$M^2 = M_x^2 + M_y^2 + M_z^2$$
$$M_- = M_x - iM_y$$
$$M_+ = M_x + iM_y.$$

Note that M^2 commutes with M_x, M_y and M_z. (Exercise - use the commutation relations $(*)$.)

Proposition 9.2. $M_z M_- = M_-(M_z - 1)$
$M_z M_+ = M_+(M_z + 1).$

Proof.

$$M_z M_- = M_z(M_x - iM_y)$$
$$= M_z M_x - iM_z M_y$$
$$= M_x M_z + iM_y - iM_z M_y$$
$$= (M_x - iM_y)(M_z - 1)$$
$$\therefore M_z M_- = M_-(M_z - 1)$$
$$M_z M_+ = M_z(M_x + iM_y)$$
$$= M_x M_z + iM_y + iM_z M_y$$
$$= (M_x + iM_y)(M_z + 1)$$
$$\therefore M_z M_+ = M_+(M_z + 1).$$

$//$

Now suppose that $M_z a^{(m)} = m a^{(m)}$. That is, suppose that $a^{(m)}$ is an eigenvector of $a^{(m)}$ with eigenvalue m. Then

$$M_z M_- a^{(m)} = M_-(M_z - 1)a^{(m)}$$
$$= M_-(m - 1)a^{(m)}$$
$$= (m - 1)M_- a^{(m)}.$$

Thus $M_- a^{(m)}$ becomes a new eigenvector of M_z with eigenvalue $(m - 1)$.

We can define

$$ca^{(m-1)} = M_-a^{(m)}$$

with the constant c to be determined.

Assume that $a^{(m)}$ is normalized for each m so that $1 = \sum_n a_n^{(m)} a_n^{(m)*}$. Then

$$1 = \frac{1}{cc^*}\sum_n [M_-a^{(m)}]_n [M_-a^{(m)}]_n^*.$$

Hence

$$1 = \frac{1}{cc^*}\sum_n [M_-a^{(m)}]_n^* [M_-a^{(m)}]_n$$

$$= \frac{1}{cc^*}\sum_n [a^{(m)^*}]_n [M_-^* M_-a^{(m)}]_n$$

$$= \frac{1}{cc^*}\sum_n [a^{(m)^*}]_n [M_+ M_-a^{(m)}]_n.$$

Now

$$M_+M_- = (M_x + iM_y)(M_x - iM_y)$$

$$= M_x^2 + M_y^2 + i(M_yM_x - M_xM_y)$$

$$= M_x^2 + M_y^2 + i(-iM_z)$$

$$= M_x^2 + M_y^2 + M_z$$

$$M_+M_- = M^2 - M_z^2 + M_z.$$

We may choose $a^{(m)}$'s so that

$$M^2 a^{(m)} = ka^{(m)}$$

(since M^2 commutes with the other operators) for a fixed k, and ask for the resulting range of m-values.

$$c^*c = \sum_n a_n^{(m)*}(M_+M_-)a_n^{(m)}$$

$$\begin{pmatrix} M_-a^{(m)} &=& ca^{(m-1)} \\ M_+a^{(m-1)} &=& c^*a^{(m)} \end{pmatrix}$$

$$\Rightarrow \boxed{|c|^2 = k - m^2 + m}.$$

Thus $|c| = (k - m(m-1))^{1/2}$. From this it follows that $k = j(j+1)$ and $2j$ is an integer.

Proof. Let $m = -j$ be the lowest state - meaning that $M_- a^{(m)} = 0$. This implies $c = 0$ for $m = -j$ whence $0 = k - (-j)(-j-1)$ whence $k = j(j+1)$.

By the same argument, if the largest value of m is j' then $k = j'(j'+1)$. Thus $j = j'$ and $2j$ is integer. //

Example. $j = 1/2$.

Then there are two states. Let

$$a^{(1/2)} = \begin{pmatrix} 1 \\ 0 \end{pmatrix}, \ a^{(-1/2)} = \begin{pmatrix} 0 \\ 1 \end{pmatrix}$$

$$c = [j(j+1) - m(m-1)]^{1/2}$$

$$= \left[\frac{1}{2}\left(\frac{3}{2}\right) - \frac{1}{2}\left(-\frac{1}{2}\right) \right]^{1/2} = [4/4]^{1/2} = 1$$

$$\therefore \ c = 1.$$

Thus

$$\left. \begin{array}{l} M_- \begin{pmatrix} 1 \\ 0 \end{pmatrix} = \begin{pmatrix} 0 \\ 1 \end{pmatrix} \\ M_- \begin{pmatrix} 0 \\ 1 \end{pmatrix} = \begin{pmatrix} 0 \\ 0 \end{pmatrix} \end{array} \right\} \Rightarrow M_- = \begin{pmatrix} 0 & 0 \\ 1 & 0 \end{pmatrix}$$

$$\left. \begin{array}{l} M_z \begin{pmatrix} 1 \\ 0 \end{pmatrix} = \frac{1}{2} \begin{pmatrix} 1 \\ 0 \end{pmatrix} \\ M_z \begin{pmatrix} 0 \\ 1 \end{pmatrix} = -\frac{1}{2} \begin{pmatrix} 0 \\ 1 \end{pmatrix} \end{array} \right\} \Rightarrow M_z = \frac{1}{2} \begin{pmatrix} 1 & 0 \\ 0 & -1 \end{pmatrix}.$$

Similarly, $M_+ = \begin{pmatrix} 0 & 1 \\ 0 & 0 \end{pmatrix}$ and

$$\left. \begin{array}{lll} M_x & = & \frac{1}{2} \begin{pmatrix} 0 & 1 \\ 1 & 0 \end{pmatrix} = \frac{1}{2} \sigma_x \\ M_y & = & \frac{1}{2} \begin{pmatrix} 0 & i \\ -i & 0 \end{pmatrix} = \frac{1}{2} \sigma_y \\ M_z & = & \frac{1}{2} \begin{pmatrix} 1 & 0 \\ 0 & -1 \end{pmatrix} = \frac{1}{2} \sigma_z \end{array} \right\} \quad \boxed{\begin{array}{l} M_- = M_x - iM_y \\ M_+ = M_x + iM_y \end{array}}$$

where σ_x, σ_y, σ_z are the Pauli matrices satisfying

$$\sigma_x^2 = \sigma_y^2 = \sigma_z^2 = 1$$

$$\sigma_x \sigma_y = -\sigma_y \sigma_x = i\sigma_z$$

Quick Review of Wave Mechanics.

$$\psi = Ae^{i(kx-wt)}$$

$$\frac{\partial \psi}{\partial t} = -iw\psi \qquad \boxed{E = h\nu = \hbar w}$$

$$(\hbar = h/2\pi)$$

$$i\hbar\frac{\partial \psi}{\partial t} = E\psi \qquad \boxed{p = \hbar k}$$

$$-i\hbar\frac{\partial \psi}{\partial x} = \hbar k\psi = p\psi.$$

De Broglie suggested (1923) that particles such as electrons could be regarded as having wave-like properties, and his fundamental associations were

$$E = \hbar w$$

$$p = \hbar k$$

for energy E and momentum p. These quantities can be combined in a simple wave-function $\psi = e^{i(kx-wt)}$.

This led Schrödinger in the same year to postulate the equation

$$\boxed{i\hbar\frac{\partial \psi}{\partial t} = H\psi}$$

where $H = p^2/2m + V$ represents total energy, kinetic plus potential. And one associated the **operator**

$$\frac{\hbar}{i}\frac{\partial}{\partial x}$$

to momentum. So

$$H = \frac{(-i\hbar)^2}{2m}\frac{\partial^2}{\partial x^2} + V = -\frac{\hbar^2}{2m}\frac{\partial^2}{\partial x^2} + V.$$

At about the same time, Heisenberg invented matrix mechanics, based on non-commutative algebra. Heisenberg's mechanics is based on an algebra where position x and momentum p no longer commuted, but rather

$$xp - px = \hbar i.$$

The formal connection of this algebra and the Schrödinger mechanics is contained in the association of **operators** (such as $(\hbar/i)\,\partial/\partial x$ to p) for physical quantities.

Note that

$$\frac{\hbar}{i}\frac{\partial}{\partial x}(xf) - x\frac{\hbar}{i}\frac{\partial}{\partial x}(f)$$
$$= \frac{\hbar}{i}\left[f + x\frac{\partial f}{\partial x} - x\frac{\partial f}{\partial x}\right] = \frac{\hbar}{i}\,f.$$

Thus $xp - px = \hbar i$ in the Schrödinger theory as well.

One consequence is that position and momentum are not simultaneously observable. Observability is associated with a Hermitian operator, and its eigenbehavior.

It took some years, and debate before Max Born made the interpretation that ψ measures a "complex probability" for the event - with $\psi^*\psi$ as the real probability.

To return to the subject of rotations, let us consider the role of angular momentum in quantum mechanics. Let

$$\vec{r} = ix + jy + kz$$
$$\vec{p} = ip_x + jp_z + kp_z$$

where i, j, k denotes the standard basis for $\mathbf{R}^3$. The classical angular momentum is given by $\vec{L} = \vec{r} \times \vec{p}$.

$$\vec{L} = \vec{r} \times \vec{p} = \begin{vmatrix} \vec{i} & \vec{j} & \vec{k} \\ x & y & z \\ p_x & p_y & p_z \end{vmatrix}.$$

In quantum mechanics, $\vec{p} == i\hbar\vec{\nabla}$. Thus

$$L_x \longleftrightarrow \frac{\hbar}{i}\left(y\frac{\partial}{\partial z} - z\frac{\partial}{\partial y}\right)$$
$$L_y \longleftrightarrow -\frac{\hbar}{i}\left(x\frac{\partial}{\partial z} - z\frac{\partial}{\partial x}\right)$$
$$L_z \longleftrightarrow \frac{\hbar}{i}\left(x\frac{\partial}{\partial y} - y\frac{\partial}{\partial x}\right)$$
$$L^2 = |\vec{L} \cdot \vec{L}| = L_x^2 + L_y^2 + L_z^2.$$

Let $[a, b] = ab - ba$. Then

$$\left.\begin{array}{rcl} [L_x, L_y] & = & i\hbar L_z \\ [L_y, L_z] & = & i\hbar L_x \\ [L_z, L_x] & = & i\hbar L_y \end{array}\right\} \vec{L} \times \vec{L} = i\hbar \vec{L}.$$

$[L^2, L_z] = 0$, so L^2 commutes with $\vec{L}$.

The formalism for angular momentum in quantum mechanics is identical to the formalism for unitary representation of rotational invariance. The operators for angular momentum correspond to rotational invariance of ψ.

Digression on Wave Mechanics.

Note that $\sin(X + Y) + \sin(X - Y) = 2\sin(X)\cos(Y)$.

$$\boxed{\sin\left(\frac{2\pi}{\lambda}(x - ct)\right)} \quad \begin{array}{rcl} \lambda & = & \text{wave-length} \\ c & = & \text{wave-velocity} \end{array}$$

$$\sin\left(\frac{2\pi}{\lambda}(x - ct)\right) = \sin(kx - wt)$$

$$k = 2\pi/\lambda, \qquad w = 2\pi c/\lambda = 2\pi\nu.$$

Let

$$u = a\left[\sin\underbrace{(kx - wt)}_{X+Y} + \sin\underbrace{(k'x - w't)}_{X-Y}\right]$$

$$\Rightarrow X = \left(\frac{k + k'}{2}\right)x - \left(\frac{w + w'}{2}\right)t$$

$$Y = \left(\frac{k - k'}{2}\right)x - \left(\frac{w - w'}{2}\right)t$$

$$u = a\sin\left[\left(\frac{k + k'}{2}\right)x - \left(\frac{w + w'}{2}\right)t\right]\cos\left[\left(\frac{k - k'}{2}\right)x - \left(\frac{w - w'}{2}\right)t\right].$$

If $k - k' = \delta k$, $w - w' = \delta w$ then $u \approx \left[a\cos\left(\frac{\delta k}{2}x - \frac{\delta w}{2}t\right)\right]\sin(kx - wt)$

$$V_g = \text{group velocity} = \delta w/\delta k$$

$$\therefore V_g = \frac{d(c/\lambda)}{d(1/\lambda)} = \frac{dc/d\lambda}{\lambda d(1/\lambda)} + c = -\lambda\frac{dc}{d\lambda} + c.$$

De Broglie postulated $E = h\nu = \hbar w$ and, by analogy with photon,

$p = E/c = \dfrac{h\nu}{c} = \dfrac{h}{\lambda} = \dfrac{\hbar 2\pi}{\lambda} = \hbar k$. With these conventions we have

$$V_g = \frac{d(c/\lambda)}{d(1/\lambda)} = \frac{d\nu}{d(1/\lambda)} = \frac{d(h\nu)}{d(h/\lambda)}$$
$$\therefore V_g = dE/dp$$

Compare this with the classical situation:

$$p = mv$$
$$E = \frac{1}{2}mv^2 + U \quad (U = \text{ potential})$$
$$\frac{dE}{dp} = \frac{1}{m}\frac{dE}{dv} = \frac{1}{m}(mv + 0) = v.$$

De Broglie's identification of energy $E = \hbar w$ and momentum $p = \hbar k$ creates a correspondence of classical velocity with the group velocity of wave packets.

Digression on Relativity.

In this digression I want to show how the pattern

$$\left\{ \begin{array}{l} \sigma_x^2 = \sigma_y^2 = \sigma_z^2 = 1 \\[4pt] \sigma_x\sigma_y = \sqrt{-1}\,\sigma_z \\[4pt] \sigma_y\sigma_z = \sqrt{-1}\,\sigma_x \\[4pt] \sigma_z\sigma_x = \sqrt{-1}\,\sigma_y \end{array} \right\}$$

arises in special relativity.

In special relativity there are two basic postulates:

1. Physical laws take the same form in two reference frames that are moving at constant velocity with respect to one another.

2. The speed of light as observed by persons in two inertial frames (i.e. moving at constant velocity with respect to one another) is the same.

The first postulate of special relativity is almost a definition of physical law. The second postulate is far less obvious. It is justified by experiment (the Michelson-Morley experiment) and the desire to have Maxwell's equations for electromagnetism satisfy the first postulate. [In the Maxwell theory there is a traveling wave solution that has velocity corresponding to the velocity of light. This led to the identification of light/radio waves and the waves of this model. If the Maxwell equations indeed satisfy 1., then 2. follows.]

Now let's consider events in special relativity. For a given observer, an event may be specified by **two times**: $[t_2, t_1] = e$. The two times are

$$t_1 = \text{time signal sent}$$

$$t_2 = \text{time signal received.}$$

In this view, the observer sends out a light signal at time t_1. The signal reflects from the event (as in radar), and is received by the same observer (transmitted along his/her world-line) at time t_2.

I shall assume (by convention) that **the speed of light equals unity**. Then the world paths of light are indicated by $45°$ lines (to the observer's (vertical) time line t). Thus the arrow pathways in the diagram above indicate the emission of a signal at time t_1, its reception (reflection) at the event at time t, and its reception (after reflection) by the observer at time t_2. With $c = \text{light speed} = 1$, we have that

$$t_2 - t_1 = 2x$$

$$t_2 + t_1 = 2t$$

where x and t are the **space** and **time** coordinates of the event e. Thus $[t_2, t_1] = [t + x, t - x]$ gives the translation from light-cone (radar) coordinates to spacetime coordinates.

Regard the ordered pair $[a, b]$ additively so that: $[a, b] + [c, d] = [a + c, b + d]$ and $k[a, b] = [ka, kb]$ for a scalar k. Then

$$[t_2, t_1] = t[1, 1] + x[1, -1].$$

Let 1 abbreviate $[1, 1]$ and σ abbreviate $[1, -1]$. Then

$$e = [t_2, t_1] = t1 + x\sigma.$$

This choice and translation of coordinates will be useful to us shortly.

Now let's determine how the light-cone coordinates of an event are related for two reference systems that move at constant velocity with respect to one another. In our representation, the world-line for a second inertial observer appears as shown below.

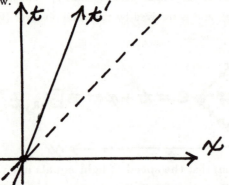

A coordinate (x, t) on the line labelled t' denotes the position and time as measured by $\mathcal{O}$ of a "stationary" observer $\mathcal{O}'$ in the second coordinate system. Note that the relative velocity of this observer is $v = x/t$. Since the velocity is constant, the world-line appears straight. (Note also that a plot of distance as measured by $\mathcal{O}'$ would not necessarily superimpose on the $\mathcal{O}'$ diagram at 90° to the line t'.)

Thus we can draw the system $\mathcal{O}$ and $\mathcal{O}'$ in the same diagram, and consider the pictures of signals sent from $\mathcal{O}$ to $\mathcal{O}'$:

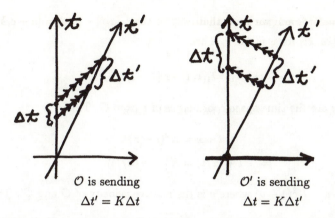

$\mathcal{O}$ is sending $\mathcal{O}'$ is sending

$\Delta t' = K\Delta t$ $\Delta t = K\Delta t'$

The diagrams above tell the story. If $\mathcal{O}$ sends two signals to $\mathcal{O}'$ at a time interval Δt, then $\mathcal{O}'$ will receive them at a time interval $K\Delta t$. The constant K is defined by

$$\frac{\Delta \text{ (time received)}}{\Delta \text{ (time sent)}} = K \, .$$

Here we use the first principle of relativity to deduce that $\mathcal{O} \to \mathcal{O}'$ ($\mathcal{O}$ sending to $\mathcal{O}'$) and $\mathcal{O}' \to \mathcal{O}$ have the **same constant** K. In fact, K is a function of the relative velocity of $\mathcal{O}$ and $\mathcal{O}'$. (See [BO]. This is Bondi's K-**calculus**.)

Let's find the value of K.

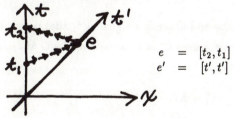

$$e = [t_2, t_1]$$
$$e' = [t', t']$$

Consider a point e on $\mathcal{O}'$'s time-line. Then $e = [t_2, t_1]$ from $\mathcal{O}$'s viewpoint and $e' = [t', t']$ represents the same event from the point of view of $\mathcal{O}'$ (just a time, no distance).

Then we have

$$t' = Kt_1$$

$$t_2 = Kt'$$

(by relativity and the assumption that $\mathcal{O}$ and $\mathcal{O}'$ have synchronized clocks at the origin).

Thus $t_2 = K^2 t_1$ and we know that

$$t_2 = t + x$$
$$t_1 = t - x$$

where t and x are the time-space coordinates of e from $\mathcal{O}$. Thus

$$t + x = K^2(t - x)$$
$$\Rightarrow \quad 1 + (x/t) = K^2(1 - (x/t))$$

Here $1 + v = K^2(1 - v)$ where v is the relative velocity of $\mathcal{O}$ and $\mathcal{O}'$. Thus

$$K = \sqrt{\frac{1 + v}{1 - v}}.$$

We now use K to determine the general transformation of coordinates:

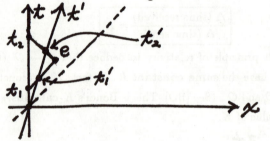

The event e is now off $\mathcal{O}''$s time-line and $\mathcal{O}$ sends a signal at t_1 that passes $\mathcal{O}'$ at t_1' ($\mathcal{O}'$ measurement), reflects from e, passes $\mathcal{O}'$ at t_2', is received by $\mathcal{O}$ at t_1. Then we have (using the same assumptions)

$$t_1' = K t_1$$
$$t_2 = K t_2'$$

Hence

$$[t_2', t_1'] = [K^{-1} t_2, K t_1].$$

This is the Lorentz transformation in light-cone coordinates.

With the help of the velocity formula for K and the identifications

$$
\begin{aligned}
t_2 &= t + x & \quad t_2' &= t' + x' \\
t_1 &= t - x & \quad t_1' &= t' - x'
\end{aligned}
$$

it is an easy matter to transform this version of the Lorentz transformation into the usual coordinates. I prefer the following slightly more algebraic approach [LK17], [LK30].

1: Regard $[K^{-1}t_2, Kt_1]$ as a special case of a product structure given by the formula

$$[A, B] * [C, D] = [AC, BD].$$

Thus $[K^{-1}, K]*[t_2, t_1] = [K^{-1}t_2, Kt_1] = [t'_2, t'_1]$ so that the Lorentz transformation is expressed through this product. I call this the **iterant algebra** [LK17].

2: Rewrite

$$[K^{-1}, K] = [X + T, -X + T]$$
$$= T1 + X\sigma.$$

Then

$$T + X\sigma = \frac{1 + (X/T)\sigma}{\sqrt{1 - (X/T)^2}}$$
$$(T^2 - X^2 = K^{-1}K = 1 \Rightarrow \sqrt{T^2 - X^2} = 1)$$
$$\Rightarrow T\sqrt{1 - (X/T)^2} = 1$$

and $v = -(X/T)$ because

$$K = \sqrt{\frac{1 + v}{1 - v}}$$

$$\Rightarrow T + X = \sqrt{\frac{1 - v}{1 + v}}$$

$$T - X = \sqrt{\frac{1 + v}{1 - v}}$$

$$\Rightarrow \frac{1 + (X/T)}{1 - (X/T)} = \frac{\sqrt{(1 - v)/(1 + v)}}{\sqrt{(1 + v)/(1 - v)}} = \frac{1 - v}{1 + v}$$

$$\therefore \boxed{[K^{-1}, K] = \frac{1 - v\sigma}{\sqrt{1 - v^2}}}$$

Using this iterant algebra, we have

$$1^2 = [1, 1] * [1, 1] = [1, 1] = 1$$
$$\sigma^2 = \sigma * \sigma = [1, -1] * [1, -1] = [1, 1] = 1$$
$$1 * \sigma = \sigma * 1 = \sigma.$$

Thus

$$[t_2', t_1'] = [K^{-1}, K] * [t_2, t_1]$$

$$\Updownarrow$$

$$t' + x'\sigma = \left(\frac{1 - v\sigma}{\sqrt{1 - v^2}}\right) + \left(\frac{x - vt}{\sqrt{1 - v^2}}\right)\sigma.$$

Thus

$$\boxed{\begin{array}{rcl} t' & = & (t - xv)/\sqrt{1 - v^2} \\ x' & = & (x - vt)/\sqrt{1 - v^2} \end{array}}$$

This is the classical form of the Lorentz transformation with light speed equal to unity.

Remark. We are accustomed to deriving the Lorentz transformation and then noting that $t'^2 - x'^2 = t^2 - x^2$ and that the group of Lorentz transformations is the group of linear transformations leaving the form $t^2 - x^2$ invariant. Here we knew that $t_2' t_2' = K^{-1} t_2 K t_1 = t_2 t_1$ early on, and hence that

$$(t' + x')(t' - x') = (t + x)(t - x)$$

$$\therefore \ t'^2 - x'^2 = t^2 - x^2.$$

This digression on relativity connects with our previous discussion of spin and angular momentum as follows: **The algebra of the Pauli matrices arises naturally in relation to the Lorentz group.** To see this we must consider what we have done so far in the context of a Euclidean space of spatial directions and displacements. So let us suppose that spacetime has dimension $n + 1$ with n dimensions of space. Let σ_0 denote the temporal direction and let $\sigma_1, \sigma_2, \ldots, \sigma_n$ be a basis of unit spatial directions. Thus we identify

$$\sigma_1 = (1, 0, 0, \ldots, 0)$$

$$\sigma_2 = (0, 1, 0, \ldots, 0)$$

$$\vdots$$

$$\sigma_n = (0, 0, 0, \ldots, 1).$$

Let $\sigma = a_1 \sigma_1 + a_2 \sigma_2 + \ldots + a_n \sigma_n$ denote a Euclidean unit spatial direction so that $a_1^2 + a_2^2 + \ldots + a_n^2 = 1$.

We are **given** that

1) $\sigma_i^2 = 1 \qquad i = 1, \ldots, n$

2) $\sigma^2 = 1$

by our assumption that $t\sigma_0 + x\sigma_i$ $(i = 1, \ldots, n)$ and $t\sigma_0 + x\sigma$ represent points in spacetime with these given directions. (Any given, spatial direction produces a two-dimensional spacetime that must obey the laws we have set out for such structures.)

Now multiply out the formula for σ^2:

$$\sigma^2 = (a_1\sigma_1 + \ldots + a_n\sigma_n)(a_1\sigma_1 + \ldots + a_n\sigma_n)$$

$$\sigma^2 = \sum_{i=1}^{n} a_i^2\sigma_i^2 + \sum_{i \neq j} a_i a_j \sigma_i \sigma_j$$

$$= \sum_{i=1}^{n} a_i^2 + \sum_{i<j} a_i a_j (\sigma_i\sigma_j + \sigma_j\sigma_i)$$

$$\therefore \sigma^2 = 1 + \sum_{i<j} a_i a_j (\sigma_i\sigma_j + \sigma_j\sigma_i).$$

Since $\sigma^2 = 1$ for all choices of $(a_1, a_2, \ldots, a_n)$ with $a_1^2 + \ldots + a_n^2 = 1$, this calculation implies that

$$\sigma_i\sigma_j = -\sigma_j\sigma_i \text{ for } i \neq j.$$

Let's look at the case of 4-dimensional spacetime, and then we have

$$e = t\sigma_0 + x\sigma_1 + y\sigma_2 + z\sigma_3.$$

If we take $\sigma_1, \sigma_2, \sigma_3$ as Pauli matrices then,

$$\sigma_1 = \begin{pmatrix} 1 & 0 \\ 0 & -1 \end{pmatrix}, \quad \sigma_2 = \begin{pmatrix} 0 & i \\ -i & 0 \end{pmatrix}, \quad \sigma_3 = \begin{pmatrix} 0 & 1 \\ 1 & 0 \end{pmatrix}$$

The spacetime point $e = t\sigma_0 + x\sigma_1 + y\sigma_2 + z\sigma_3$ is then represented by a 2×2 Hermitian matrix as:

$$e = t\begin{pmatrix} 1 & 0 \\ 0 & 1 \end{pmatrix} + x\begin{pmatrix} 1 & 0 \\ 0 & -1 \end{pmatrix} + y\begin{pmatrix} 0 & i \\ -i & 0 \end{pmatrix} + z\begin{pmatrix} 0 & 1 \\ 1 & 0 \end{pmatrix}$$

$$e = \begin{pmatrix} t+x & iy+z \\ -iy+z & t-x \end{pmatrix}.$$

Note that the determinant of e is the spacetime metric:

$$\text{Det}(e) = \begin{vmatrix} t+x & iy+z \\ -iy+z & t-x \end{vmatrix}$$

$$\text{Det}(e) = t^2 - x^2 - y^2 - z^2.$$

This is a (well-known) that leads to a very perspicuous description of the full group of Lorentz transformations.

That is, we wish to consider the construction of linear transformations of $\mathbf{R}^4 = \{(t, x, y, z)\}$ that leave the quadratic form $t^2 - x^2 - y^2 - z^2$ invariant. Let's use the Hermitian representation,

$$e = \begin{pmatrix} A & W \\ \overline{W} & B \end{pmatrix}$$

where A and B are real numbers, and W is complex, with conjugate $\overline{W}$. Let P be any complex 2×2 matrix with $\text{Det}(P) = 1$. The group of such matrices is called $SL(2, \mathbf{C})$. Then $(PeP^*)^* = Pe^*P^* = PeP^*$. Hence PeP^* is Hermitian. Furthermore,

$$\text{Det}(PeP^*) = \text{Det}(P)\text{Det}(e)\text{Det}(P^*)$$

$$= \text{Det}(P)\text{Det}(e)\text{Det}(P)$$

$$= \text{Det}(e).$$

Thus, if $\mathcal{L}$ denotes the group of Lorentz transformations, then we have exhibited a $2-1$ map $\pi : SL(2, \mathbf{C}) \to \mathcal{L}$.

This larger group of Lorentz transformations includes our directional transforms

$$[A, B] \mapsto [K^{-1}A, KB].$$

Let's see how:

$$[A, B] = [T + X, T - X] \leftrightarrow \begin{bmatrix} A & 0 \\ 0 & B \end{bmatrix}$$

$$\therefore [K^{-1}A, KB] \leftrightarrow \begin{bmatrix} K^{-1}A & 0 \\ 0 & KB \end{bmatrix} = \begin{bmatrix} 1/\sqrt{K} & 0 \\ 0 & \sqrt{K} \end{bmatrix} \begin{bmatrix} A & 0 \\ 0 & B \end{bmatrix} \begin{bmatrix} 1/\sqrt{K} & 0 \\ 0 & \sqrt{K} \end{bmatrix}.$$

Quaternionic Representation.

The Pauli matrices and the quaternions are intimately related. In the quaternions we have three independent orthogonal unit vectors i, j, h with $i^2 = j^2 = h^2 = ijk = -1$ where the vectors $1, i, j, k$ form a vector space basis for 4-dimensional real space $\mathbf{R}^4$. This gives $\mathbf{R}^4$ the structure of an associative division algebra. The translation between Pauli matrices and quaternions arises via:

$$i \leftrightarrow -\sqrt{-1}\sigma_1 = -\sqrt{-1}\begin{pmatrix} 1 & 0 \\ 0 & -1 \end{pmatrix} = \begin{pmatrix} -\sqrt{-1} & 0 \\ 0 & \sqrt{-1} \end{pmatrix}$$

$$j \leftrightarrow -\sqrt{-1}\sigma_2 = -\sqrt{-1}\begin{pmatrix} 0 & \sqrt{-1} \\ -\sqrt{-1} & 0 \end{pmatrix} = \begin{pmatrix} 0 & 1 \\ -1 & 0 \end{pmatrix}$$

$$k \leftrightarrow -\sqrt{-1}\sigma_3 = -\sqrt{-1}\begin{pmatrix} 0 & 1 \\ 1 & 0 \end{pmatrix} = \begin{pmatrix} 0 & -\sqrt{-1} \\ -\sqrt{-1} & 0 \end{pmatrix}$$

$$1 \leftrightarrow 1 = \begin{pmatrix} 1 & 0 \\ 0 & 1 \end{pmatrix}$$

$$\therefore \ a + bi + cj + dk \leftrightarrow \begin{pmatrix} a - b\sqrt{-1} & c - d\sqrt{-1} \\ -c - d\sqrt{-1} & a + b\sqrt{-1} \end{pmatrix} = \begin{pmatrix} Z & W \\ -\overline{W} & \overline{Z} \end{pmatrix}$$

With this identification, we see that the **unit quaternions** are isomorphic to

$$SU(2) = \left\{ \begin{pmatrix} Z & W \\ -\overline{W} & \overline{Z} \end{pmatrix} \Big| \ \begin{array}{c} Z\overline{Z} + W\overline{W} = 1 \\ Z, W \text{ complex numbers} \end{array} \right\}.$$

The upshot of this approach is that we can represent spacetime points quaternionically. It is convenient, for this, to change conventions and use

$$\sqrt{-1}T + iX + jY + kZ = p.$$

Then

$$p\overline{p} = (\sqrt{-1}T + iX + jY + kZ)(\sqrt{-1}T - iX - jY - kZ)$$
$$= -T^2 + X^2 + Y^2 + Z^2.$$

Note that we are here working in complexified quaternions with $\sqrt{-1}$ commuting with i, j and k.

Since $p\overline{p}$ represents the Lorentz metric, we can represent Lorentz transformations quaternionically via

$$\boxed{T(p) = gpg \text{ where } g = t + \sqrt{-1}v, \ \|v\|^2 - t^2 = 1}$$

$$T(p)\overline{T(p)} = qpg\overline{(gpg)} \qquad \boxed{g\overline{g} = 1}$$
$$= gpg\,\overline{g}\,\overline{p}\,\overline{g}$$
$$= gp(\sqrt{-1}t + v)(\sqrt{-1}t - v)\overline{p}\,\overline{g}$$
$$= gp(-t^2 + \|v\|^2)\overline{p}\,\overline{g}$$
$$= gp\,\overline{p}\,\overline{g}$$
$$= (p\overline{p})g\overline{g}$$
$$\therefore\; T(p)\overline{T(p)} = p\overline{p}.$$

We must verify here under what circumstances $T(p)$ is a point in spacetime. Note that

$$g = t + \sqrt{-1}v = t + v_1\sigma_1 + v_2\sigma_2 + v_3\sigma_3$$

and if $g^2 = (t^2 + v \cdot v) + \sqrt{-1}(2tv)$ actually represents our Lorentz boost, then gpg will perform the boost for $p = \sqrt{-1}\,t + q$, q in the direction v. If $q \perp v$ then $gqg = q$. Compare this discussion with the treatment of reflections in the next section.

10^0. Quaternions, Cayley Numbers and the Belt Trick.

First we will develop the quaternions by reflecting on reflections. Then come string tricks, and interlacements among the topological properties of braids and strings, and algebras.

So we begin with **the algebra of reflections in $\mathbf{R}^3$**. Let $\vec{u} \in \mathbf{R}^3$ be a unit vector. By $\mathbf{R}^3$ we mean Euclidean three-dimensional space. Thus

$$\mathbf{R}^3 = \{(x, y, z) | x, y, z \text{ real numbers}\}$$
$$\vec{u} = (u_1, u_2, u_3)$$
$$\vec{u} \text{ a unit vector means } \|\vec{u}\| = \sqrt{u_1^2 + u_2^2 + u_3^2} = 1 .$$

We have the inner product $\vec{a} \cdot \vec{b} = a_1 b_1 + a_2 b_2 + a_3 b_3$. Each unit vector $\vec{u}$ determines a plane $M\vec{u}$ that is perpendicular to it:

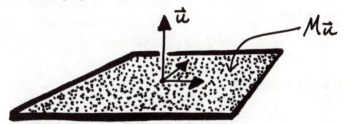

In order to use $M\vec{u}$ as a mirror, that is in order to reflect a vector $\vec{a}$ with respect to it, we first write $\vec{a} = \vec{a}_\perp + \vec{a}_\|$ where $\vec{a}_\perp$ is perpendicular to $M\vec{u}$ and $\vec{a}_\|$ is parallel to $M\vec{u}$. Then

$$R(\vec{a}) = -\vec{a}_\perp + \vec{a}_\|$$

accomplishes the reflection.

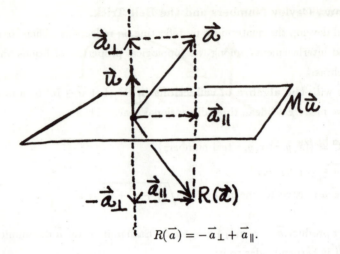

$$R(\vec{a}) = -\vec{a}_\perp + \vec{a}_\parallel.$$

And we also see that

$$\vec{a}_\perp = (\vec{a} \cdot \vec{u})\vec{u}$$

(since $\vec{u}$ is a unit vector). Thus

$$\vec{a}_\parallel = \vec{a} - (\vec{a} \cdot \vec{u})\vec{u}$$

and $\boxed{R(\vec{a}) = \vec{a} - 2(\vec{a} \cdot \vec{u})\vec{u}}$. This gives a specific formula for reflection in the plane perpendicular to the unit vector $\vec{u}$.

We wish to represent reflections by an algebra structure on the points of three-space.

As geometry is translated into algebra the geometric roles (a little geometry in the algebra) **become inverted. That which is mirrored in geometry becomes a (notational) mirror in algebra.**

In symbols:

$$R(X) = uXu.$$

Here X (which is **acted on by the mirror** in geometry) becomes a **notational**

mirror, reflecting u on either side.

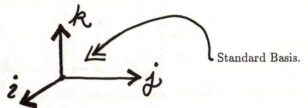

Standard Basis.

We want uXu to be the reflection of X in the plane perpendicular to u.

Therefore:

$$\left\{ \begin{array}{rcl} kkk & = & -k \\ kjk & = & j \\ kik & = & i \end{array} \right\}$$

Let's assume that this algebra is **associative** (i.e. $(ab)c = a(bc)$ for all a, b, c) and that **non-zero vectors have inverses.** Then multiplying $kkk = -k$ by k^{-1} we have

$$kk = -1$$
$$kjk = j$$
$$kik = i.$$

Therefore, multiplying the second two equations by k, we have

$$jk = -kj$$
$$ik = -ki.$$

The same argument can be applied to reflections in the planes perpendicular to i and j. Hence $i^2 = j^2 = k^2 = -1$, and, in fact, by the same argument $u^2 = -1$ for any unit vector u.

This may seem a bit surprising, since $-1 \notin \mathbf{R}^3$! In fact, we see that for this scheme to work, it is necessary to use $\mathbf{R}^4 = \{(x, y, z, t)\}$ with $(0, 0, 0, -1)$ representing the "scalar" -1.

What about ij? Is this forced by our scheme? With the given assumptions it is not, however one way to fulfill the conditions is to use:

$$ij = k$$
$$jk = i$$
$$ki = j$$

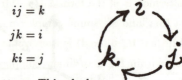

the familiar vector-cross-product pattern. This algebra

$$i^2 = j^2 = k^2 = ijk = -1$$

is **Hamilton's Quaternions.** It produces an algebra structure on $\mathbf{H} = \mathbf{R}^4 = \{xi + yj + zk + t\}$ that is associative and so that non-zero elements have inverses. In fact, writing

$$q = t + ix + jy + kz$$
$$\bar{q} = t - ix - jy - kz$$

then $q\bar{q} = \bar{q}q = t^2 + x^2 + y^2 + z^2$.

An element of the form $ia + jb + kc \in \mathbf{R}^3$ is a **pure quaternion** (analogous to a pure imaginary complex number). Let

$$X = iX_1 + jX_2 + kX_3$$
$$Y = iY_1 + jY_2 + kY_3.$$

Then the quaternionic product is given by the formula

$$XY = -X_1Y_1 - X_2Y_2 - X_3Y_3$$
$$+ \begin{vmatrix} Y_1 & X_2 \\ Y_1 & Y_2 \end{vmatrix} k - \begin{vmatrix} X_1 & X_3 \\ Y_1 & Y_3 \end{vmatrix} j + \begin{vmatrix} X_2 & X_3 \\ Y_2 & Y_3 \end{vmatrix} i$$
$$\therefore XY = -(X \cdot Y) + (X \times Y)$$

where $X \times Y$ is the vector cross product in $\mathbf{R}^3$.

Note that, as we have manufactured them, there is a 2-dimensional sphere's worth of pure quaternionic square roots of -1. That is, if $U \in \mathbf{R}^3$ then

$$UU = -U \cdot U + U \times U$$
$$\therefore U^2 = -U \cdot U$$
$$\therefore U^2 = -1 \leftrightarrow U \cdot U = 1$$
$$\therefore U^2 = -1 \leftrightarrow U \in S^2.$$

Here S^2 denotes the set of points in $\mathbf{R}^3$ at unit distance from the origin.

Any quaternion is of the form $a + bu$ where $a, b \in \mathbf{R}$ and $u \in S^2$. This gives $\mathbf{H}$ the aspect to a kind of "spun" complex numbers since $\mathbf{H}_u = \{a + bu | a, b \in \mathbf{R}, u \text{ fixed}\}$ is isomorphic to $\mathbf{C} = \{a + bi | a, b \in \mathbf{R}, i = \sqrt{-1}\}$. Multiplication within $\mathbf{H}_u$ is ordinary complex multiplication, but multiplication between different $\mathbf{H}_u$ depends upon the quaternionic multiplication formulas.

Another important property of quaternionic multiplication is that it is **norm-preserving**. That is, $\|qq'\| = \|q\| \, \|q'\|$ where $\|q\| = q\bar{q}$. This is, of course easy to see since

$$\|qq'\| = (qq')\overline{(qq')}$$
$$= (qq')(\overline{q'} \, \bar{q})$$
$$= q(q'\overline{q'})\bar{q}$$
$$= (q\bar{q})(q'\overline{q'})$$
$$\therefore \|qq'\| = \|q\| \, \|q'\|.$$

Here we have used an easily verified fact: $\overline{qq'} = \overline{q'} \, \bar{q}$ **and** the associativity of the multiplication.

In fact one way of thinking about how the quaternions were produced is that we started with the complex numbers $\mathbf{C} = \{a + bi\}$ and we tried to **add a new imaginary** j $(j^2 = -1)$ **so that** $jZ = \bar{Z}j$, $(Z = a+ib, \bar{Z} = a-ib)$ and so that the multiplication was associative. (Then $k = ij$ and $k^2 = (ij)(ij) = i(ji)j = i(\bar{i}j)j = i(-ij)j = (i(-i))(jj) = (+1)(-1) \therefore k^2 = -1$, recovering the quaternions.)

That this procedure treads on dangerous ground is illustrated when we try to repeat it! **Try to add an element** J **to H so that the new algebra is associative,** $JA = \bar{A}J \, \forall A \in \mathbf{H}$, **and** J **is invertible.** Then

$$J(AB) = \overline{(AB)}J$$
$$= (\overline{B}\,\overline{A})J$$
$$= \overline{B}(\overline{A}J)$$
$$= \overline{B}(J\overline{\overline{A}})$$
$$= \overline{B}(JA)$$
$$= (\overline{B}J)A$$
$$= (J\overline{\overline{B}})A$$
$$= (JB)A$$
$$J(AB) = J(BA)$$
$$J^{-1}(J(AB)) = J^{-1}(J(BA))$$
$$(J^{-1}J)(AB) = (J^{-1}J)(BA)$$
$$\therefore AB = BA.$$

Since this is to hold for any quaternions A and B and since $AB \neq BA$ in general in **H** we see that **associativity is too much to demand of this extension of the quaternions.**

There does exist an algebra structure on eight-dimensional space $\mathbf{H} + J\mathbf{H}$ such that $J^2 = -1$, $JZ = \overline{Z}J$, called the **Cayley numbers.** It is necessarily non-associative. How can one determine the Cayley multiplication?? We now initiate a Cayley digression.

Cayley Digression.

First a little abstract work. Let's assume that we are working with a normed algebra $\mathcal{A}$. We are given Euclidean space $\mathbf{R}^n$ having a multiplicative structure $\mathbf{R}^n \times \mathbf{R}^n \to \mathbf{R}^n$ and a notion of conjugation $Z \mapsto \overline{Z}$ so that $\overline{\overline{Z}} = Z$. (For **R** itself, $\overline{r} = r$.) Assume that the Euclidean norm is given by $\|Z\| = Z\overline{Z}$ and that the **multiplication is norm-preserving**

$$\|ZW\| = \|Z\|\,\|W\|.$$

(This no longer follows from $\|Z\| = Z\overline{Z}$ since we can't assume associativity.)

Define a **bilinear form**

$$[Z, W] = \frac{1}{2}(Z\overline{W} + \overline{Z}W).$$

(We assume that $\overline{Z + W} = \overline{Z} + \overline{W}$ also.) And finally assume that **the norm on the algebra is non-degenerate** in the sense that

$$\begin{aligned} [Z, X] &= 0 \quad \forall X \\ \Rightarrow Z &= 0 \end{aligned}$$

Lemma 10.1. $2[Z, W] = \|Z + W\| - \|Z\| - \|W\|$.

Proof. Easy.

Proposition 10.2. (J. Conway [CON1]).

(1) $[ac, bc] = [a, b][c]$

(2) $[ac, bd] + [ad, bc] = 2[a, b][c, d]$

(3) $[ac, b] = [a, b\bar{c}]$.

Here a, b, c, d are any elements of the algebra $\mathcal{A}$ and $[a] = \|a\|$.

Proof.

$$[(a + b)c] = [ac + bc] = [ac] + [bc] + 2[ac, bc]$$

$$\|$$

$$[a + b][c] = ([a] + [b] + 2[a, b])[c]$$

$$\Rightarrow [ac, bc] = [a, b][c], \text{ proving (1)}.$$

To prove (2) replace c by $(c + d)$ in (1):

$$[a, b][c + d] = \qquad [a(c + d), b(c + d)] = [ac + ad, bc + bd]$$

$$\|$$

$$[a, b]([c] + [d] + 2[c, d]).$$

Continue expanding and collect terms. For (3) note that $\bar{c} = 2[c, 1] - c$,

$(2[c, 1] - c = (c \cdot \overline{1} + \bar{c} \cdot 1) - c = \bar{c})$ and let $d = 1$ in (2). Then

$$[ac, b] = 2[a, b][c, 1] - [a, bc]$$

$$= [a, b(2[c, 1] - c)]$$

$$= [a, b\bar{c}].$$

This completes the proof of the proposition. $\qquad //$

Conway's proposition shows that we can use the properties

$$[ab] = [a][b]$$
$$[a, a] = [a]$$
$$[a, b][c] = [ac, bc]$$
$$2[a, b][c, d] = [ac, bd] + [ad, bc]$$
$$[ac, b] = [a, b\overline{c}]$$
$$([a, bc] = [a\overline{c}, b],\ [ac, b] = [c, \overline{a}b], \text{etc.})$$

plus the non-degeneracy

$$[Z, t] = [W, t] \quad \forall t \Rightarrow Z = W$$

to ferret out the particular properties of the multiplication. Just watch!

(1) $$\overline{\overline{x}} = x$$

Proof.

$$[x, t] = [1, t\overline{x}]$$
$$= [1 \cdot \overline{\overline{x}}, t]$$
$$\therefore [x, t] = [\overline{\overline{x}}, t]$$
$$\therefore x = \overline{\overline{x}}.$$

//

(2) $$\overline{ab} = \overline{b}\,\overline{a}$$

Proof.

$$[ab, t] = [a, t\overline{b}]$$
$$= [\overline{t}a, \overline{b}]$$
$$\therefore [ab, t] = [\overline{t}, \overline{b}\,\overline{a}]$$
$$[ab, t] = [1(ab), t]$$
$$= [1, t\overline{(ab)}] = [\overline{t}, \overline{ab}]$$
$$\therefore [\overline{t}, \overline{b}\,\overline{a}] = [\overline{t}, \overline{ab}]$$
$$\therefore \overline{b}\,\overline{a} = \overline{ab}$$

//

Now suppose that H is a subalgebra on which $[\,,\,]$ is non-degenerate, and that there exists $J \perp H$ ($\perp$ means $[h, J] = 0 \quad \forall h \in H$ whence $h\overline{J} + J\overline{h} = 0$ whence (assuming for a moment $\overline{J} = -J$) $J\overline{h} = hJ$. This was our previous extension idea. Here it is embodied geometrically as perpendicularity.). We assume $[J] = 1$.

Theorem 10.3. Under the above conditions $H + JH$ is a subalgebra of the given algebra $\mathcal{A}$, and $H \perp JH$. Furthermore, if $a, b, c, d \in H$ then

$$(a + Jb)(c + Jd) = (ac - d\overline{b}) + J(cb + \overline{a}d).$$

Remark. If H is $\mathbf{H}$, the quaternions, then this specifies the Cayley multiplication.

Proof. Note $\overline{J} = 2[J, 1] - J = -J$ since $J \perp H \quad (1 \in H)$.

$$[a, Jb] = [a\overline{b}, J] = 0 \qquad \forall a, b \in H$$
$$\therefore H \perp JH.$$

Also

$$[Ja, Jb] = [J][a, b] = [a, b]$$
$$\therefore JH \text{ is isometric to } H.$$

Claim 1. $aJ = J\overline{a} \quad \forall a \in H$

Proof.

$$[aJ, t] = [J, \overline{a}t]$$
$$= 2[J, \overline{a}][1, t] - [Jt, \overline{a}]$$
$$([ac, bd] = 2[a, b][c, d] - [ad, bc])$$
$$\therefore [aJ, t] = -[Jt, \overline{a}]$$
$$= -[t, \overline{J}\ \overline{a}]$$
$$= [t, J\overline{a}]$$
$$= [J\overline{a}, t].$$

Claim 2. $(Jb)c = J(cb)$

412

Proof.

$$[(Jb)c, t] = [Jb, t\bar{c}]$$
$$= [\bar{b}J, t\bar{c}]$$
$$= -[\bar{b}\,\bar{c}, tJ] \quad \text{(as above)}$$
$$= [(\bar{b}\,\bar{c})J, t]$$
$$\therefore (Jb)c = (\bar{b}\,\bar{c})J$$
$$= J(\overline{\bar{b}\,\bar{c}})$$
$$\therefore (Jb)c = J(cb).$$

//

Claim 3. $a(Jb) = J(\bar{a}b)$

Proof.

$$[a(Jb), t] = [Jb, \bar{a}t]$$
$$= -[Jt, \bar{a}b] \quad \text{(by perpendicularity)}$$
$$= [t, J(\bar{a}b)]$$
$$\therefore a(Jb) = J(\bar{a}b).$$

//

Claim 4. $(Ja)(Jb) = -b\bar{a}.$

Proof.

$$[(Ja)(Jb), t] = [Jb, (\overline{Ja})t]$$
$$= [\bar{b}J, (\overline{Ja})t]$$
$$= -[\bar{b}t, (\overline{Ja})J]$$
$$= -[(\bar{b}t)\bar{J}, \overline{Ja}]$$
$$= -[(\bar{b}t)\bar{J}, \bar{a}\,\bar{J}]$$
$$= -[\bar{b}t, \bar{a}][\bar{J}]$$
$$= -[t, b\bar{a}]$$
$$\therefore (Ja)(Jb) = -b\bar{a}.$$

This completes the proof. //

Applying this construction iteratively, beginning with the reals **R**, we find

$$\mathbf{R} \;\hookrightarrow\; \mathbf{C} \;\hookrightarrow\; \mathbf{H} \;\hookrightarrow\; \mathbf{K}$$

reals   complexes   quaternions   Cayley numbers

At each stage some element of abstract structure is lost. Order goes on the passage from **R** to **C**. Commutativity disappears on going from **C** to **H**. The Cayley numbers **K** lose associativity. Finally, the process of the theorem, when applied to **K** itself simply collapses. Thus $\mathbf{R} \subset \mathbf{C} \subset \mathbf{H} \subset \mathbf{K}$ are the only non-degenerate normed algebras. (This was originally proved in [HU]).

The situation is deeper than that. One can ask for division algebra (not necessarily normed) structures on Euclidean spaces, and again the only dimensions are 1, 2, 4 and 8. But now the result requires deep algebraic topology (see [A1] for an account).

Exercise.

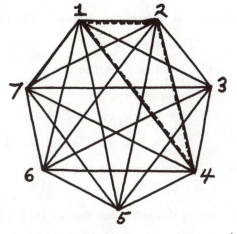

Show that there is a basis for **K**, $\{1, e_1, e_2, e_3, e_4, e_5, e_6, e_7\}$, with

$$e_i^2 = -1$$

$$e_i e_j = -e_j e_i \qquad i \neq j$$

and

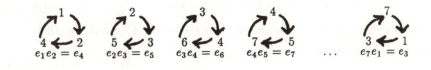

$$\boxed{e_i e_{i+1} = e_{i+3}}\ \text{(indices modulo 7)}$$

This ends our digression into the Cayley numbers.

Rotations and Reflections.

So far we have used the quaternions to represent reflections. If $R_u : \mathbf{R}^3 \to \mathbf{R}^3$ is the reflection in the plane perpendicular to the unit vector $u \in S^2$, then

$$R(p) = R_u(p) = upu$$

is the quaternion formula for the reflection.

We have verified this for the frame $\{i, j, k\}$. To see it in general, let $\{\vec{u}_1, \vec{u}_2, \vec{u}_3\}$ be a right-handed frame of orthogonal unit vectors in $\mathbf{R}^3$. Then we know (using $\vec{u}_i \cdot \vec{u}_j = 0\ \ i \neq j$, $\vec{u}_i \cdot \vec{u}_i = 1$, $\vec{u}_1 \times \vec{u}_2 = \vec{u}_3$ etc...) that $\vec{u}_1^2 = \vec{u}_2^2 = \vec{u}_3^2 = \vec{u}_1 \vec{u}_2 \vec{u}_3 = -1$ as quaternions. Hence the formula will hold in any frame, hence for any unit vector u.

Recall that we also know that the reflection is given by the direct formula

$$R(p) = p - 2(u \cdot p)u,$$

as shown at the beginning of this section. Therefore

$$p - 2(u \cdot p)u = upu$$
$$= (-u \cdot p + u \times p)u$$
$$= -(u \cdot p)u - (u \times p) \cdot u + (u \times p) \times u$$
$$\therefore p - 2(u \cdot p)u = -(u \cdot p)u + (u \times p) \times u$$
$$\therefore \boxed{p - (u \cdot p)u = (u \times p) \times u}.$$

This is a "proof by reflection" of this well-known formula about the triple vector product.

Remark. It is important to note that the vector cross product is a **non-associative** multiplication. Thus

$$i \times (i \times j) = i \times k = -j$$
$$(i \times i) \times j = 0 \times j = 0.$$

The associativity of the quaternion algebra gives rise to specific relations about vector cross products.

We now turn from reflections to rotations. **Every rotation of three dimensional space is the resultant of two reflections.** If the planes M_1, M_2 make an angle θ with each other, then the rotation resulting from the composition of reflecting in M_1 and then in M_2 is a rotation of angle 2θ about the axis of intersection of the two planes. View Figure 10.1.

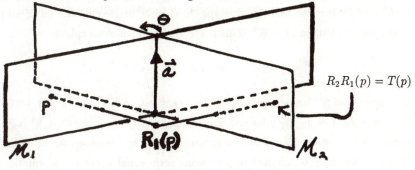

$T = R_2 R_1$ is rotation about $\vec{a}$ by 2θ

Figure 10.1

Now let's examine the algebra of the product of two reflections. Let $u, v \in S^2$. Then the vector $u \times v$ is the axis of intersection of the planes M_u, M_v of vectors perpendicular to u and to v respectively. Thus we expect $Rv \circ Ru = T$ to be a rotation about $u \times v$ with angle 2θ for $\theta = $ (angle between u and v) $= \arccos(u \cdot v)$. And we have

$$T(p) = Ru(Rv(p))$$
$$= u(vpv)u$$
$$T(p) = (uv)p(vu).$$

Let $g = uv = -(u \cdot v) + u \times v$ and note that $vu = -(v \cdot u) + v \times u = +\overline{uv} = \overline{g} = g^{-1}$. Thus $T(p) = gp\overline{g} = gpg^{-1}$ where

$$g = -\cos(\theta) - \sin(\theta)w,$$

$w = (u \times v)/\|u \times v\|$. The global minus sign is extraneous since

$$gpg^{-1} = (-g)p(-g)^{-1}.$$

Hence

416

Proposition 10.4. Let $g = (\cos\theta) + (\sin\theta)u$ where $u \in S^2$ is a unit quaternion. Then $T(p) = gpg^{-1} = gp\bar{g}$ defines a mapping $T : \mathbf{R}^3 \to \mathbf{R}^3$ ($\mathbf{R}^3$ is the set of pure quaternions), and **T is a rotation with axis u and angle of rotation about u equal to 2θ.**

Put a bit more abstractly, we have shown that there is a **2 to 1 mapping** $\pi : S^3 \to SO(3)$ given by $\pi(g)(p) = gpg^{-1}$. Here S^3 represents the set of all unit length quaternions in $\mathbf{H} = \mathbf{R}^4$. This is the three-dimensional sphere

$$S^3 = \{ai + bj + ck + t | a^2 + b^2 + c^2 + t^2 = 1\}.$$

Any element of S^3 can be represented as $g = e^{u\theta} = (\cos\theta) + (\sin\theta)u$ with $u \in S^2 \subset \mathbf{R}^3$. Since $\|qq'\| = \|q\|\,\|q'\|$ for any quaternions $q, q' \in \mathbf{H}$, we see that S^3 is closed under multiplication. Hence S^3 is a group. This is the Lie group structure on S^3. $SO(3)$ is the group of orientation preserving orthogonal linear transformations of $\mathbf{R}^3$ – hence the group of rotations of $\mathbf{R}^3$.

The mapping $\pi : S^3 \to SO(3)$ is $2 \to 1$ since $\pi(g) = \pi(-g)$. We conclude that, **topologically, $SO(3)$ is the result of identifying antipodal points in** S^3. In topologists parlance this says that $SO(3)$ is homeomorphic to $\mathbf{RP}^3$ = the real projective three space = the space of lines through the origin in $\mathbf{R}^4 \cong$ the three dimensional sphere S^3 modulo antipodal identifications.

Remark. In general, $\mathbf{RP}^n$ is the space of lines in $\mathbf{R}^{n+1}$ (through the origin) and there is a double covering $\pi : S^n \to \mathbf{RP}^n$. Historically, the term projective space comes from projective geometry. Thus $\mathbf{RP}^2 = S^2/$ (antipodal identifications) is homeomorphic to **a disk D^2 whose boundary points are antipodally identified.**

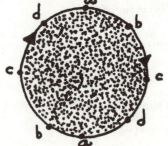

RP²

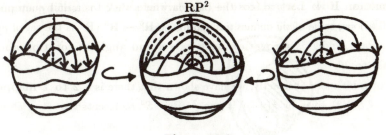

Figure 10.2

This is a strange space, but arises naturally in projective geometry as the **completion of a plane by adding a circle of points at infinity** so that travel outward along any ray through the origin will go through ∞ and return along the negative ray.

"∞"

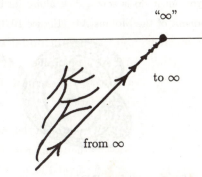

to ∞

from ∞

Exercise. Show directly that $SO(3) \cong D^2/\sim$ where D^3 represents a three-dimensional ball, and $\sim$ denotes antipodal identification of boundary points. [HINT: Let D^3 have radius π. Associate to $\vec{v} \in D^3$ a rotation (right-handed) about $\vec{v}$ of $\|\vec{v}\|$ radians.]

Let's return for a moment to the projective plane **RP²**. This is a closed surface, like the surface of a sphere or of a torus, but (it can be proved) there is no way to embed this surface into three dimensional space without introducing singularities. If we start to sew the boundary of a disk to itself by antipodal identifications, some of the difficulties show themselves:

$\mathcal{M}$

Begin the process by identifying two intervals results in a Möbius strip $\mathcal{M}$ as above. In the projective plane, if you go through infinity and come back, you come back with a twist! In projective three-space, the same process will turn your handedness. This consideration is probably the origin of the Möbius band. The Möbius band is a creature captured from infinity, and brought to live in bounded space.

To complete $\mathcal{M}$ to $\mathbf{RP}^2$, all you must do is sew a disk along its boundary circle to the single boundary component of the Möbius $\mathcal{M}$. (Figure 10.3).

D^2  $\mathcal{M}$

Instructions for producing $\mathbf{RP}^3$
Figure 10.3

This description, $\mathbf{RP}^3 = \mathcal{M} \cup D^2$, gives us a very good picture of the topology of $\mathbf{RP}^3$. In particular it lets us spot a loop α on $\mathbf{RP}^2$ that can't be contracted, but $\alpha^2 = $ "the result of going around on α twice" **is** contractible. The loop α is the core of the band $\mathcal{M}$. α^2 is the boundary of $\mathcal{M}$ (you go around $\mathcal{M}$ twice when traversing the boundary). The curve α^2 is contractible, because it is identified with the boundary of the sewing disk D^2. (α is a generator of $\pi_1(\mathbf{RP}^2)$ and $\pi_1(\mathbf{RP}^2) \cong \mathbf{Z}_2$.)

The same effect happens in **RP**^{3}, and hence in $SO(3)$. **There is a loop** α **on** $SO(3)$ **that is not contractible, but** α^2 **is contractible.** This loop can be taken to be $\alpha(t) = r(2\pi t, \overrightarrow{n})$ where $r(\theta, \overrightarrow{n})$ is a rotation by angle θ about the axis $\overrightarrow{n}$ (call it the north pole). Note that each $\alpha(t)$ is a rotation and that $\alpha(0) = \alpha(1) =$ identity.

Given this loop α of rotations we can examine its geometry and topology by creating an image of the action of $\alpha(t)$ as t ranges from 0 to 1. To accomplish this, let $\hat{\alpha} : S^2 \times [0,1] \to S^2 \times [0,1]$ be defined by

$$\hat{\alpha}(p,t) = (\alpha(t)p, t).$$

If you visualize $S^2 \times [0,1]$ as a 3-dimensional annulus with a large S^2 (2-dimensional sphere) on the outside, and a smaller S^2 on the inside (See Figure 10.4.) then I can illustrate

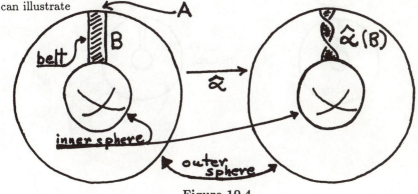

Figure 10.4

the action of $\hat{\alpha} : S^2 \times [0,1] \to S^2 \times [0,1]$ by considering the image of a band $B \subset S^2 \times [0,1]$ so that $B = A \times [0,1]$ where A is an arc on S^2 passing through the north pole. Since $\alpha(t)$ is a rotation about the north pole by $2\pi t$, this band receives a 2π twist under $\hat{\alpha}$. And the fact that $\hat{\alpha}^2$ can be deformed to the identity map implies that $\hat{\alpha}^2(B)(=$ band with 4π-twist) **can be deformed in** $S^2 \times I$, keeping the ends of the band fixed, **so that all twist is removed from the band.**

This is a "theoretical prediction" based on the topological structure of $SO(3)$. In fact our prediction can be "experimentally verified" as shown in Figure 10.5.

This is known to topologists as the **belt trick** and to physicists as the **Dirac string trick**.

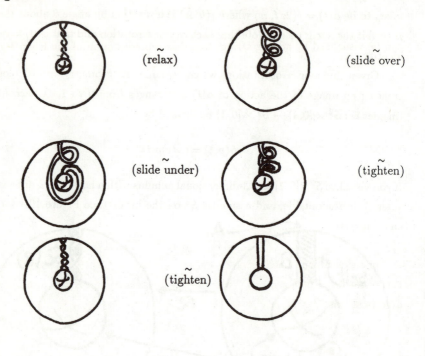

Figure 10.5

Another way of putting the string trick is this. Imagine a sphere suspended in the middle of a room attached by many strings to the floor, ceiling and walls. Turning the ball by 2π will entangle the strings, and turning it by 4π will apparently entangle them further. However with the 4π turn, all the strings can be disentangled without removing their moorings on the walls or the sphere, and without moving the sphere!

A particle with spin is something like this ball attached to its surroundings by string. It's amplitude changes under a 2π rotation, and is restored under the 4π rotation. Compare with Feynman [FE1].

It is interesting to note that the structure of the quaternion group

$$i^2 = j^2 = k^2 = ijk = -1$$

can be visualized by adding strings to a disk-representation of the Klein 4-group

$$I^2 = J^2 = K^2 = 1, \quad IJ = JI = K.$$

To see this let's take a disk

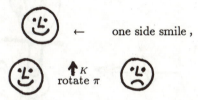

and one side frown. Let I denote a rotation of π about an axis perpendicular to this page. Let J be a π rotation about a horizontal axis and K a π rotation about a vertical axis.

Then, applying these operations to the disk we have

and it is easy to see that $I^2 = J^2 = K^2 = 1$, $IJ = JI = K$ in the sense that these equalities mean identical results for the disk.

If we now add strings to the disk, creating a puppet, then a 2π rotation leaves tangled strings. Let i, j, k denote the same spatial rotations as applied to our puppet.

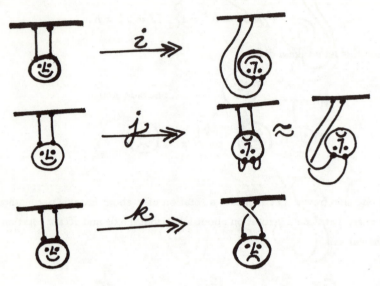

Figure 10.6

As we see, k^2 is no longer 1 since there is a 2π twist on the strings. The same goes for i^2, j^2. Thus we may identify a $(\pm 2\pi)$ twist on the strings as -1 and set $i^2 = j^2 = k^2 = -1$ since by the belt trick $(i^2)^2 = (j^2)^2 = (k^2)^2 = 1$.

If a $\pm 2\pi$ twist is -1, then we can also identify a $\pm \pi$ twist with $\pm\sqrt{-1}$! And this is being done with all three entities i, j, k. Note that $\bar{i} = -i$, $\bar{j} = -j$, $\bar{k} = -k$ are all obtained either by reversing the rotation, switching crossings, or adding a $\pm 2\pi$ twist on the strings.

The other quaternion relations are present as well:

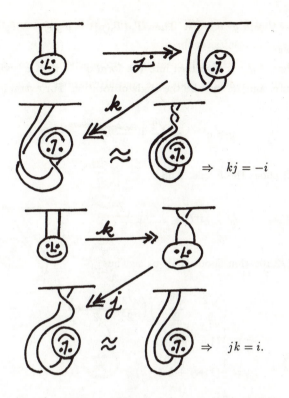

$$\Rightarrow \quad kj = -i$$

$$\Rightarrow \quad jk = i.$$

Thus the puppet actually reproduces the quaternion relations. In its way this diagrammatic situation is analogous to our diagram algebra where a different string trick produces the relations in the Temperley-Lieb algebra.

Another comment about possible extensions of the string trick arises in connection with the Cayley numbers. As we have seen, the Cayley multiplicative structure is given by starting with $i, j, k, -1$ and adding J so that $Jz = \overline{z}J$, $J^2 = -1$. **Is there a string trick interpretation of the Cayley numbers?**

Return to Proposition 10.4.

Let's return to Proposition 10.4. This shows how to represent rotations of $\mathbf{R}^3 = \{xi + yj + zk | x, y, z \text{ real}\}$ via quaternions. For $g = e^{u\theta} = (\cos\theta) + (\sin\theta)u$, u a unit length quaternion in $\mathbf{R}^3$, the mapping $T : \mathbf{R}^3 \to \mathbf{R}^3$, $T(p) = gp\overline{g}$, is a rotation about the axis u by 2θ. One of the charms of this representation is that it lets us compute the axis and rotation angle of the composite of two rotations. For, given two rotations $g = e^{u\theta}$ and $g' = e^{u'\theta'}$. Then $T_{g'}(T_g(p)) = g'gp\overline{g}\overline{g'} = (g'g)p\overline{(g'g)}$. Thus $T_{g'} \circ T_g = T_{g'g}$.

This means that if $g'g$ is rewritten into the form $e^{u''\theta''}$ then u'' will be the axis of the composite, and $2\theta''$ will be the angle of rotation. For example,

$$g = e^{k\pi/4} = \frac{\sqrt{2}}{2} + \frac{\sqrt{2}}{2} k$$

is a rotation of $\pi/2$ about the k axis. And

$$g' = e^{j\pi/4} = \frac{\sqrt{2}}{2} + \frac{\sqrt{2}}{2} j$$

corresponds to a $\pi/2$ rotation about the j-axis. Thus

$$g'g = \left(\frac{\sqrt{2}}{2} + \frac{\sqrt{2}}{2} j\right)\left(\frac{\sqrt{2}}{2} + \frac{\sqrt{2}}{2} k\right)$$
$$= \frac{1}{2}(1+j)(1+k)$$
$$= \frac{1}{2}(1+j+k+jk)$$
$$= \frac{1}{2}(1+j+k+i)$$
$$\therefore g'g = \frac{1}{2} + \frac{\sqrt{3}}{2}\left(\frac{i+j+k}{\sqrt{3}}\right)$$
$$\therefore g'g = e^{((i+j+k)/\sqrt{3})\pi/3}.$$

Hence the composite of two $\pi/2$ rotations about the orthogonal axes j and k is a rotation of $2\pi/3$ around the diagonal axis. Figure 10.7 illustrates this directly via a drawing of a rotated cube.

This ends our chapter on the quaternions, Cayley numbers and knot and string interpretations. We showed how these structures arise naturally via looking for a model of the algebraic structure of reflections in space. In this model, the geometry is mirrored in the algebra: the geometric point being mirrored becomes notational mirror for the representation of the mirror plane: $R(p) = upu$, u a unit vector perpendicular to the mirror plane.

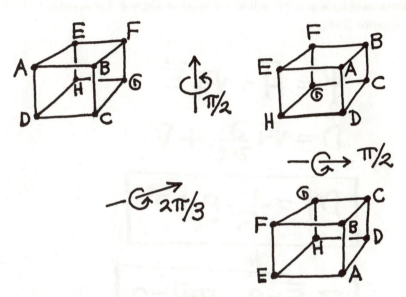

Figure 10.7

Epilog. This chapter has been a long excursion, from reflections to normed algebras, quaternions, string tricks and back to reflections. We have touched upon the quaternions and their role in spin and special relativity (section 9^0 Part II). The role of the Cayley numbers in physics is an open question, but the quaternions, invented long before relativity, make a perfect fit with relativity, electromagnetism and the mathematics of spin.

Exercise. [SI]. Let $D = \sqrt{-1}\frac{\partial}{\partial t} + \vec{\nabla}$ where $\vec{\nabla} = i\frac{\partial}{\partial x} + j\frac{\partial}{\partial y} + k\frac{\partial}{\partial z}$. Here i, j, k generate the quaternions, and $\sqrt{-1}$ is an extra root of -1 that commutes with i, j and k. Let $\psi = H + \sqrt{-1}\,E$ where $E = E_x i + E_y j + E_z k$, $H = H_x i + H_y j + H_z k$ are fields on $\mathbf{R}^3$. (i) Show that the equation $D\psi = 0$, interpreted in complexified quaternions is identical to the vacuum Maxwell equations. (ii) Determine the non-vacuum form of Maxwell's equations in this language. (iii) Prove the relativistic invariance of the Maxwell equations by using the quaternions. (iv) Investigate the nonlinear field equations ([HAY], [STEW]) $D\psi = \frac{1}{2}\psi\sqrt{-1}\psi^*$ where $\psi^* = H - \sqrt{-1}E$. (v) Let $K = K_1 + \sqrt{-1}K_2$ be a constant where K_1 and K_2 are linear combinations of i, j and k. Show that the equation $D\psi = K\psi$ is equivalent to the Dirac equation [DIR].

$$\Psi = \vec{H} + \sqrt{-1}\,\vec{E}$$

$$D = \sqrt{-1}\,\frac{\partial}{\partial t} + \nabla$$

$$\boxed{D\Psi = \frac{1}{\sqrt{-1}}\,\rho + \vec{J}}$$

$$\|\|\|$$

$$\boxed{\begin{array}{l} \nabla \cdot \vec{E} = \rho \;,\; \nabla \cdot \vec{H} = 0 \\[4pt] \dfrac{\partial \vec{H}}{\partial t} = -\nabla \times \vec{E} \\[4pt] \dfrac{\partial \vec{E}}{\partial t} + \vec{J} = \nabla \times \vec{H} \end{array}}$$

11^0. The Quaternion Demonstrator.

As we have discussed in section 10^0 of Part II, The Dirac String Trick ([FE1], [BAT]) illustrates an elusive physical/mathematical property of the spin of an electron. An observer who travels around an electron (in a proverbial thought experiment) will find that the wave function of the electron has switched its phase for each full turn of the observer. See [AH] for a discussion of this effect. Here we repeat a bit of the lore of section 10^0, and continue on to an application of the string trick as a quaternion demonstrator.

It is a fact of life in quantum mechanics that a rotation of 2π does not bring you back to where you started. Nevertheless, in the case of the spin of the electron (more generally for spin $1/2$), a 4π rotation does return the observer to the original state. These facts, understandable in terms of the need for unitary representations of rotations of three space to the space of wave functions, seem very mysterious when stated directly in the language of motions in three dimensional space.

In order to visualize these matters, Dirac found a simple demonstration of a topological analog of this phenomenon using only strings, or a belt. As the **Belt Trick** this device is well-known to generations of topologists. In fact, a closely related motion, the **Plate Trick** (or Philippine Wine Dance) has been known for hundreds of years. (See [BE], p.122.)

Note that the trick shows that a 4π twist in the belt can be removed by a topological ambient isotopy that leaves the ends of the belt fixed. We have illustrated the fixity of the ends of the belt by attaching them to two spheres. The belt moves in the complement of the spheres. This trick is easy to perform with a real belt, using an accomplice to hold one end, and one's own two hands to hold the other end as the belt is passed around one end.

It is also well-known that the belt trick illustrates the fact that the fundamental group of the space of orientation preserving rotations of three dimensional space $(SO(3))$ has order two. This is, in turn, equivalent to the fact that $SO(3)$ is double covered by the unit quaternions, $SU(2)$.

It is the purpose of this section to explore the belt trick and the group $SU(2)$. I will show how a natural generalization of the belt trick gives rise to a topological/mechanical model for the quaternion group, and how subgroups of $SU(2)$ are

visualized by attaching a belt to an object in three space. This generalizes the belt trick to a theorem, (the **Belt Theorem** - see below) about symmetry and $SU(2)$. The Belt Theorem is a new result.

Finally, we show how to make a computer graphics model of the belt trick. This model is actually a direct visualization of the homotopy of the generator of $\pi_1(SO(3))$ with its inverse.

The Quaternion Demonstrator.

In order to create a mechanical/topological demonstration of the quaternion group, follow these instructions.

1. Obtain a cardboard box, a small square of cardboard and a strip of paper about a foot long and an inch wide.

2. Tape one end of the strip to the cardboard box, and tape the other end of the strip to the square of cardboard.

Your **demonstrator** should appear as shown below.

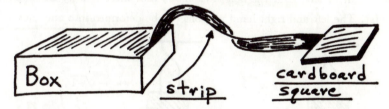

Before we construct the quaternions it is useful to do the belt trick and then to construct a square root of minus one. After a few square roots of minus one have appeared (one for each principal direction in space), we shall find that the quaternions have been created.

Belt Trick.

To demonstrate the belt trick, set the cardboard box of the demonstrator on a table, pick up the card, and begin with an untwisted belt (The paper strip shall be called the belt.). Twist the card by 720 degrees, and demonstrate the belt trick as we did in Section 10^0.

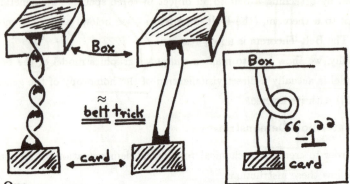

Minus One.

Since the belt trick disappears a twist of 720 degrees, it follows that **a twist of 180 degrees is the analog of the square root of minus one.** Four applications of the 180 degree twist bring you back to the beginning (with a little help from topology). If this is so, then **a twist of 360 degrees must be the analog of minus one.** Of course you may ask of the 360 degree twist - which one, right or left? But the belt trick shows that there is **only one 360 degree twist.** The left and right hand versions can be deformed into one another.

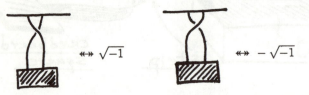

The presence of a 360 degree twist on the belt indicates multiplication by minus one.

Many Square Roots of Minus One.

There are many square roots of minus one. Just rotate the card around any axis by 180 degrees. Doing this four times will induce the belt trick and take you back to start. In particular, consider the rotations about three perpendicular axes as shown below.

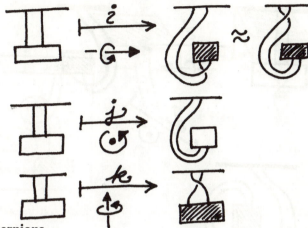

The Quaternions.

Having labelled these three basic roots of minus one i, j, k we find that

$$i^2 = j^2 = k^2 = ijk = -1.$$

These are the defining relations for the quaternion group.

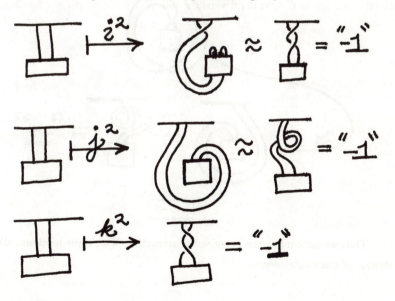

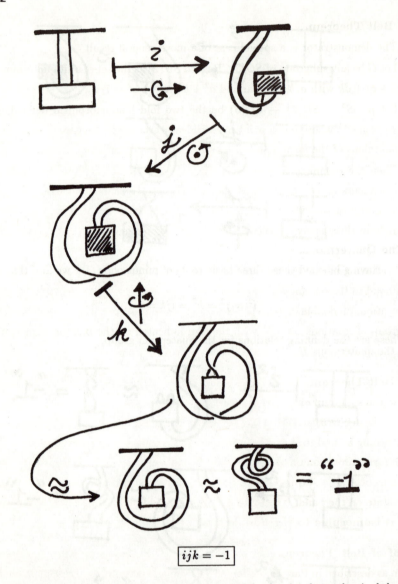

$$\boxed{ijk = -1}$$

Thus we have made the promised construction of a simple mechanical demonstrator of the quaternions.

The Belt Theorem.

The demonstrator is a special case of a more general result.

Let G be any subgroup of $SO(3)$. Regard $SU(2)$ as the set of unit quaternions: $a + bi + cj + dk$ with a, b, c, d real and $a^2 + b^2 + c^2 + d^2 = 1$.

Let $p : S^3 = SU(2) \to SO(3)$ be the two fold homomorphism defined by $p(g)(v) = gvg^{-1}$ where v is a point in Euclidean three space – regarded as the set of quaternions of the form $ai + bj + ck$ with a, b, c real numbers.

Then the preimage $\widehat{G} = p^{-1}(G)$ is a subgroup of $SU(2)$ that has twice as many elements as G. $\widehat{G}$ is called the **double group** of G. For example, the double group of the Klein Four Group, K, (i.e. the group of symmetries of a rectangle in three space) is the quaternion group, H, generated by i, j and k in $SU(2)$.

We have seen that the quaternions arise from the Klein Four Group, by attaching a band to the rectangle whose symmetries generate K. With this attachment, the rotations all double in value (2π is non-trivial and of order two), and the resulting group of rotations coupled with the sign indicated on the band is isomorphic with the quaternions H.

The Belt Theorem. Let A be an object embedded in three space, and let $G(A)$ denote the subgroup of $SO(3)$ of rotations that preserve A. $G(A) = \{s \text{ in } SO(3) | s(A) = A\}$. Then the double group $\widehat{G}(A) \subset SU(2)$ is realized by attaching a band to A and following the prescription described above. That is, we perform a rotation on A and catalog the state of the band topologically. Two rotations of A are $\widehat{}$-**equivalent** if they are identical on A, **and** they produce the same state on the band. Then **the set of** $\widehat{}$-**equivalence classes of rotations of A is isomorphic to the double group** $\widehat{G}(A)$.

Proof of Belt Theorem. Let $\pi : SU(2) \to SO(3)$ be the covering homomorphism as described in Part II, section 10^0. Given an element $\hat{g} \in \widehat{G} \subset SU(2)$ where G is the symmetry group of an object $\mathcal{O} \subset \mathbf{R}^3$, let $g = \pi(\hat{g})$ and let β denote the band that is attached to $\mathcal{O}$. We assume that the other end of β is "attached to infinity" for the purpose of this discussion. That is, $\mathcal{O}$ is equipped with an infinite band with one end attached to $\mathcal{O}$. Twisting $\mathcal{O}$ sends some curling onto the band.

434

Each position of $\mathcal{O}$ in space curls the band, but by the belt trick we may compare 2π and 4π rotations of $\mathcal{O}$ and find that they differ by a 2π twist in the band "at infinity". That is, we can make the local configurations of $(\mathcal{O}, \beta)$ the same for g and g followed by 2π, but the stationary part of the band some distance away from $\mathcal{O}$ will have an extra twist in the second case.

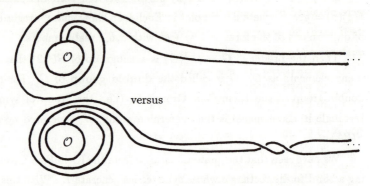

As a result, we get a doubling of the set of spatial configurations of $\mathcal{O}$ to the set of configurations of $(\mathcal{O}, \beta)$. If S is the original set of configurations, let $\widehat{S}$ be the set of configurations for $(\mathcal{O}, \beta)$. Then S is in 1-1 correspondence with G, hence $\widehat{S}$ is in 1-1 correspondence with $\widehat{G}$. It is easy to see that the latter correspondence is an isomorphism of groups.   //

Programming the Belt Trick.

The program illustrates a belt that stretches between two concentric spheres in three dimensional space. A belt will be drawn that is the image under rotations of an untwisted belt that originally stretches from the north pole of the larger sphere to the north pole of the smaller sphere.

Call this belt the **original belt**. Upon choosing an axis of rotation, the program draws a belt that is obtained from the original belt via rotation by $2\pi t$ around the axis of the t-**shell**. The t-shell (with $0 < t < 1$) is a sphere concentric to the two boundary spheres at $t = 0$ and $t = 1$. Thus, at the top ($t = 0$), the belt is unmoved. It then rotates around the axis and is unmoved at the bottom ($t = 1$) – the inner concentric sphere.

For each choice of axis we get one image belt. The axis with $u_1 = 0$, $u_2 = 1$,

$u_3 = 0$ gives a belt with a right-handed 2π twist. As we turn the axis through π in the $u_1 - u_2$ plane, the belt undergoes a deformation that moves it around the inner sphere and finally returns it to its original position with a reversed twist of 2π.

The set of pictures of the deformation (stereo pairs after the program listing) were obtained by running the program for just such a set of axes. The program itself uses the quaternions to accomplish the rotation. Thus we are using the quaternions to illustrate themselves!

```
10 PRINT"DIRAC STRING TRICK!"
15 PRINT"COPYRIGHT(1989)L.H.KAUFFMAN - KNOTS INC."
20 PRINT"INPUT AXIS U1i+U2j+U3k"
25 PI=3.141592653886
27 U1=0:U2=1:U3=0:'DEFAULT AXIS
30 INPUT"U1";U1:INPUT"U2";U2:INPUT"U3";U3
40 U=SQR(U1*U1+U2*U2+U3*U3)
50 U1=U1/U:U2=U2/U:U3=U3/U
60 PRINT"THE TWISTED BELT!":CLS
65 'SOME CIRCLES
70 FOR I=0 to 100
75 A=COS(2*PI*I/100):B=SIN(2*PI*I/100)
80 PSET(120+100*A,125-100*B)
90 PSET(120+20*A,125-20*B)
95 PSET(370+100*A,125-100*B)
97 PSET(370+20*A,125-20*B)
100 NEXT I
105 'DRAW BELT
110 FOR I=0 TO 50
120 FOR J=0 TO 20
130 X = -10+J
140 R = 100-I*(8/5)
150 Y+SQR(R*R-X*X)
160 Z=0
170 P1=X:P2=Y:P3=Z
180 IF I=0 OR I=50 THEN 210
190 TH=(PI/50)*I
200 GOSUB 2000:X=Y:Y=Z:Z=W
210 PSET(120+X,125-Y)
215 PSET(370+X+.2*Z,125-Y)
220 NEXT J
230 NEXT I
```

```
 235 A$=INKEY$:IF A$="" THEN 235
 237 GOTO 10
 240 END
1000 'THE QUATERNION SUBROUTINE!
1020 'INPUT A+Bi+Cj+Dk
1040 'INPUT E+Fi+Gj+Hk
1060 X=A*E-B*F-C*G-D*H
1070 Y=B*E+A*F-D*G+C*H
1080 Z=C*E+D*F+A*G-B*H
1090 W=D*E-C*F+B*G+A*H
1100 RETURN
2000 'ROTATION ROUTINE
2020 'INPUT AXIS U1i+U2j+U3k
2030 'INPUT ANGLE TH=(1/2)rotation angle
2040 'INPUT POINT P1i+P2j+P3k
2050 P=COS(TH):Q=SIN(TH)
2060 A=P:B=Q*U1:C=Q*U2:D=Q*U3
2070 E=0:F=P1:G=P2:H=P3
2080 GOSUB 1000:'QMULT
2090 E=P:F=-B:G=-C:H=-D
2100 A=X:B=Y:C=Z:D=W
2110 GOSUB 1000:'QMULT
2120 RETURN
```

The Quaternionic Arm.

In an extraordinary conversation about quaternions, quantum mechanics, quantum psychology [O], quantum logic and the martial art Wing Chun, the author and Eddie Oshins discovered that the basic principle of the quaternion demonstrator can be done with only a single human arm. Hold your (right) arm directly outward perpendicular to your body with the palm up. Turn your arm by 180 degrees. This is i. Now bend your elbow until your hand points to your body. This is j. Now push your hand out in a stroke that first moves your palm downward - until your arm is once again fully extended from your body with your palm up. This is k. The entire movement produces an arm that is twisted by 360 degrees. Whence $ijk = -1$.

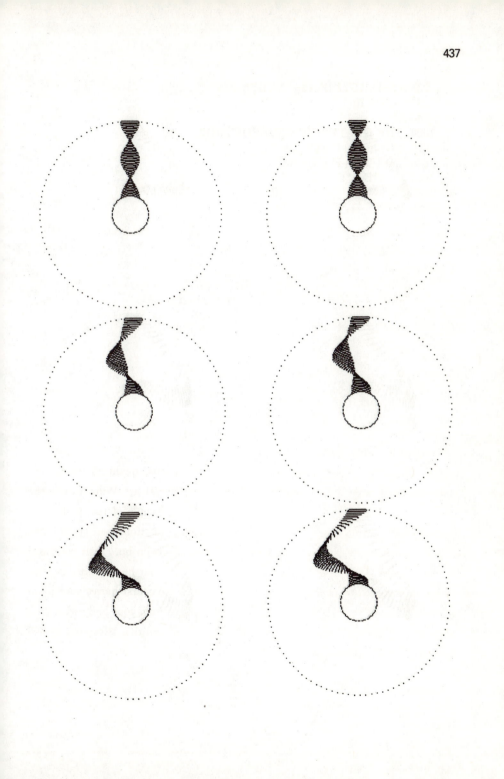

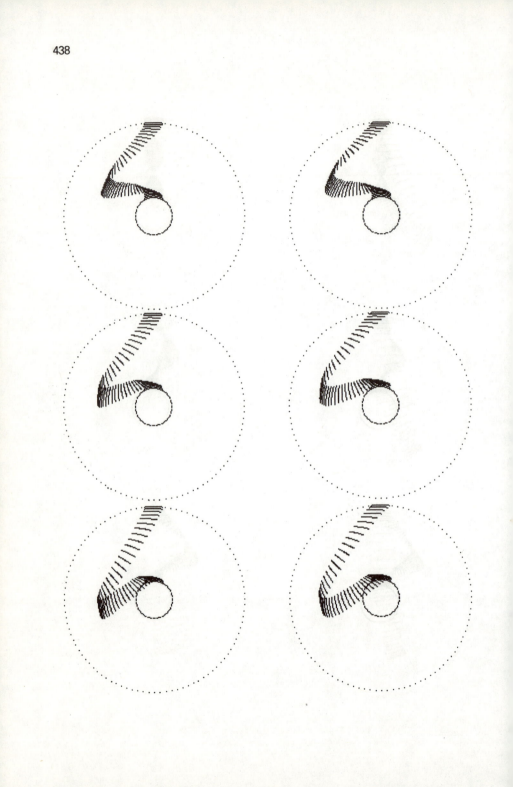

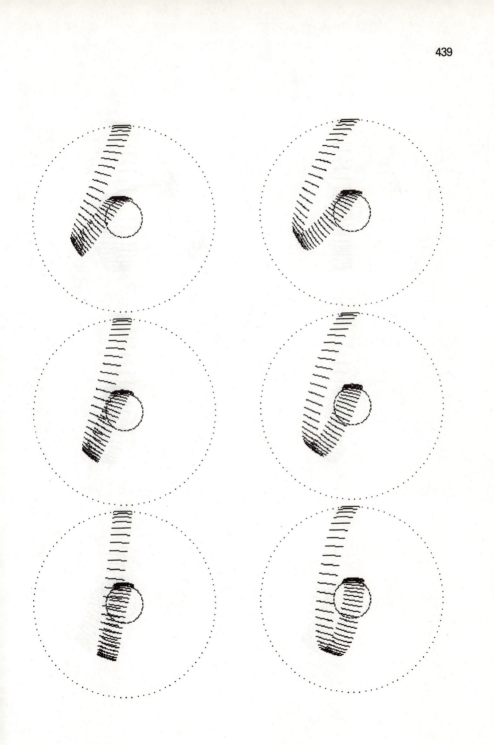

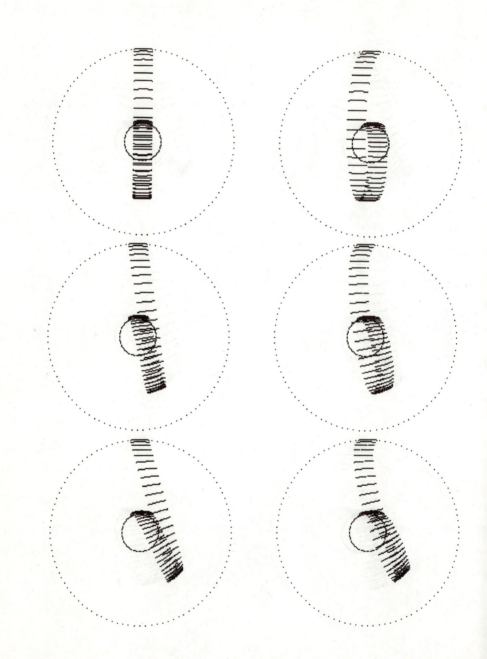

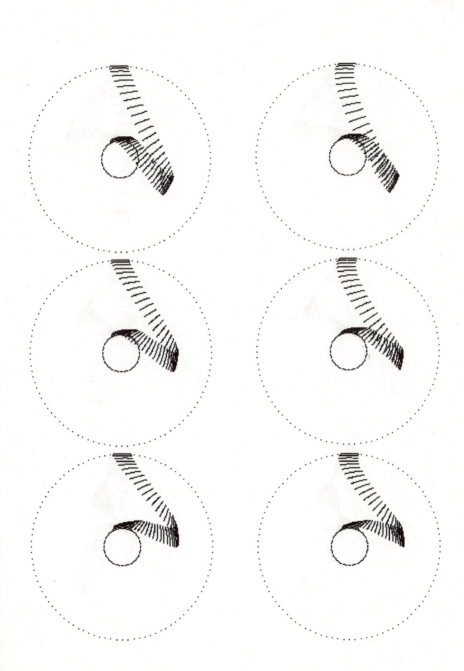

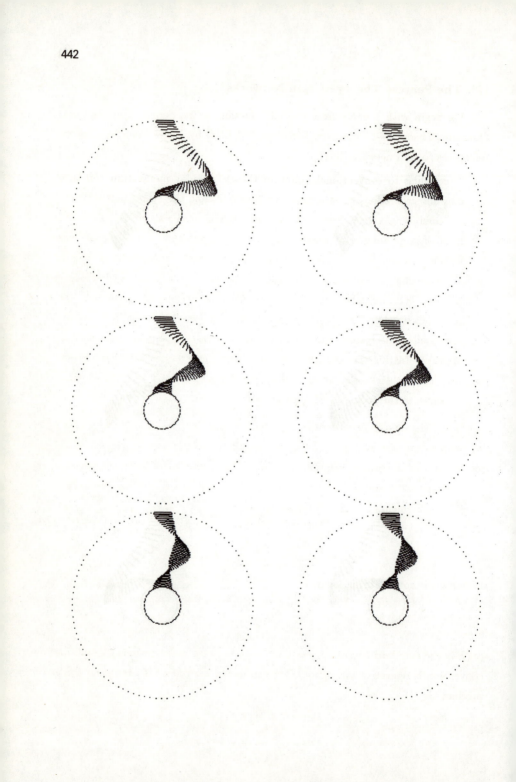

12^0. The Penrose Theory of Spin Networks.

We begin with a quote from the introduction to the paper "Combinatorial Quantum Theory and Quantized Directions" [PEN4] by Roger Penrose. For the references to spin nets see [HW].

"According to conventional quantum theory, angular momentum can take only integral values (measured in units of $\hbar/2$) and the (probabilistic) rules for combining angular moments are of combinatorial nature. ... In the limit of large angular momenta, we may ... expect that the quantum rules for angular momentum will determine the geometry of directions in space. We may imagine these directions to be determined by a number of spinning bodies. The angles between their axes can then be defined in terms of the probabilities that their total angular momentua (i.e. their "spins") will be increased or decreased when, say, an electron is thrown from one body to another. In this way the geometry of directions may be built up, and the problem is then to see whether the geometry so obtained agrees with what we know of the geometry of space and time."

The Penrose spin networks are designed to facilitate calculations about angular momentum and $SL(2)$. Perhaps the most direct route into this formalism is to give part of the group-theoretic motivation, then set up the formal structure.

Recall that a spinor is a vector in two complex variables, denoted by ψ^A $(A = 1, 2)$. Elements $U \in SL(2)$ act on spinors via

$$(U\psi)^A = \sum_B U_B^A \psi^B.$$

Defining the conjugate spinor by

$$\psi_A^* = \epsilon_{AB} \psi^B$$

(ϵ_{AB} is the standard epsilon: $\epsilon_{12} = 1 = -\epsilon_{21}$, $\epsilon_{11} = \epsilon_{22} = 0$. Einstein summation convention is operative throughout.), we have the natural $SL(2)$ invariant inner product

$$\psi\psi^* = \psi^A \epsilon_{AB} \psi^B.$$

In order to diagram this inner product, let

$$\psi^A \longleftrightarrow \boxed{}^{\,A}$$

Then it is natural to lower the index via

$$\psi^*_B \longleftrightarrow B \,\boxed{}$$

so that

$$\psi\psi^* \longleftrightarrow \boxed{}\,\boxed{}$$

with the cap $A \frown B$ representing the epsilon matrix ϵ_{AB}.

Unfortunately (perhaps fortunately!) this notation is not topologically invariant since if $\frown \longleftrightarrow \epsilon_{AB}$ and $\smile \longleftrightarrow \epsilon^{AB}$ then

$$\longleftrightarrow \epsilon_{AB}\epsilon^{BC} = -\delta^C_A \longleftrightarrow -$$

where δ^C_A denotes the Kronecker delta. We remedy this situation by replacing

$$\left\{ \begin{array}{ccc} A \quad B & \longleftrightarrow & \sqrt{-1}\epsilon_{AB} \\[2em] C \quad D & \longleftrightarrow & \sqrt{-1}\epsilon^{CD} \end{array} \right\}$$

the epsilons by their multiples by the square root of -1.

This gives a topologically invariant diagrammatic theory, but translations are required in carrying $SL(2)$-invariant tensors into the new framework. For example, the new loop value is

$$A \bigcirc B = (\sqrt{-1})^2 + (-\sqrt{-1})^2 = -2,$$

whence the name negative dimensional tensors.

Even with this convention, we still need to add minus signs for each crossing in order to have

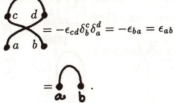

$$= -\epsilon_{cd}\delta_b^c\delta_a^d = -\epsilon_{ba} = \epsilon_{ab}$$

$$= \quad \bigcap_{a \quad b} .$$

Thus $\displaystyle \bigtimes_{c \quad d}^{a \quad b} = -\delta_d^a\delta_c^b$ by definition. With these conventions, we have the **basic binor identity**:

$$\bigtimes + \bigcup_{\bigcap} + \big)\big(= 0.$$

Proof.

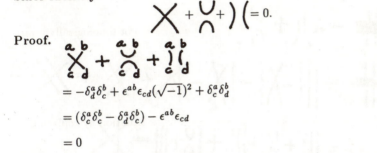

$$= -\delta_d^a\delta_c^b + \epsilon^{ab}\epsilon_{cd}(\sqrt{-1})^2 + \delta_c^a\delta_d^b$$

$$= (\delta_c^a\delta_c^b - \delta_d^a\delta_c^b) - \epsilon^{ab}\epsilon_{cd}$$

$$= 0 \qquad\qquad\qquad\qquad\qquad\qquad //$$

As aficionados of the bracket (Part I) we recognize the possibility of a topological generalization via

$$\bigtimes = A \bigcup_{\bigcap} + A^{-1}\big)\big($$

$$\bigcirc = -A^2 - A^{-2},$$

but this is postponed to the next section.

The next important spin network ingredient is the **antisymmetrizer**. This is a diagram sum associated to a bundle of lines, and it is denoted by

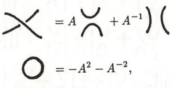

where N denotes the number of lines. The antisymmetrizer is defined by the formula

$$\begin{array}{c}\blacksquare\!\!\uparrow^N\end{array} = \sum_{\sigma \in S_N} \text{sgn}(\sigma)\; \boxed{\sigma}$$

where σ runs over all permutations in S_N (the set of permutations on N distinct objects), and the σ in the box denotes the diagrammatic representation of this permutation as a braid projection. Thus

$$\boxed{\,}^{N}\!\!\! = N!\,|^{N}\!\!\! ,\ and$$

$$\boxed{\,} = ||-X$$

$$= ||+\overset{\cup}{\underset{\cap}{X}}+)(\qquad [X+\overset{\cup}{\underset{\cap}{X}}+)(=\phi]$$

$$\boxed{\,} = 2)(+\overset{\cup}{\underset{\cap}{X}}.$$

$$\boxed{\,} = |||-X|-|X-\overset{}{X}\!\!|+\overset{}{X}\!\!\times+\times\!\!\times$$

$$= |||+\overset{\cup}{\underset{\cap}{X}}|+|||+|\overset{\cup}{\underset{\cap}{X}}+|||-\overset{}{X}\!\!\!$$

$$+\overset{\cup}{\underset{\cap}{X}}\!\!+\overset{\cup}{\underset{\cap}{X}}\!\!|+|\overset{\cup}{\underset{\cap}{X}}+|||$$

$$+\overset{\cup}{\underset{\cap}{X}}\!\!+\overset{\cup}{\underset{\cap}{X}}\!\!|+|\overset{\cup}{\underset{\cap}{X}}+|||$$

$$= 5|||+3\overset{\cup}{\underset{\cap}{X}}|+3|\overset{\cup}{\underset{\cap}{X}}+\overset{\cup}{\underset{\cap}{X}}+\overset{\cup}{\underset{\cap}{X}}-\overset{}{X}\!\!\!$$

$$= 5|||+3\overset{\cup}{\underset{\cap}{X}}|+3|\overset{\cup}{\underset{\cap}{X}}+\overset{\cup}{\underset{\cap}{X}}+\overset{\cup}{\underset{\cap}{X}}$$

$$+\overset{}{X}\!\!\times+\times\!\!\times \qquad [\overset{}{X}\!\!\times=\overset{\cup}{\underset{\cap}{X}}]$$

$$\boxed{\,} = 6|||+4\overset{\cup}{\underset{\cap}{X}}+4|\overset{\cup}{\underset{\cap}{X}}+2\overset{\cup}{\underset{\cap}{X}}+2\overset{\cup}{\underset{\cap}{X}}.$$

These antisymmetrizers are the basic ingredients for making spin network calculations of Clebsch-Gordon coefficients, $3j$ and $6j$ symbols, and other apparatus of angular momentum. In this framework, the 3-vertex is defined as follows:

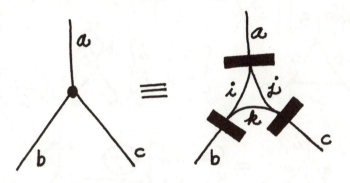

Here a, b, c are positive integers satisfying the condition that the equations $i + j = a$, $i + k = b$, $j + k = c$ can be solved in non-negative integers. Think of spins being apportioned in this way in interactions along the lines. This vertex gives the spin-network analog of the quantum mechanics of particles of spins b and c interacting to produce spin a.

Formally, the $6j$ symbols are defined in terms of the re-coupling formula

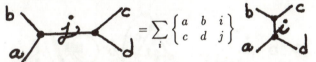

To obtain a formula for these recoupling coefficients, one can proceed diagrammatically as follows.

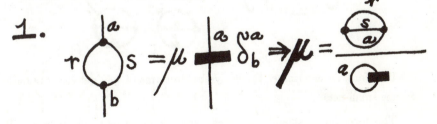

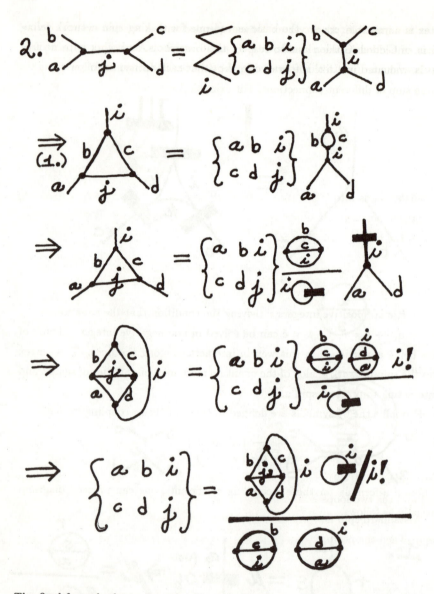

The final formula determines the $6j$ symbol in terms of spin network evaluations of some small nets.

This formulation should make clear what we mean by a spin network. It is a graph with trivalent vertices and spin assignments on the lines so that each 3-

vertex is admissible. Each vertex comes equipped with a specific cyclic ordering and is embedded in the plane to respect this ordering. A network with no free ends is evaluated just like the bracket except that each antisymmetrizer connotes a large sum of different connections. For example,

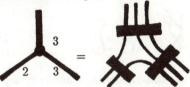

and this expands into $(2!)(3!)^2 = 72$ local configurations. The rule for expanding the antisymmetrizers takes care of the signs. Each closed loop in a given network receives the value (-2). Thus for

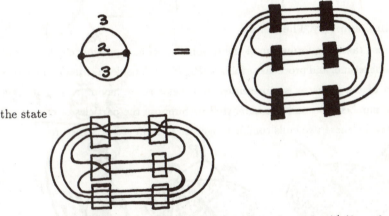

the state

contributes $(-1)^4(-2)^2$ since the total permutation sign is $(-1)^4$ (from four cross-ings) and the two loops each contribute (-2).

A standard spin network convention is to ignore self-crossings of the trivalent graph. Then the norm, $\|G\|$, of a closed spin network G is given by the formula

$$\|G\| = \sum_{\sigma}(-1)^{\chi(\sigma)}(-2)^{\|\sigma\|}$$

where σ runs over all states in the sense (of replacing the bars of the antisymmetriz-ers by specific connections), and $\chi(\sigma)$ is the total number of crossings (within the bars) in the state. $\|\sigma\|$ is the number of closed loops.

450

Just as with the bracket, we can also evaluate by using the binor expansion

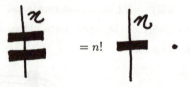

$$-2 = $$

and $\quad \bigotimes = -\bigcirc\!\!\!\!-\ -\ \bigotimes = -(-2)^2 - (-2) = -4 + 2 = -2.$

Note also, that the antisymmetrizers naturally satisfy the formula

$$= n! \quad \bullet$$

The Spin-Geometry Theorem.

Penrose takes the formalism of the spin networks as a possible combinatorial basis for spacetime, and proceeds as follows [PEN4]: "Consider a situation in which a portion of the universe is represented by a network   having among its free ends, one numbered m and one numbered n. Suppose the particles or structures represented by these two ends combine together to form a new structure.

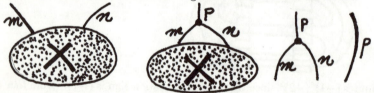

We wish to know, for any allowable p, what is the probability that the angular momentum number of this new structure be p." Penrose takes this probability to be given by the formula

where the norm of a network with free ends is the norm of two copies of that net plugged into one another. (analog of $\psi\psi^*$).

"Consider now the situation below involving two bodies with large angular momenta M and N.

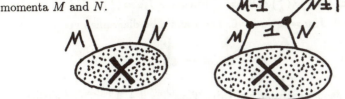

We define the **angle** between the axes of the bodies in terms of relative probability of occurrence of $N + 1$ and $N - 1$ respectively, in the "experiment" shown above. ... To eliminate the possibility that part of the probability be due to ignorance, we envisage a repetition of the experiment as shown below.

If the probability given in the second experiment is essentially unaffected by the result of the first experiment then we say that the angle θ between the axes of the bodies is well-defined and is determined by this probability. It then turns out that

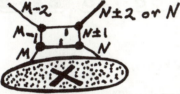

$$\cong \frac{1}{2}\cos(\theta)$$

and from this ... **it is possible to show that the angles obtained in this way satisfy the same laws as do angles in a three-dimensional Euclidean space.**" This is the Spin-Geometry Theorem.

Calculating Vector Coupling Coefficients.

Here I show how to translate a standard problem about angular momenta into the spin network language. Let J_1, J_2, J_3 be the standard generators for the Lie algebra of $SU(2)$ (compare section 9^0, Part II).

$$J_1 = \frac{1}{2}\begin{pmatrix} 0 & 1 \\ 1 & 0 \end{pmatrix}, \; J_2 = \frac{1}{2}\begin{pmatrix} 0 & -i \\ i & 0 \end{pmatrix}, \; J_3 = \frac{1}{2}\begin{pmatrix} 1 & 0 \\ 0 & -1 \end{pmatrix}.$$

Let $J_+ = J_1 + iJ_2 = \left(\begin{smallmatrix} 0 & 1 \\ 0 & 0 \end{smallmatrix}\right)$ and $J_- = J_1 - iJ_2 = \left(\begin{smallmatrix} 0 & 0 \\ 1 & 0 \end{smallmatrix}\right)$ be the raising and lowering operators, so that for $u = \left(\begin{smallmatrix} 1 \\ 0 \end{smallmatrix}\right)$ and $d = \left(\begin{smallmatrix} 0 \\ 1 \end{smallmatrix}\right)$ we have $J_3 u = (\frac{1}{2})u$, $J_3 d = (-\frac{1}{2})d$ and $J_+ d = u$, $J_+ u = 0$, $J_- u = d$, $J_- d = 0$. Let u and d be diagramed via

Let denote **symmetrization**. Thus . Then, letting

$$\tilde{J} = (J \otimes 1 \otimes 1 \otimes \ldots \otimes 1) + (1 \otimes J \otimes 1 \otimes \ldots) + \ldots,$$

we let

and get

$$J_3(V(k,\ell)) = \left(\frac{k-\ell}{2}\right)V(k,\ell)$$

in the corresponding tensor product. This yields an irreducible representation of $s\ell(2)$ with eigenvalues $\{-m/2, \ldots, m/2\}$ and $m = (k+\ell)/2$. We write

with $m = \frac{r-s}{2}$, $j = \frac{r+s}{2}$. Then $J_3|jm\rangle = m|jm\rangle$, and the raising and lowering operators move these guys around.

We wish to compute the vector coupling coefficient, or Wigner coefficient **giving the amplitude for three spins to combine to zero**. This is denoted by

$$\begin{bmatrix} j_1 & j_2 & j_3 \\ m_1 & m_2 & m_3 \end{bmatrix}.$$

In order to compute this amplitude, **we form the product state** $|j_1 m_1\rangle|j_2 m_2\rangle|j_3 m_3\rangle$ **and apply antisymmetrizers to reduce it to a scalar.**

Question. Why do we apply epsilons $(\underset{ab}{\Pi} = \epsilon_{ab})$ (i.e. antisymmetrizers) to reduce these tensors to scalars?

Answer. To combine to zero involves spin conservation. Thus the possibilities are

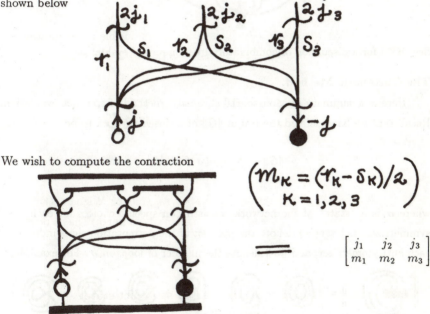

$\uparrow$(time's arrow) $+\dfrac{1}{2}$ $\bullet$ $\bullet -\dfrac{1}{2}$ and $-\dfrac{1}{2}$ $\bullet$ $\bullet +\dfrac{1}{2}$.

And remember, the difference between these two is the difference in phase of the wave function. Thus this change in sign

$$\Pi_{ab} = -\Pi_{ba}$$

corresponds to the change in phase when you walk around a spin $(\frac{1}{2})$ particle (Compare with sections 9^0, 10^0 and 11^0 of Part II.).

The diagram representing the structure of three spins of total spin zero is as shown below

We wish to compute the contraction

$$\left(m_K = (r_K - \delta_K)/2 \atop K = 1, 2, 3 \right)$$

$$= \begin{bmatrix} j_1 & j_2 & j_3 \\ m_1 & m_2 & m_3 \end{bmatrix}$$

In order to translate this into a spin network, we replace Π by $\sqrt{-1}\,\Pi = \cup$ and Π by $\sqrt{-1}\,\Pi = \cap$, and **symmetrizers become anti-symmetrizers** (by the convention of minus signs for crossings). Thus

$$\begin{bmatrix} j_1 & j_2 & j_3 \\ m_1 & m_2 & m_3 \end{bmatrix} =$$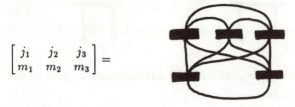

This reduces the calculation of the vector coupling coefficient to the evaluation of the spin net

$$\left(j_K = \frac{r_K + s_K}{2} \atop K = 1, 2, 3 \right)$$ 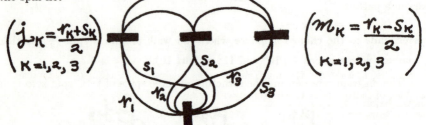 $$\left(m_K = \frac{r_K - s_K}{2} \atop K = 1, 2, 3 \right)$$

See [HW] for a discussion of combinatorial techniques of evaluation.

The Chromatic Method.

Here is a summary of Moussouris' chromatic method of spin net evaluation. Recall that we have defined the **norm** $\|G\|$ of a closed network to be

$$\|G\| = \sum_{\vec{\sigma}} \operatorname{sgn}(\vec{\sigma})(-2)^{\|\vec{\sigma}\|}$$

where $\vec{\sigma}$ is a "state" of the network consisting in specific choices for each anti-symmetrizer. Let $\operatorname{sgn}(\vec{\sigma})$ denote the product of the permutation signs induced at each antisymmetrizer, and $\|\vec{\sigma}\|$ denote the number of loops in $\vec{\sigma}$. For example,

$$= (-2)^3 - (-2)^2 - (-2)^2 + (-2) + (-2) - (-2)^2$$
$$= -8 - 4 - 4 - 2 - 2 - 4$$
$$= -24.$$

Note that in this language, we can think of the network and its state expansion as synonymous.

In order to facilitate computations it is convenient to define a polynomial $P_G(\delta)$ as follows

$$P_G(\delta) = \left(\sum_{\vec{\sigma}} \mathrm{sgn}(\vec{\sigma}) \delta^{\|\vec{\sigma}\|} \right) \bigg/ \Pi e!$$

where Πe ! denotes the product of the factorial of the multiplicities assigned to the edges of G. Now note that in these strand nets the edges enter antisymmetrizer bars in a pattern

where a, b, c, d denote the multiplicities of the lines. The term $\Pi e!$ denotes the product of factorials of these multiplicities. Of course the standard spin network norm occurs for $\delta = -2$.

If $\delta = N$ (a positive integer), we can regard the strands as labelled from an index set of "colors" $\mathcal{I} = \{1, 2, \ldots, N\}$. Then a non-zero state corresponds to an N coloring of the (individual strands) so that for each strand entering a bar there is exactly one strand of the same color leaving the bar. (Note that N must be greater than or equal to the maximum multiplicity entering an antisymmetrizer in order to use this set up.)

Thus for a given choice of permutations at the bars, we can count cycles in the resulting graph, and stipulate that **cycles sharing a bar have different colors.**

For a state $\vec{\sigma}$, let $c_i(\vec{\sigma})$ denote the number of cycles in $\vec{\sigma}$ of **type** i (assuming that we have classified the possible cycles into **types** so that cycles in a given type can receive the same color). The possibility of this type classification rests on the observation: **All terms in the state summation for $P_G(\delta)$ with a given set of cycle numbers $\{c_i\}$ have the same sign.**

Proof. The sign is the parity of the number of crossings that occur at the bars. The total number of crossings between different cycles is even. (Recall that for spin nets we do not count crossings away from bars.) Hence the total parity of the

456

bar-crossings equals the parity of the "spurious" intersections of different cycles away from the bars. The number of spurious intersections is

$$\sum_{e_i \cap e_j} e_i \cdot e_j - \sum_{\text{self-intersecting cycles}} c_k^2.$$

Hence the overall parity is determined by $\{c_k\}$.   //

Note that for each allowed coloring of the cycles, the number of terms in the sum is $\Pi e!$ (as defined above). Thus we have proved the

Theorem 12.1. The loop polynomial $P_G(\delta)$ of the strand net G is given by the formula (for $\delta = N$ a positive integer)

$$P_G(\delta) = \sigma \sum_{\{c_i\}} \epsilon K(\{c_i\}, \delta)$$

where

$$\sigma = (-1)^{\sum_{e_i \cap e_j} (e_i \cdot e_j)}$$

$$\epsilon = (-1)^{\sum_{\text{self intersections}} c_k \cdot c_k}$$

and $K(\{c_i\}, N)$ is equal to the number of allowed colorings of the configuration of $\{c_i\}$ cycles. The sum is over all non-negative cycle numbers $\{c_i\}$ such that for each edge $e_i = \sum c_j$ where the c_j catalog those cycles that pass through e_i.

Remark. Knowing $P_G(\delta)$ as a polynomial in δ via its evaluations at positive integers, allows the evaluation at $\delta = -2$. We get an integer for $P_G(-2)$ since any polynomial that is integer valued at positive integers is integer valued at negative integers as well. In fact we will use the following:

$$\binom{N}{K} = \frac{N(N-1)\ldots(N-K+1)}{K!}$$

$$\Rightarrow \binom{-N}{K} = \frac{(-N)(-N-1)\ldots(-N-K+1)}{K!} \in \mathbf{Z}.$$

Note that $(-N)! = (-1)^N (2N-1)!/(N-1)!$.

Example. Here is an application of the theorem. Let G be the network shown below.

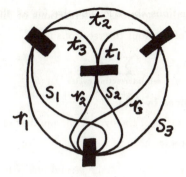

This net has 9 edges, 4 bars and 6 cycle types as shown below

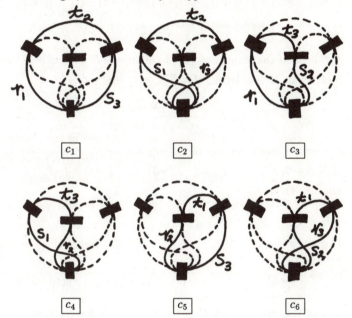

Thus, we have

$$t_1 = c_5 + c_6 \qquad s_1 = c_2 + c_4 \qquad r_1 = c_1 + c_3$$
$$t_2 = c_1 + c_2 \qquad s_2 = c_3 + c_6 \qquad r_2 = c_4 + c_5$$
$$t_3 = c_3 + c_4 \qquad s_3 = c_1 + c_5 \qquad r_3 = c_2 + c_6$$

and, letting $\pi = s_1 r_2 + s_1 r_3 + r_3 s_2 + c_2 + c_4 + c_6$, we have $(-1)^\pi$ is the sign of the corresponding state. The equations above let us solve for all the c_i in terms of $z = c_1$:

$$c_1 = z \qquad\qquad c_5 = s_3 - z$$
$$c_2 = t_2 - z \qquad\qquad c_6 = t_1 - s_3 + z$$
$$c_3 = r_1 - z$$
$$c_4 = t_3 - r_1 + z$$

Thus

$$\pi = s_1 r_2 + s_1 r_3 + r_3 s_2 + (t_2 - z) + (t_3 - r_1 + z) + (t_1 - s_3 + z)$$
$$\pi = s_1 r_2 + s_1 r_3 + r_3 s_2 + t_1 + t_2 + t_3 - s_3 - r_1 + z.$$

Here $K(\{c_i\}, N)$ equals the number of ways to distribute N objects in G bins with c_i objects in each bin. Hence, letting $J = \sum c_i = t_1 + t_2 + t_3$,

$$K(\{c_i\}, N) = \frac{N!/(N - J)!}{c_1! c_2! c_3! c_4! c_5! c_6!}.$$

The chromatic sum is over the range of z giving non-negative values for all c_i.

Thus, at $N = -2$, $N!/(N - J)! = (-2)(-3)\ldots(-J-1) = (-1)^J (J+1)!$. Let $q = \pi - z$ (above). Then

$$(\pi e!) P_G(-2) =$$
$$(-1)^q (J+1)! \sum_z \frac{(-1)^z}{z!(t_2 - z)!(r_1 - z)!(t_3 - r_1 + z)!(s_3 - z)!(t_1 - s_3 + z)!}$$

Exercise. Recall that the network G of this example is derived from the recoupling network for spins $\begin{bmatrix} j_1 & j_2 & j_3 \\ m_1 & m_2 & m_3 \end{bmatrix}$ as described in the first part of this section. Re-express the evaluation in terms of the j's and m's and compare your results with the calculations of Clebsch-Gordon coefficients in any text on angular momentum.

13^0. Q-Spin Networks and the Magic Weave.

The last section has been a quick trip through the formalism and interpretation of the Penrose spin networks. At the time they were invented, these networks seemed a curious place to **begin** a theory. Their structure was obtained by a process of **descent** from the apparatus of quantum mechanical angular momentum.

Yet the basis of the spin network theory is the extraordinarily simple binor identity

$$\times + \smile_\frown +)(\; = 0,$$

and we know this to be a special case of the bracket polynomial identity

$$\asymp \; = A \smile_\frown + A^{-1})($$

(brackets removed for convenience). In the theory of the bracket polynomial the loop value is $\delta = -A^2 - A^{-2}$, hence binors occur for $A = -1$. We have already explained in Part I (section 9^0) how the bracket is related to the quantum group $SL(2)q$ for $\sqrt{q} = A$.

Thus, it should be clear to the reader that there is a possibility for generalizing the Penrose spin nets to q-spin nets based upon the bracket identity. In this section I shall take the first steps in this direction, and outline what I believe will be the main line of this development. Even in their early stages, the q-spin networks are very interesting. For example, we shall see that the natural generalization of the anti-symmetrizer leads to a "magic weave" that realizes the special projection operators in the Temperley-Lieb algebra (see section 16^0 of Part I). We then use the q-spin nets to construct Turaev-Viro invariants of 3-manifolds. ([TU2], [LK19], [LK20].)

Before beginning the technicalities of q-spin nets, I want to share foundational thoughts about the design of such a theory and its relationship with quantum mechanics and relativity. This occurs under the heading – **network design**. The reader interested in going directly into the nets can skip across to the sub-heading **q-spin nets**.

Network Design – Distinction as Pregeometry.

What does it mean to produce a combinatorial model of spacetime? One thinks of networks, graphs, formal relationships, a fugue of interlinked constructions. Yet what will be fundamental to such an enterprise is a point of view, or a place to stand, from which the time space unfolds. Before constructing, before speaking and even before logic stands a single concept, and that is the concept of **distinction.** I take this concept first in ordinary language where it has a multitude of uses and a curious circularity. For there can be no seeing without a seer, no speaking without a speaker. We, who would discuss this concept of distinction, become distinguished through the conversation of distinction that is us. Distinction is a concept that requires its own understanding. Fortunately, there are examples: Draw a distinction; draw a circle in the plane; delineate the boundary of this room; indicate the elements of the set of prime integers. Speech brings forth an architecture of distinctions and a framework for clarity and for action.

The specific models we use for space and spacetime are extraordinary towers of distinctions. Think of creating the positive integers from the null set (Yea, think of creating the null set from the void!), the rationals from the integers, the reals from limits of sequences of rationals, the coordinate spaces, the metrics and all the semantics of interpretation - just to arrive at Minkowski spacetime and the Lorentz group! It is no wonder that there is a desire for other ways to tell this story. (Compare [F12], [KVA], [LK17], [SB2].)

There are other ways. At the risk of creating a very long digression, I want to indicate how **the Lorentz group arises naturally in relation to any distinction.** We implicitly use the Lorentz group in the ordinary world of speech, thought and action. Recall from section 9^0 of Part II that in light cone coordinates for the Minkowski plane the Lorentz transformation has the form $\mathcal{L}[A, B] = [KA, K^{-1}B]$ where K is a non-zero constant derived from the relative velocity of the two inertial frames. In fact, it is useful to remember that for time-space coordinates (t, x), the corresponding light-cone coordinates are $[t + x, t - x]$ (for light speed $= 1$). Thus $(t - x)$ can be interpreted as the time of emission of a light signal and $(t + x)$ as the time of reception of that signal after it has reflected from an event at the point (t, x).

Writing $[t+x, t-x]$ as $[t+x, t-x] = t \cdot 1 + x\sigma$ where $1 = [1,1]$ and $\sigma = [1,-1]$ we recover standard coordinates and the Pauli algebra. That is, it is natural to define

$$[X,Y][Z,W] = [XZ,YW]$$
$$[X,Y][Z,W] = [X+Z, Y+W]$$

in the light-cone coordinates. Then $\mathcal{L}[A,B] = [K, K^{-1}][A,B]$, and the Lorentz transformation is a product in this algebra. In this algebra structure $\sigma^2 = [1,-1][1,-1] = [1,1] = 1$, and in order to obtain a uniform algebraic description over arbitrary space directions, we conclude that $\sigma^2 = 1$ for any unit direction. This entails a Clifford algebra structure on the space of directions [LK17]. For example, if $\sigma_x, \sigma_y, \sigma_z$ denote unit perpendicular directions in three space with $\sigma_x\sigma_y = -\sigma_y\sigma_x$, $\sigma_x\sigma_z = -\sigma_z\sigma_x$, $\sigma_y\sigma_z = -\sigma_z\sigma_y$ and $\sigma_x^2 = \sigma_y^2 = \sigma_x^2 = 1$, then $(a\sigma_x + b\sigma_y + c\sigma_z)^2 = a^2 + b^2 + c^2$ for scalars a, b, c commuting with $\sigma_x, \sigma_y, \sigma_z$. Hence, any unit direction has square 1.

By this line of thought we are led to the minimal Pauli representation of a spacetime point as a Hermitian 2×2 matrix H:

$$H = \begin{bmatrix} T+X & Y+\sqrt{-1}Z \\ Y-\sqrt{-1}Z & T-X \end{bmatrix} = T + X\sigma_x + Y\sigma_y + Z\sigma_z$$

and $\|H\|^2 = \text{Det}(H) = T^2 - X^2 - Y^2 - Z^2$.

All this comes from nothing but light cone coordinates and the principle of relativity. I will descend **underneath** this structure and look at the distinctions out of which it is built. (See [LK30].)

Therefore, begin again. Let a distinction be given

with sides labelled A and B. Call this distinction $[A, B]$. Consider A and B as **evaluations** of the two sides of the distinction. (In our spacetime scenario the distinction has the form [**after, before**], and numerical times are the evaluations.) Now suppose another observer gives evaluations $[A', B']$ with $A' = \rho A$, $B' = \lambda B$ for non-zero ρ and λ. Then

$$[\rho, \lambda] = \sqrt{\rho\lambda}[K, K^{-1}]$$

where $K = \sqrt{\rho/\lambda}$.

$$\boxed{[A', B'] = \sqrt{\rho\lambda}[KA, K^{-1}B]} \; .$$

Thus, **up to a scale factor the numerical transformation of any distinction is a Lorentz transformation.** In this sense the Lorentz group operates in all contexts of evaluation. The Lorentz group arises at once along with the concept of (valued) distinction.

We should look more closely to see if the Pauli algebra itself is right in front of us in the first distinction. Therefore, consider once more a distinction with sides labelled A and B: I will suppose only the most primitive of evaluations - namely $+1$ and -1. Thus the space of evaluations consists in the primitive distinction $-1/+1$ or unmarked/marked. By choosing a mark we can represent -1 as an un-marked side and +1 as a **marked** side.

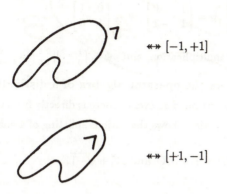

$$\leftrightarrow\; [-1, +1]$$

$$\leftrightarrow\; [+1, -1]$$

In any case there are now two basic operations:

(1) **Change** the status of a side.

(2) **Interchange** the relative status of two sides.

I represent **change** by two operations $p[A, B] = [+A, -B]$ (change inside) and $q[A, B] = [-A, +B]$ (change outside). The interchange is given by the formula $\Phi[A, B] = [B, A]$. We have $pq[A, B] = [-A, -B] = -[A, B]$. Hence $pq = -1$, or $q = -p$. Thus -1 denotes the operation changing the status of two sides. For convenience, I let $\bar{p}$ denote $-p$. Of course 1 denotes the identity operation. We then have $1, p, \Phi$ and $p\Phi$ with $p\Phi[A, B] = [+B, -A]$. Hence

$$(p\Phi)^2 = -1.$$

The distinction (observed) gives rise of its own accord to a square root of negative one, and three elements of square one $(1, p, \Phi)$. By introducing an extra commuting $\sqrt{-1}$ (If it exists, why not replicate it?) we get $\sigma_x = p$, $\sigma_y = \Phi$, $\sigma_z = \sqrt{-1}p\Phi$ and the event $T + Xp + Y\Phi + Z\sqrt{-1}p\Phi$ corresponds directly to the Hermitian matrix

$$H = \begin{bmatrix} T+X & Y+\sqrt{-1}Z \\ Y-\sqrt{-1}Z & T-X \end{bmatrix}$$

$$= T\begin{bmatrix} 1 & 0 \\ 0 & 1 \end{bmatrix} + X\begin{bmatrix} 1 & 0 \\ 0 & -1 \end{bmatrix} + Y\begin{bmatrix} 0 & 1 \\ 1 & 0 \end{bmatrix} + \sqrt{-1}Z\begin{bmatrix} 0 & 1 \\ -1 & 0 \end{bmatrix}.$$

In fact, these 2×2 matrices

$$1 = \begin{bmatrix} 1 & 0 \\ 0 & 1 \end{bmatrix}, p = \begin{bmatrix} 1 & 0 \\ 1 & -1 \end{bmatrix}, \Phi = \begin{bmatrix} 0 & 1 \\ 1 & 0 \end{bmatrix}$$

represent our operations by left multiplication, and $p\Phi = \begin{bmatrix} 1 & 0 \\ 0 & -1 \end{bmatrix}\begin{bmatrix} 0 & 1 \\ 1 & 0 \end{bmatrix} = \begin{bmatrix} 0 & 1 \\ -1 & 0 \end{bmatrix}$.

Spacetime arises directly from the operator algebra of a distinction.

The basic Hermitian representation of an event proceeds directly from the idea of a distinction. Also, this point of view shows the curious splitting of 4-space into

$$[T + X, T - X] = T + \sigma X \quad \text{and} \quad Y + \sqrt{-1}Z.$$

In the form

$$T + Xp + (Y + Z\sqrt{-1}p)\Phi$$

we see that **operationally** the split is between the actions that consider **one side** or **both sides** (**change** and **interchange**) of the distinction.

The important point about this line of reasoning is that, technicalities aside, spacetime is an elementary concept, proceeding outward from the very roots of our being in the world. Resting in this, it will be interesting to see how spacetime arises in the q-spin nets. Since we are building it in from the beginning, it is already there.

In particular, the formula

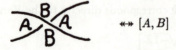

now appears in a new light as a knot theoretic way to write an ordered pair $[A, A^{-1}]$. (And $[A, A^{-1}]$ is a possible Lorentz transformation in light cone coordinates.) Here the crossing indicates the distinction $[A, B]$:

$$\underset{\text{B}}{\overset{\text{B}}{A \times A}} \quad \leftrightarrow \quad [A, B]$$

It is not the case at this level that the splices $\overset{\smile}{\wedge}$ and $)($ have algebra structure related to the Pauli algebra. However, this **does** occur in the FKT [LK2] model

of the Conway-Alexander polynomial (see also [LK3], [LK9]). In that model, we
have

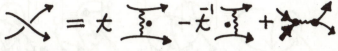

where the dots indicate the presence of state markers. Here a state of a link shadow
consists in a choice of pointers from regions to vertices, with two adjacent regions
selected without pointers (the **stars**):

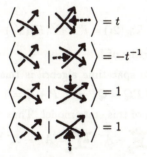

In a state, each vertex receives exactly one pointer. The vertex weights are

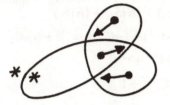

(For a negative crossing t and t^{-1} are interchanged.)

It is easy to prove (by direct comparison of state models) that

$$\nabla_K(2i) = i^{w(K)/2}\langle K\rangle(\sqrt{i})$$

where $\nabla_K = \sum_S \langle K|S\rangle$ is the Conway-Alexander polynomial using this state model,
$w(K)$ is the writhe of K, and $\langle K\rangle$ is the bracket polynomial (Part I, section 3^0).

Example.

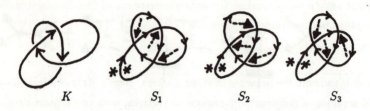

$$\nabla_K = \langle K|S_1\rangle + \langle K|S_2\rangle + \langle K|S_3\rangle$$
$$= (t)(-t^{-1}) + (-t^{-1})(-t^{-1}) + (t)(t)$$
$$= -1 + t^{-2} + t^2$$
$$\nabla_K = (t - t^{-1})^2 + 1$$

$$\nabla_K = z^2 + 1$$
$$\nabla_K(2i) = -3.$$

Exercise. Prove that

$$\nabla_K(2i) = i^{w(K)/2}\langle K\rangle(\sqrt{i})$$

where $w(K)$ denotes the writhe of K.

The point in regard to spacetime algebra is this: Let $P = $ ⧖ , $Q = $ ⧗ . Then $P^2 = P$, $Q^2 = Q$ and $PQ = QP = 0$, $P + Q = 1$ by the exclusion principles embedded in the formalism of this state model. Thus

$$A \text{⧖} - A^{-1}\text{⧗} = AP - A^{-1}Q \equiv [\mathsf{A}, \mathsf{A}^{-1}]$$

coincides with our spacetime algebra if we take P and Q to be the light cone generators

$$P = \frac{1+\sigma}{2}, \quad Q = \frac{1-\sigma}{2}$$

$(\sigma^2 = 1)$.

In summary, we have seen that elementary concepts of spacetime occur directly represented in the formalisms of the bracket (Jones) and Conway-Alexander polynomials. It will be useful to keep this phenomenon in mind as we construct the q-spin networks.

q-Spin Networks.

The first task is to generalize the antisymmetrizers of the standard theory. These are projection operators in the sense that

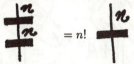

$$= n! \quad$$

and they kill off the non-identity generators of the Temperley-Lieb algebra $e_1, \ldots, e_{n-1}$:

$$\underbrace{\bigcup_{\bigcap} \| \cdots \|}_{e_1}, \ldots, \underbrace{\| \cdots \| \bigcup_{\bigcap}}_{e_{n-1}}, \text{ since } \qquad = 0 \text{ by antisymmetry.}$$

These remarks mean that the expanded forms (using the binor identity) of the antisymmetrizers are special projection operators in the Temperley-Lieb algebra. For example

$$\qquad = \| \| - X = 2 \| \| + \bigvee_{\bigcap}$$

Similarly,

$$\qquad = 6 \| \| \| + 4 \bigvee_{\bigcap} \| + 4 \| \bigvee_{\bigcap} + 2 \bigvee_{\bigcap} \bigvee + 2 \bigvee_{\bigcap}$$

We shall now see that a suitable generalization produces these operators in the full Temperley-Lieb algebra.

We use the bracket identity $X = A \bigvee_{\bigcap} + A^{-1}\,)($ in place of the binor identity and use loop value $\delta = -A^2 - A^{-2}$. A q-**spin network** is nothing more than a link diagram with special nodes that are designated as anti-symmetrizers (to be defined below). Thus any q-spin network computes its own bracket polynomial.

Define the q-**symmetrizer** by the formula

$$\boxed{}^{n} = \sum_{\sigma \in S_n} (A^{-3})^{T(\sigma)} \boxed{\hat{\sigma}}$$

where $T(\sigma)$ is the minimal number of transpositions needed to return σ to the identity, and $\hat{\sigma}$ is a minimal braid representing σ with all negative crossings, that is with all crossings of the form shown below with respect to the braid direction

Example 1.

$$\text{╫} = \| + \bar{A}^3 \text{╳} = \| + \bar{A}^3 \left[A \underset{\cap}{\cup} + \bar{A}^1 \right) \text{)(}$$

Hence $\text{╫} = (1 + A^{-4})[\ \| \ -\delta^{-1} \underset{\cap}{\cup} \ ]$. Here $\delta = -A^2 - A^{-2}$. We recognize $1 - \delta^{-1}e_1$ as the element f_1 of section 16^0, Part I. It is the first Temperley-Lieb projector.

Define the **quantum factorial** $[n]!$ by the formula

$$\boxed{[n]! = \sum_{\sigma \in S_n} (A^{-4})^{T(\sigma)}}.$$

Exercise.

$$[n]! = \sum_{\sigma \in S_n} (A^{-4})^{T(\sigma)}$$

$$\Rightarrow [n]! = \prod_{k=1}^{n} \left(\frac{1 - A^{-4k}}{1 - A^{-4}} \right).$$

In the example, note that $[2]! = 1 + A^{-4}$. Thus

$$\text{╫} = [2]! f_1.$$

Recall from section 16^0, Part I that the Temperley-Lieb projectors are defined inductively via the equations

$$\left\{ \begin{array}{l} f_0 = 1 \\ f_{n+1} = f_n - \mu_{n+1} f_n e_{n+1} f_n \\ \\ \mu_1 = 1/\delta \\ \mu_{n+1} = 1/(\delta - \mu_n) \end{array} \right\}.$$

Thus

$$f_1 = 1 - \delta^{-1} e_1$$

$$f_2 = 1 - \mu_2(e_1 + e_2) + \mu_1\mu_2(e_1 e_2 + e_2 e_1)$$

$$\mu_1 = 1/\delta$$

$$\mu_2 = \delta/(\delta^2 - 1).$$

Theorem 13.1. The Temperley-Lieb elements f_n with $\delta = -A^2 - A^{-2}$ are equivalent to the q-symmetrizers. In particular, we have the formula

$$f_{(n-1)} = \frac{1}{[n]!} \quad \text{}$$

Proof. f_n is characterized by $f_n^2 = f_n$, $e_i f_n = f_n e_i = 0 \quad i = 1, \dots, n-1$. That the q-symmetrizer annihilates $e_1, \dots, e_{n-1}$ follows as illustrated in the example below for $n = 2$:

$= \cup + A^{-3}$ $= \cup + (A^{-3})(-A^3) \cup = 0.$

This completes the proof. //

I have emphasized the construction of these generalized antisymmetrizers because they clarify knot theoretically the central ingredient for Lickorish's construction of three manifold invariants from the bracket - as described in section 16^0, Part I. Thus these 3-manifold invariants are, in fact, q-spin network evaluations!

Example.

Hence

Exercise. Verify this formula by direct bracket expansion.

Now we can define the 3-vertex for q-spin nets just as before:

470

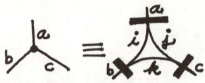

$$i + j = a$$
$$j + k = c$$
$$i + k = b$$

with q-symmetrizers in place of antisymmetrizers.

The recoupling theory goes directly over, and this context can be used to define a theory of q-angular momentum and to create invariants of three manifolds. (See [LK19], [LK20]) for our work.

Exercise.

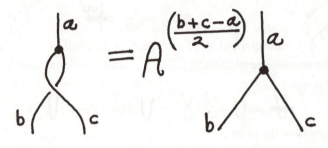

Exercise.

$$b \bigcirc c = \phi \quad \text{if} \quad a \neq a'$$

$$b \bigcirc c = \left[\frac{\boxed{a \, b \, c}}{\boxed{a}} \right] \,\vdash$$

Exercise. Given that

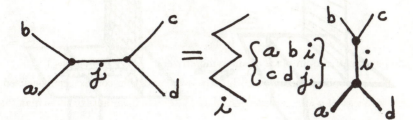

Show that (Elliot-Biedenharn Identity)

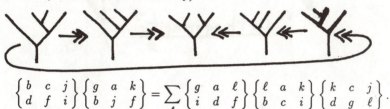

$$\begin{Bmatrix} b & c & j \\ d & f & i \end{Bmatrix} \begin{Bmatrix} g & a & k \\ b & j & f \end{Bmatrix} = \sum_\ell \begin{Bmatrix} g & a & \ell \\ i & d & f \end{Bmatrix} \begin{Bmatrix} \ell & a & k \\ b & c & i \end{Bmatrix} \begin{Bmatrix} k & c & j \\ d & g & \ell \end{Bmatrix}.$$

Exercise.

$$\sum_i \begin{Bmatrix} a & b & i \\ c & d & j \end{Bmatrix} \begin{Bmatrix} d & a & \kappa \\ b & c & i \end{Bmatrix} = \delta_{j\kappa}$$

(Orthogonality).

Exercise. Read [TU2].

In this paper Turaev and Viro construct invariants of 3-manifolds via a state summation involving the $q - 6j$ symbols. In [LK20] we (joint work with Sostenes Lins) show how to transfer their idea to an invariant based on the re-coupling theory of q-spin networks. The underlying toplogy for both approaches rests in the **Matveev moves** [MAT] shown below.

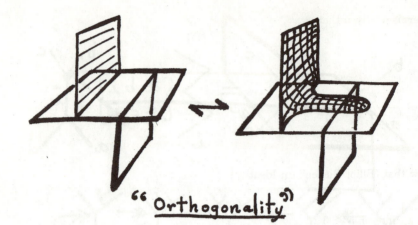

"_Orthogonality_"

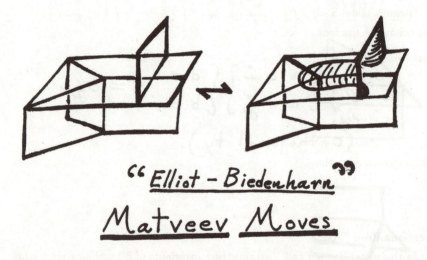

"_Elliot - Biedenharn_"
Matveev Moves

These are moves on special spines for three-manifolds that generate piecewise linear homeomorphisms of three-manifolds. In such a spine, a typical vertex appears as shown below with four adjacent one-cells, and six adjacent two-cells. Each one-cell abuts to three two-cells.

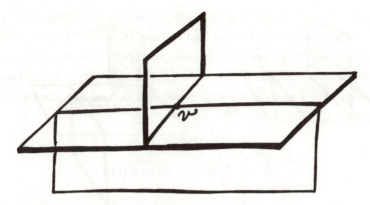

For an integer $r \geq 3$ the **color set** is $C(r) = \{0, 1, 2, \ldots, r-2\}$. A **state at level** r of the three-manifold M is an assignment of colors from $C(r)$ to each of the two-dimensional faces of the spine of M.

Given a state S of M, assign to each vertex the tetrahedral symbol whose edge colors are the face colors at that vertex as shown below:

Assign to each edge the $q - 3j$ symbol associated with its triple of colors

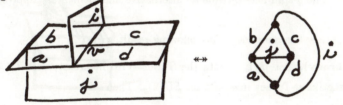

and assign to each face the Chebyshev polynomial $\Delta_i = (x^{i+1} - x^{-(i+1)})/(x - x^{-1})$ for $x = -A^2$.

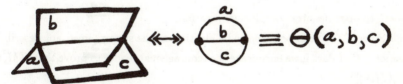

whose index is the color of that face.

Then, for $A = e^{i\pi/2r}$, let $I(M,r)$ denote the sum over all states for level r, of the product of vertex evaluations and the face evaluations divided by the edge evaluations **for those evaluations that are admissible and non-zero.** Thus, if any edge has zero evaluation in a state S, then this state is dropped from the summation. For a given r, the edge evaluations are non-zero exactly when $a + b + c \leq 2r - 4$.

$$I(M,r) = \sum_S \frac{\prod_v \text{TET}(v, S) \prod_f \Delta_{S(f)}}{\prod_e \theta(S_a(e), S_b(e), S_c(e))}$$

Here $\text{TET}(v, s)$ denotes the tetrahedral evaluation associated with a vertex v, for the state S. $S(f)$ is the color assigned to a face f and $S_a(e)$, $S_b(e)$, $S_c(e)$ are the triplet of colors associated with an edge in the spine.

This is our [LK20] version of the Turaev-Viro invariant. It follows via the orthogonality and Elliot-Biedenharn identities that $I(M,r)$ is invariant under the Matveev moves. The q-spin nets provide an alternative and elementary approach to this subject.

In [TU5] Turaev has announced the following result.

Theorem (Turaev). Let $|M|_q$ denote the Turaev-Viro invariant, and $Z_q(M)$ denote the Reshetikhin-Turaev invariant for $SU(2)_q$. Then

$$|M|_q = Z_q(M)\overline{Z_q(M)}$$

where the bar denotes complex conjugation.

Since $Z_q(M)$ can be taken to be identical to the surgery invariant described in section 16^0 of Part I, we encourage the (courageous) reader to prove that $I(M,r) = |Z_q(M,r)|^2$.

Exercise. Read [HAS]. These authors suggest a relationship between quantum gravity and the semi-classical limit of $6j$ symbols - hence the title "Spin networks are simplicial quantum gravity". Investigate this paper in relation to q-spin networks. Compare with the ideas in [MOU] and [CR3].

14^0. Knots and Strings – Knotted Strings.

It is natural to speculate about knotted strings. At this writing, such thoughts can only be speculation. Nevertheless, there are some pretty combinations of ideas in this domain, and they reflect on both knot theory and physics. (See [ROB], [ZA], [Z2].)

To begin with, the string as geometric entity is a surface embedded in space-time. Suppose, for the purpose of discussion, that this spacetime is of dimension four. (I shall speculate a little later about associating knots with super strings in 10-space.) A surface embedded in dimension four can itself be knotted – for example, there exist knotted two dimensional spheres in Euclidean four space. However, the most interesting matter for topology is the consideration of a string interaction vertex embedded in spacetime. Abstractly the vertex is a sphere with four punctures, and is pictured as shown below

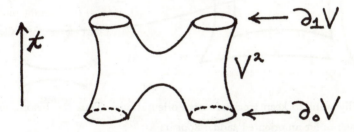

Thus we view the four-vertex as a process where two one-dimensional strings interact, tracing out a surface of genus zero with two punctures at $t = 0$ and two punctures at $t = 1$. Let V^2 denote this surface. Now, letting spacetime be $\mathbf{E}^4 = \mathbf{R}^3 \times \mathbf{R} = \{(x, t)\}$, we consider embedding $F : V^2 \to \mathbf{E}^4$ such that $\partial_0 V \hookrightarrow \mathbf{R}^3 \times 0$ and $\partial_1 V \hookrightarrow \mathbf{R}^3 \times 1$. Here $\partial_0 V$ and $\partial_1 V$ are indicated in the diagram above, each of these partial boundaries consists in two circles. The embedding is assumed to be smooth, and we can take this to mean that

(i) All but finitely many levels $F(V^2) \cap \{(x, t) \mid t \text{ fixed}\} = F_t(V)$ are embeddings of a collection of circles (hence a link) in $\mathbf{R}_t^3 = \{(x, t) \mid x \in \mathbf{R}^3\}$.

(ii) The critical levels involve singularities that are either **births** (minima), **deaths** (maxima) or **saddle points**.

476

In a birth at level t_0 there is a sudden appearance of a point at level t_0. The point becomes an unknotted circle in the levels immediately above t_0. At a maximum or death point a circle collapses to a point and disappears from higher levels.

At a saddle point, two curves touch and rejoin as illustrated below.

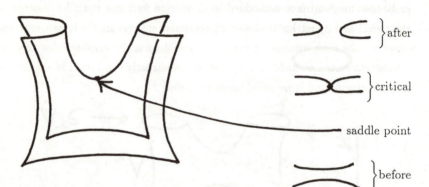

We usually want to keep track of curve orientations. Thus the before, critical, after sequence for an oriented saddle appears as:

before   critical   after

The requirement on the production of a standard vertex is that we

(i) start with two knots

(ii) allow a sequence of births, deaths, and saddles that ends in two knots.

(iii) the genus of the surface so produced is zero.

(As long as the surface is connected, the last requirement is equivalent to the condition that $s = b + d + 2$ where s denotes the number of saddles, b denotes the number of births and d denotes the number of deaths.)

For example, for two unknotted strings the vertex look like

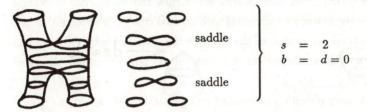

$$s = 2$$
$$b = d = 0$$

Now the remarkable thing is that two knotted strings can interact at such an embedded four vertex and produce a pair of entirely different knots. For example, the trefoil and its mirror image annihilate each other, producing two unknots.

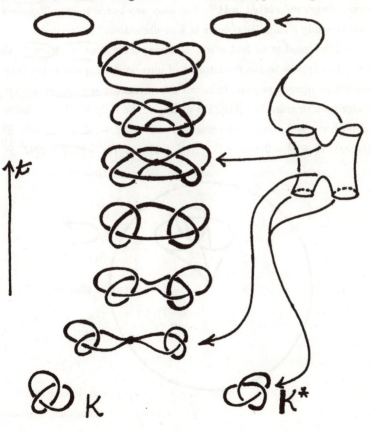

In general a knot and its mirror image annihilate in this way. Some knots, called slice knots (see [FOX1]) can interact with an unknot and disappear. In general there is a big phenomena of interactions of such knotted strings. The corresponding topological questions are central to the study of the interface between three and four dimensions. This mathematical structure will be inherent in any theory of knotted strings.

Now the primary objection to knotted strings is that the superstring is necessarily living in a space of 10 dimensions, and there is no known theory for associating an embedding of the string in spacetime that is detailed enough to account for knotting. Furthermore, one dimensional curves and two dimensional surfaces are always unknotted in $\mathbf{R}^{10}$. The easy way out is to simply declare that it is useful to study string like objects in four-dimensional spacetime.

The difficult road is to find a way to incorporate knottedness with the superstring. Moving now to the department of unsupported speculation, here is an idea. Recall that many knots and links can be represented as neighborhood boundaries of algebraic singularities [MI]. For example, the right-handed trefoil knot is the neighborhood boundary of the singular variety $Z_1^2 + Z_2^3 = 0$ in $\mathbf{C}^2$. That is $K = \{(z_1, z_2) \mid z_1^2 + z_2^3 = 0 \text{ and } |z_1|^2 + |z_2|^2 = 1\} = \mathcal{L}(K) = V(z_1^2 + z_2^3) \cap S^3$.

In this algebraic context, we can "suspend" K to a manifold of dimension 7 sitting inside the unit sphere in $\mathbf{R}^{10}$.

$$\Sigma^7(K) = \mathcal{V}(z_1^2 + z_2^3 + (z_3^2 + z_4^2 + z_5^2)) \cap S^9$$

$$\Sigma^7(K) \subset S^9 \subset \mathbf{R}^{10}.$$

It is tantalizing to regard $\Sigma^7(K) \subset \mathbf{R}^{10}$ as the "string". One would have to rewrite the theory to take into account the vibrational modes of this 7-manifold associated with the knot. The manifold $\Sigma^7(K)$ has a natural action of the orthogonal group $0(3)$ (acting on (z_3, z_4, z_5)) and $\Sigma^7(K)/0(3) \cong D^4$ with $K \subset S^3 \subset D^4$ representing the inclusions of the orbit types. Thus we can recover spacetime (as D^4) and the knot from $\Sigma^7(K)$. This construction is actually quite general (see [LK]) and any knot $K \subset S^3$ has an associated $0(3)$-manifold $\Sigma^7(K) \subset \mathbf{R}^{10}$. Thus this link-manifold [LK] construction could be a way to interface knot theory and the theory of superstrings.

The manifolds $\Sigma^7(K)$ are quite interesting. For example $\Sigma^7(3,5)$ is the Milnor exotic 7-sphere [MI]. Here $(3,5)$ denotes a torus knot of type $(3,5)$ - that is a knot winding three times around the meridian of a torus while it winds five times around the longitude. Thus we are asking about the vibrational modes of exotic differentiable spheres in $\mathbf{R}^{10}$.

This mention of exoticity brings to mind another relationship with super-string theory. In [WIT4] Witten has investigated global anomalies in string theory that are associated with exotic differentiable structures on 10 and 11 dimensional manifolds. In [LK22] we investigate the relationship of these anomalies with specific constructions for exotic manifolds. These constructions include the constructions of link manifolds and algebraic varieties just mentioned, but in their 11-dimensional incarnations.

For dimension 10 the situation is different since exotic 10-spheres do not bound parallelizable manifolds. In this regard it is worth pointing out that such **very exotic n-spheres** are classified by $\pi_{n+k}(S^k)/\mathrm{Im}(J)$ where $\pi_{n+k}(S^k)$ denotes a (stable) homotopy group of the sphere S^k and $\mathrm{Im}(J)$ denotes the image of the J-homomorphism

$$J : \pi_n(SO(k)) \to \pi_{n+k}(S^k)$$

The very exotic spheres live in the darkest realms of homotopy theory. A full discussion of this story appears in [KV]. The J-homomorphism brings us back the Dirac string trick, for the simplest case of it is

$$J : \pi_1(SO(3)) \to \pi_4(S^3).$$

The construction itself is exactly parallel to our belt twisting geometry of section 11^0, and here it constructs the generator of $\pi_4(S^3) \cong \mathbf{Z}_2$. To see this, we must define J. Let $g : S^n \to SO(k)$ represent an element of $\pi_n(SO(k))$. Then we get $g^* : S^n \times S^{k-1} \to S^{k-1}$ via $g^*(a,b) = g(a)b$. Associated with any map $H : X \times Y \to Z$ there is a map $\widehat{H} : X * Y \to S(Z)$ where $X * Y$ denotes the space of all lines joining points of X to points of Y, and $S(Z)$ denotes the suspension of Z obtained by joining Z to two points. In general, $S^n * S^{k-1} \cong S^{n+k}$ and $S(S^{k-1}) \cong S^k$. Thus g^* induces a map $\widehat{g^*} : S^{n+k} \to S^k$ representing an element in $\pi_{n+k}(S^k)$. This association of g with $\widehat{g^*}$ is the J-homomorphism.

In the case of $\pi_1(SO(3))$ we have $g : S^1 \to SO(3)$ representing the loop of rotations around the north pole of S^2. The map

$$g^* : S^1 \times S^2 \to S^2$$

induces $\widehat{g^*} : S^1 * S^2 \to S(S^2)$ hence $\widehat{g^*} : S^4 \to S^3$. This element generates $\pi_4(S^3)$. In the belt trick, we used the same idea to a different end. $g : S^1 \to SO(3)$ gave rise to $g : I \to SO(3)$ where I denotes the unit interval and $g(0) = g(1)$. Then $g^* : I \times S^2 \to I \times S^2$ via $g^*(t,p) = (g(t)p, t)$ and we used g^* to twist a belt stretched between $S^2 \times 0$ and $S^2 \times 1$. The fact that the $720°$ twisted belt can be undone in $I \times S^2$ without moving the ends is actually a visualization of $\pi_4(S^3) \cong \mathbf{Z}_2$.

Elements of $\pi_{n+k}(S^k)$ in the image of the J-homomorphism can correspond to exotic spheres that bound parallelizable manifolds (such as the Milnor spheres in dimensions 7 and 11). Thus there is a train of relationships among the concepts of knotted strings, high dimensional anomalies, the belt trick and the intricacies of the homotopy groups of spheres.

Exercise 14.1. Show that any knot K can interact with its mirror image K^* through an embedded genus zero string vertex in $3+1$ spacetime to produce two unknots.

Exercise 14.2. The **projective plane**, $\mathbf{P}^2$, is by definition the quotient space of the two-dimensional sphere S^2 $(S^2 = \{(x, y, z) \in \mathbf{R}^3 \mid x^2 + y^2 + z^2 = 1\})$ by the antipodal map $T : S^2 \to S^2$, $T(x, y, z) = (-x, -y, -z)$. Topologically this is equivalent to forming $\mathbf{P}^2$ from a disk D^2 by identifying antipodal points on the boundary of the disk. Examine the decomposition of $\mathbf{P}^2$ indicated by the figure below:

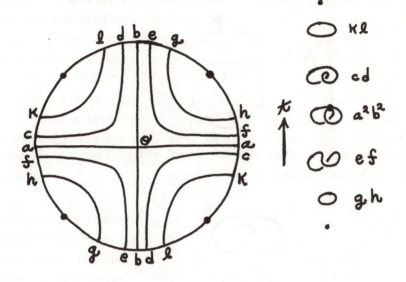

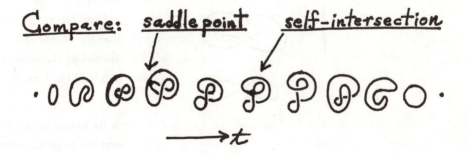

In this figure, points that are identified with one another on the disk boundary are indicated by the same letter. Therefore we see that the two arcs with endpoints c, d form a circle in $\mathbf{P}^2$, as do the arcs with endpoints e, f. Similarly, the arcs labelled a, a and b, b each form circles. Thus we see that except for the points p and q the figure indicates a decomposition of $\mathbf{P}^2$ into circle all disjoint from one another except for the a and b circles at θ. We can therefore imagine $\mathbf{P}^2$ as described by a birth of the point p; p grows to become a circle c_1, c_1 undergoes a saddlepoint singularity to become yet another circle c_2; c_2 dies at q.

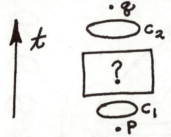

But how does a circle go through a saddle point with itself to produce a single circle? The answer lies in a twist:

With this idea, we can produce an embedding of $\mathbf{P}^2$ in $\mathbf{R}^4$ as follows.

In this figure we mean each level to indicate the intersection $\mathbf{P}^2 \cap (\mathbf{R}^3 \times t)$ for $0 \leq t \leq 1$. At $t = 0$ there is a birth (minimum). As t goes from $1/4$ to $3/8$ the curve undergoes an ambient isotopy the curled form at $t = 3/8$. Note that the trace of this ambient isotopy in $\mathbf{R}^4$ is free from singularities. The saddle point occurs at $t = 1/2$.

Corresponding to this embedding of $\mathbf{P}^2$ in $\mathbf{R}^4$ there is an **immersion** of $\mathbf{P}^2$ in $\mathbf{R}^3$. An immersion of a surface into $\mathbf{R}^3$ is a mapping that may have self-intersections of its image, but these are all locally transverse surface intersections modeled on the form of intersection of the $x - y$ plane and the $x - z$ plane in $\mathbf{R}^3$. We can exhibit an immersion by giving a series of levels, each showing an immersion of circles in a plane (possibly with singularities). The entire stack of levels then describes a surface in $\mathbf{R}^3$. Instead of using ambient isotopy as we move from plane to plane, we use **regular homotopy**. Regular homotopy is generated by the projections of the Reidemeister II and III moves. Thus

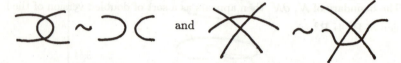 and

are the basic regular homotopies. The Whitney trick is a fundamental example of a regular homotopy:

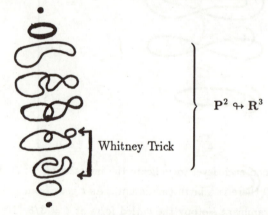

With the help of the Whitney trick we can obtain an immersion of $\mathbf{P}^2$ in $\mathbf{R}^3$:

$\mathbf{P}^2 \looparrowright \mathbf{R}^3$

Whitney Trick

Problem. You might think that the immersion above could be covered by an embedding of $\mathbf{P}^2$ into $\mathbf{R}^4$ so that each level is obtained from its ancestors by either minima, maxima, saddle points or **regular isotopy**. Prove that this is not so.

The immersion shown above for $\mathbf{P}^2 \looparrowright \mathbf{R}^3$ has a normal bundle. This is the result of taking a neighborhood of $\mathbf{P}^2 \looparrowright \mathbf{R}^3$ consisting of small normals to the surface in $\mathbf{R}^3$. We can construct this normal bundle $\mathcal{N}$ by replacing each level

curve by a thickened version as in

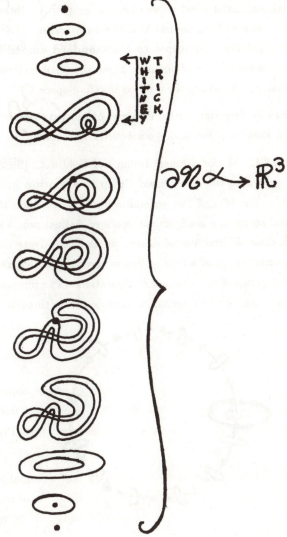

The boundary of $\mathcal{N}$, $\partial\mathcal{N}$, then appears as a sort of doubled version of the immersion of $\mathbf{P}^2 \looparrowright \mathbf{R}^3$.

WHITNEY TRICK

$\partial\mathcal{N} \looparrowright \mathbf{R}^3$

Problem. Show that the boundary of $\mathcal{N}$, $\partial\mathcal{N}$, is homeomorphic to a two sphere S^2. Thus the induced immersion $\partial\mathcal{N} \looparrowright \mathbf{R}^3$ is an immersion of S^2 in $\mathbf{R}^3$. Since the normal bundle has a symmetry (positive to negative normal directions) it is possible to **evert** this immersion $\alpha : S^2 \looparrowright \mathbf{R}^3$. In other words, you can easily turn this sphere inside out by exchanging positive and negative normals. It is not so obvious that the standard embedding of $S^2 \subset \mathbf{R}^3$ can be turned inside out through immersions and regular homotopy. ([FRA]) Nevertheless this is so. The eversion follows from the fact that **the immersion** $\alpha : S^2 \looparrowright \mathbf{R}^3$ **described above is regularly homotopic to the standard embedding of S^2.** Prove this last statement by a set of pictures. Think about sphere eversion in the context of knotted strings by lifting the eversion into four-space.

Exercise 14.3. Investigate other examples of knotted string interactions, by finding pairs of knots that can undergo a 4-vertex interaction.

Exercise 14.4. (Motion Groups) Dahm and Goldsmith (See [G] and references therein.) generalize the Artin Braid Group to a **motion group** for knots and links in $\mathbf{R}^3$. The idea of this generalization is based upon the braid group as fundamental group of a configuration space of distinct points in the plane. Just as we can think of braiding of points, we can investigate closed paths in the space of configurations of a knot or link in three space. The motion group is the fundamental group of this space of configurations. The simplest example of such a motion is obtained when one unlinked circle passes through another, as shown below:

This is the exact analog of a point moving around another point.

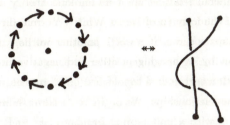

It turns out that the motion group for a collection of n disjoint circles in $\mathbf{R}^3$ is isomorphic to the subgroup of the group $\text{Aut}(n) = \text{Aut}(F(X_1,\dots,X_n))$ (the group of automorphisms of the free group on generators $X_1,\dots,X_n$) that is generated by (in the case of $n = 2$ - two strings) T, E and S where $T(x_1) = x_1^{-1}$, $T(x_2) = x_2$, $E(x_1) = x_2$, $E(x_2) = x_1$, $S(x_1) = x_2 x_1 x_2^{-1}$, $S(x_2) = x_2$. Imbo and Brownstein [IM2] calculate that the motion group for two unknotted strings in $\mathbf{R}^3$ is

$$\langle T, E, S \mid T^2 = E^2 = (TE)^4 = TST^{-1}S^{-1} = (TESE)^2 = 1\rangle.$$

This has implications for exotic statistics ([IM1], [LK28]).

Problem: Prove this result and extend it to n strings.

Exercise. Read [JE] and make sense of these ideas (elementary particles as linked and knotted quantized flux) in the light of modern developments. (Compare with [RA].)

15^0. DNA and Quantum Field Theory.

One of the most successful relationships between knot theory and DNA research has been the use of the formula of James White [W], relating the linking, twisting and writhing of a space curve. It is worth pointing out how these matters appear in a new light after linking and higher order invariants are re-interpreted a lá Witten [WIT2] as path integrals in a topological quantum field theory. This section is an essay on these relationships. We begin by recalling White's formula.

Given a space curve C with a unit normal framing $v, v^\perp$ and unit tangent t (v and $v^\perp$ are perpendicular to each other and to t, forming a differentiably varying frame, $\langle v, v^\perp, t\rangle$, at each point of C.) Let C_v be the curve traced out by the tip of ϵv for $0 < \epsilon \ll 1$. Let $Lk = Lk(C, C_v)$ be the linking number of C with this displacement C_v. Define the **total twist**, Tw, of the framed curve C by the formula $Tw = \frac{1}{2\pi} \int v^\perp \cdot dv$. Given $(x,y) \in C \times C$, let $e(x,y) = (y-x)/|y-x|$ for $x \neq y$ and note that $e(x,y) \to t/|t|$ (for t the unit tangent vector to C at x) as x approaches y. This makes e well-defined on all of $C \times C$. Thus we have $e : C \times C \to S^2$. Let $d\Sigma$ denote the area element on S^2 and define the (spatial) **writhe** of the curve C by the formula

$$Wr = \frac{1}{4\pi} \int_{C \times C} e^* d\Sigma = \frac{1}{4\pi} \int_{z \in S^2} Cr(z) dz.$$

Here $Cr(z) = \underset{p \in e^{-1}(z)}{\Sigma} J(p)$ where $J(p) = \pm 1$ according to the sign of the Jacobian of e.

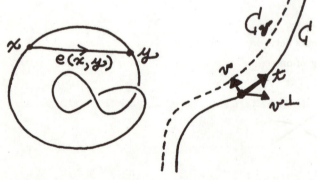

It is easy to see, from this description that the writhe coincides with the flat writhe (sum of crossing signs) for a curve that is (like a knot diagram) nearly embedded in a single plane.

With these definitions, White's theorem reads

$$Lk = Tw + Wr.$$

This equation is fully valid for differentiable curves in three space. Note that the writhe only depends upon the curve itself. It is independent of the framing. By combining two quantities (twist and writhe) that depend upon metric considerations, we obtain the linking number - a topological invariant of the pair (C, C_v).

The planar version of White's theorem is worth discussing. Here we have C and C_v forming a pair of parallel curves as in the example below:

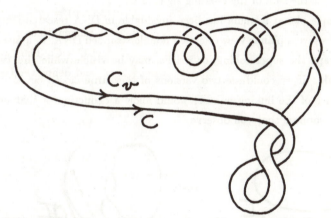

The twisting occurs between the two curves, and is calculated as the sum of $\pm(1/2)$ for each crossing of one curve with the other – in the form

The flat crossings also contribute to the total linking number, but we see that they

also catalogue the flat writhe of C. Thus

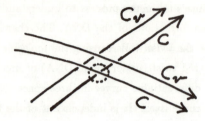

is regarded as a contribution of $(+1)$ to $w(C)$. Thus, we get

$$Lk(C, C_v) = Tw(C, C_v) + Wr(C)$$

where $Wr(C)$ is the sum of the crossing signs of C.

This simple planar formula has been invaluable in DNA research since photomicrographs give a planar view of closed double-stranded DNA. In a given electron micrograph the super-coiling, or writhe, may be visible while **the twist is unobservable.** If we could see two versions of the **same** closed circular DNA, then invariance of the linking number would allow a deduction of that number! For example, suppose that you observe

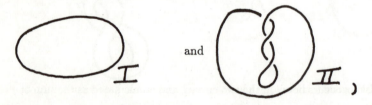

both representing the same DNA. In the first case $Wr_I = 0$. In the second case $Wr_{II} = -3$. Thus

$$LK_I = Wr_I + Tw_I = Tw_I$$
$$LK_{II} = Wr_{II} + Tw_{II} = -3 + Tw_{II}.$$

Since $Lk_I = Lk_{II}$, we conclude that $Tw_{II} = Tw_I + 3$. Since I is relaxed, one can make an estimate based on the DNA geometry of the total twist. Let's say that $Tw_I = 500$. Then, we know that $Tw_{II} = 503$.

In this way the formula of White enables the DNA researcher to deduce quantities not directly observable, and in the process to understand some of the topological and energetical transformations of the DNA. The chemical environment of the DNA can influence the twist, thus causing the molecule to supercoil either in compensation for lowered twist (underwound DNA) or increase of twist (overwound DNA).

On top of this, there is the intricacy of the geometry and topology of recombination and the action of enzymes such as topoisomerase – that change the linking number of the strands ([BCW]). These enzymes perform the familiar switching operation 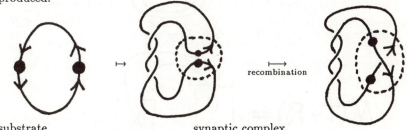 on interlinked strands of DNA by breaking one strand, and allowing the other strand to pass through it. In recombination a **substrate** with special sites on it twists into a **synaptic complex** with the sites for recombination adjacent to one another. Then in the simplest form of recombination the two sites are replaced by a crossover, and a new (possibly linked) double-stranded form is produced.

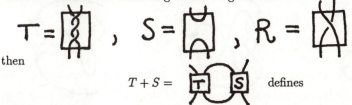

substrate   synaptic complex

(In the biological terminology a link is called a catenane.) Of course, many different sorts of knots and links can be produced by such a process. In its simplest form, this movement from substrate to synaptic complex to product of the recombination takes the form of the combining of two tangles. Thus if

$$T = \boxed{} \quad , \quad S = \boxed{} \quad , \quad R = \boxed{}$$

then

$$T + S = \boxed{T} \boxed{S} \quad \text{defines}$$

tangle addition, and any four strand tangle τ has a **numerator** $N(\tau)$ and a

denominator $D(\tau)$:

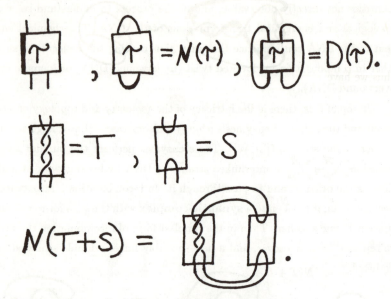

$$N(T+S) =$$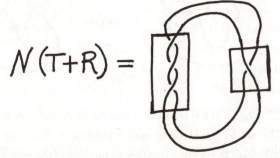

This is the synaptic complex. The product of the recombination is

$$N(T+R) =$$

A **second round** of the recombination constitutes $N(T + R + R)$:

 $\longmapsto$  $\longmapsto$

substrate   synaptic complex $N(T + R + R)$

And a **third round** yields $N(T + R + R + R)$.

Thus we have

$$N(T + S) \cong \quad \bigcirc \quad \text{Unknot}$$

$$N(T + R) \cong \quad \text{Hopf Link}$$

$$N(T + R + R) \cong \quad \text{Figure Eight Knot}$$

$$N(T + R + R + R) \cong \quad \text{Whitehead Link}$$

In experiments carried out with $T_n 3$ Resolvase [CSS] N. Cozarelli and S. Spengler found that just these products (Hopf Link, Figure Eight Knot, Whitehead Link) were being produced in the first three rounds of site specific recombination. In making a tangle-theoretic model of this process, D. W. Sumners [S] assumed that **T and R are enzyme determined constants independent of the variable geometry of the substrate S.** With this assumption, he was able to prove, using knot theory, that in order to obtain the products as shown above for the first three rounds, and assuming that $S = $, then

$$T = \qquad \text{and} \qquad R = \qquad .$$

This is a remarkable scientific application of knot theory to **deduce the underlying mechanism** of the biochemical interaction.

How did this result work? The key lay in Sumners' reduction of the problem to one involving only so-called **rational** knots and links, and then the construction of a tangle calculus involving this class of links. The rational tangles are characterized topologically by values in the extended rational numbers $Q^* = Q \cup \{1/0 = \infty\}$. An element in Q^* has the form β/α where $\alpha \in N \cup \{0\}$, (N is the natural numbers), and $\beta \in \mathbf{Z}$ with $gcd(\alpha, \beta) = 1$. Rational tangles themselves are obtained by iterating operations similar to the recombination process itself. Thus

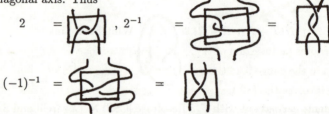

The image of a tangle is obtained by turning it 180° around the left-top to right-bottom diagonal axis. Thus

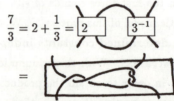

Rational numbers correspond to tangles via the continued fraction expansion. Thus

$$\frac{7}{3} = 2 + \frac{1}{3} = \boxed{2} \bigcirc \boxed{3^{-1}}$$

$$=$$

Since two rational tangles are topologically equivalent if and only if they receive the same fraction in Q^* (See [ES].), it is possible to calculate possibilities for site specific recombination in this category. Here we have an arena in which molecular operations, knot theoretic operations and the **topological information** carried out by a knot or link are in good accord!

This brings us directly to the general question: **What is the nature of the topological information carried by a knot or link?** For biology this information manifests itself in the history of a recombinant process, or in the architecture of the constituents of a cell. In this book, we have seen that such information is naturally enfolded in a quantum statistical framework. The most succinct statement of the generalized skein relations occurs for us in section 17^0 of Part I, where - as a consequence of the generalized path integral

$$Z_K = \int dA e^{(ik/4\pi)\mathcal{L}} \langle K|A \rangle$$

where $\mathcal{L}$ is the Chern-Simons Lagrangian, and $\langle K|A \rangle$ is the Wilson loop

$$\langle K|A \rangle = \mathrm{tr}\left(\mathbf{P} \, \exp\left(i \oint_K A \right) \right),$$

we find

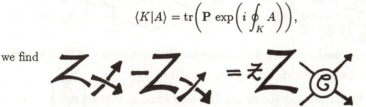

where $\mathcal{C}$ denotes an insertion of the Casimir operator for the representation of the Lie algebra in the gauge field theory. The basic change of information in switching a crossing involves the full framework of a topological quantum field theory. Yet conceptually this viewpoint is an enormous advance because it is a context within which one can begin to understand the nature of the information in the knot.

How does this viewpoint influence molecular biology? It may be too early to tell. The topological models used so far in DNA research are naive oversimplifications of the biology itself. We do not yet know how to relate the topology with the information structure and coding of DNA sequencing, let alone with the full complex of cellular architecture and interaction. Furthermore, biology occurs in the physical context of fields and the ever present quantum mechanical underpinnings of the molecules themselves.

In this regard, the Witten functional integral gives a powerful metaphor linking biology and quantum field theory. For remember the nature of the Wilson loop $\langle K|A \rangle = \mathrm{tr}\left(\mathbf{P}\exp\left(i \oint_K A\right)\right)$. This quantity is to be regarded as the limit as the number of "detectors" goes to infinity of a product of matrices plucked from the gauge potential A (defined on the three dimensional space). Each detector takes a matrix of the form $(1 + iAdx)$ at a spatial point x.

detector

I like to think of each detector as a little codon with an "antenna" and legs (—◯—) for plugging into nearby codons. Put a collection of these together into a knot

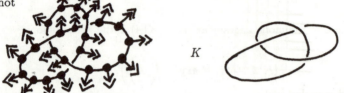

and you have an element, ready to measure the field and form $\langle K|A \rangle$. The interconnection is the product that computes the Wilson loop.

Just so, it may be that the circularly closed loops of knotted and catenaned DNA are indeed Wilson loops – receivers of information from the enveloping biological field. If this is so, then knot theory, quantum field theory and molecular biology are fundamentally one scientific subject. Information is never carried only in a discrete mode. Why not regard DNA as the basic transceiver of the morphogenetic field? Each knotted form is a receptor for different sorts of field information, and the knots are an alphabet in the language of the field.

If these ideas seem strange in the biological context, they are not at all unusual for physics. For example, Lee Smolin and Carlo Rovelli ([RS], [LS2]) have proposed a theory of quantum gravity in which knots represent (via Wilson loop integration) a basis of quantum mechanical states in the theory.

Nevertheless, we should pursue the metaphor within the biological context. In this context I take a morphogenetic field to be a field composed of physical fields, but also fields of form and recursion. A simple example will help to elucidate this point of view. Consider John H. Conway's cellular automaton the "Game of Life". This is a two dimensional automaton that plays itself out on a rectangular lattice of arbitrary size. Squares in the lattice are either marked ▓ or unmarked ░. The rules of evolution are:

1. ■ → ☐ at less than two neighbors.

2. ■ → ☐ at more than three neighbors.

3. ☐ → ■ at exactly three neighbors.

The rules are applied simultaneously on the board at each time step. As a result, very simple initial configurations such as

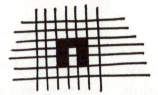

have long and complex lifetimes. At each step in the process a given square is (with its neighbor configuration) a detector of the morphogenetic field that is the rule structure for the automaton. The field is present everywhere on the board, it is a field of form, a non-numerical field. The particular local shapes on the board govern the evolutionary possibilities inherent in the field.

Another example arises naturally in Cymatics [JEN]. In Cymatics one studies the forms produced in materials under the influence of vibration - such as the nodal patterns of sand on a vibrating plate, or the form of a liquid surface under vibration. To an observer, the system appears to decompose into an organism of correlated parts, but these parts are all parts only to that observer, and are in reality, orchestrated into the whole through the vibratory field. Each apparent part is (to the extent that "it" is seen as a part) a locus of reception for information from the field, and a participant in the field of form and local influence generated in the material through the vibration.

One more example will help. This is the instantiation of autopoesis as described by Varela, Maturana and Uribe in [VMU]. Here one has a cellular space occupied by codons, 0, marked codons, ☐0, and catalysts, ∗. The marked codons can interlink: ... ☐0☐0☐0 ... . The catalyst facilitates marking via the combination of two codons to a marked codon: 0 0 ∗ → ☐0 . Marked codons may spontaneously decay

$$☐0 → 0 \ 0$$

or lose or gain their linkages. Codons can pass through linkage walls:

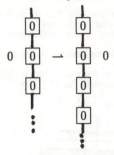

but catalysts can **not** pass through a linkage.

With these rules, the system is subjected to random perturbations in a soup of codons, marked codons and catalysts. Once closed, loops of marked codons enclosing a catalyst evolve they tend to maintain a stable form

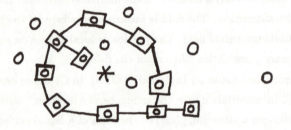

since the catalyst encourages the restoration of breaks in the loop, and has difficulty in escaping since it cannot cross the boundary that the loop creates.

In this example, we see in a microcosm how the cell may maintain its own structure through a sequence of productions that involve the very parts out of which it is constructed. The closing of the loop is essential for the maintenance of the form. It is through the closure of the loop that the cell has stability and hence a value in the system. The sensitivity of the closed loop is that of a unity that maintains its own form.

These examples give context to the possibility that there really is a deep relationship between the Jones polynomial – seen as a functional integral for topological quantum field theory in $SU(2)$ gauge – and the role of knotted DNA in molecular biology. $SU(2)$ can be related to local rotational structures (compare [BAT]) for bodies in three dimensional space (as we have discussed in relation to the Dirac string trick) and hence there is the possibility of interpreting the Wilson loop in geometric terms that relate directly to biology. This is an idea for an idea.

To make one foray towards a concrete instantiation of this idea, consider the gauge theory of deformable bodies as described in [WIL]. Here one wants to determine the net rotation that results when a deformable body goes through a given sequence of unoriented shapes, in the absence of external forces. That is, given a path in the space of unlocated shapes (This is what a swimmer, falling cat or moving cell has under its control.), what is the corresponding path in the configuration space of located shapes? How does it actually move? In order to make axial comparisons between the unlocated and located shapes we have to give each unlocated shape a standard orientation, and then assign elements of $SO(3)$ relating the actual path to the abstract path. This ends up in a formulation that is precisely an $SO(3)$ gauge theory. For a given sequence of standard shapes $S_0(t)$, one wants a corresponding sequence of oriented shapes $S(t)$ that correspond to the real path of motion. These two paths are related by an equation

$$S(t) = R(t)S_0(t)$$

where $R(t)$ is a path in $SO(3)$. Computing $R(t)$ infinitesimally corresponds to solving the equation

$$\frac{dR}{dt} = R(t)A(t) \qquad \left(A(t) = R^{-1}(t)\frac{dR}{dt} \right).$$

Once A is known, the full rotation $R(t)$ can be expressed as a path ordered exponential

$$R(t) = \mathbf{P} \exp\left(\int_{t_0}^{t} A(t)dt \right).$$

Wilczek defines a gauge potential over **the space of unlocated shapes** S_0 (also denoted by A) via

$$A_{\dot{S}_0}[S_0(t)] \equiv A(t).$$

That is, A is defined on the tangent space to S_0 - it is a vector field with a component for every direction in shape space - and takes values in the Lie algebra of $SO(3)$. As usual, the curvature tensor (or field strength)

$$F_{ij} = \frac{\partial A_i}{\partial w_j} - \frac{\partial A_j}{\partial w_i} + [A_i, A_j]$$

(w_i are a basis of tangent vectors to S_0) determines the rotations due to infinitesimal deformations of S_0.

In specific cases, the gauge potential can be determined from physical assumptions about the system in question, and the relevant conservation laws. I have described this view of gauge theories for deformable bodies because it shows how close the formalisms of gauge theory, Wilson loops, and path ordered integration are to actual descriptions of the movement of animals, cells and their constituents. We can move to $SU(2)$ gauge by keeping track of orientation-entanglement relations (as in the Dirac string trick), and averages over Wilson loop evaluations then become natural for describing the statistics of motions. This delineates a broad context for gauge field theory in relation to molecular biology.

16^0. Knots in Dynamical Systems – The Lorenz Attractor.

The illustration on the cover of this book indicates a **Lorenz knot**; that is, it represents a periodic orbit in the system of ordinary differential equations in three dimensional space discovered by Lorenz [LOR]. This system is, quite specifically,

$$X' = \sigma(Y - X)$$
$$Y' = X(r - Z) - Y$$
$$Z' = XY - bZ$$

where σ, r and b are constants. (Lorenz used $\sigma = 10$, $b = 8/3$ and $r = 28$.) The constant r is called the **Reynolds number** of the Lorenz system.

The Lorenz system is the most well-known example of a so-called chaotic dynamical system. It is extremely sensitive to initial conditions, and for low Reynolds number any periodic orbits (if they indeed exist) are unstable. Thus a generic orbit of the system wanders unpredictably, weaving its way back and forth between two unstable attracting centers. The overall shape of this orbit is stable and this set is often called the Lorenz attractor. Thus one may speak of orbits in the Lorenz attractor.

A qualitative analysis of the dynamics of the Lorenz attractor by Birman and Williams [B3] shows that periodic orbits (in fact orbits generally) should lie on a singular surface that these authors dub the "knot holder". See Figure 16.2 for the general aspect of the Lorenz attractor - as a stereo pair, and Figure 16.1 for a picture of the knot holder.

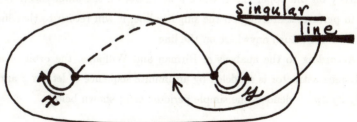

Knot Holder
Figure 16.1

502

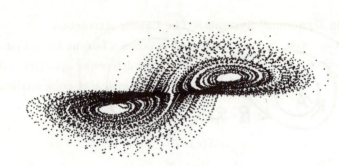

Lorenz Attractor

Figure 16.2

The knot-holder can be regarded as the result of gluing part of the boundary of an annulus to a line joining the inner and outer boundaries as shown below:

In Figure 16.1, we have indicated two basic circulations - x and y. The label x denotes a journey around the left-hand hole in the knot-holder, while y denotes a journey around the right-hand hole. Starting from the singular line, x connotes a journey on the upper sheet, while y will move on the lower sheet. If a point is told to perform x starting on the singular line, it will return to that line, and at first turn can return anywhere on the line.

According to the analysis of Birman and Williams, the orbit of a point in the Lorenz attractor is modeled by an infinite sequence of letters x and y. Thus $xy\ xy\ xy\ xy \ldots$ denotes the simple periodic orbit shown below.

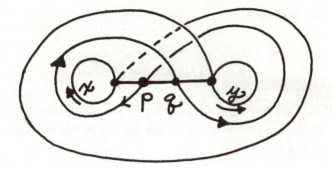

$$w = xy \; xy \; xy \; xy \ldots$$

Already, many features of the general situation are apparent here. Note that this orbit intersects the singular line in points p and q and that from p the description is $yx \; yx \; yx \ldots$. Thus we can write

$$w_p = xy \; xy \; xy \ldots$$

$$w_q = yx \; yx \; yx \ldots$$

Note that the list w_p, w_q is in lexicographic order if we take $x < y$ and that $p < q$ in the usual ordering of the unit interval homeomorphic with the singular line.

Are there any other orbits with this description $xy \; xy \; xy \ldots$? Consider the experiment of doing x and then returning to a point $q < p$:

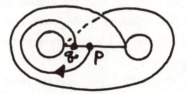

It is then not possible to perform y, since this would force the orbit to cross itself transversely (contradicting the well-definedness of the Lorenz vector field).

Therefore for $q < p$, we see that

$$w_p = xx*$$

$$w_q = x*$$

and $w_p \leq w_q$. If $w_p = w_q$ then $w_p = w_q = xxxx\ldots$ – an infinite spiraling orbit – not realistic for the Lorenz system at $r = 28$, but possible in this combinatorics. Otherwise, we find (e.g.)

$$w_p = xxxxy\ldots$$

$$w_q = xxxy\ldots$$

and $w_p < w_q$. Incidentally, we shall usually regard $xxx\ldots$ as a simple periodic orbit encircling the left hole in the knot-holder.

Now, suppose we want a periodic orbit of the form

$$w = (xyxyy)(xyxyy)(xyxyy)\ldots.$$

Then this will cross the singular line in **five** points and these points will have descriptions corresponding to the five cyclic permutations of the word $xyxyy$:

1. $xyxyy$: $w_1 = xyxyy\ xyxyy\ldots$

4. $yxyyx$: $w_4 = yxyyx\ yxyyx\ldots$

2. $xyyxy$: $w_2 = xyyxy\ xyyxy\ldots$

5. $yyxyx$: $w_5 = yyxyx\ yyxyx\ldots$

3. $yxyxy$: $w_3 = yxyxy\ yxyxy\ldots$

I have listed in order the cyclic permutations of $xyxyy$ and labelled each word with the integers 1,2,3,4,5 to indicate the lexicographic order of the words. The orbit must then pass through points p_1, p_2, p_3, p_4, p_5 on the singular line of the knot-holder so that w_i is the orbit description starting from p_i. We need (as discussed above for the simplest example) that $p_1 < p_2 < p_3 < p_4 < p_5$ since, in lexicographic order, $w_1 < w_2 < w_3 < w_4 < w_5$. See Figure 16.3 for the drawing on the knot-holder.

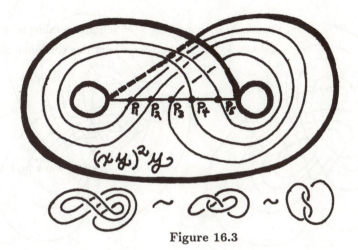

Figure 16.3

These considerations give rise to a simple algorithm for drawing a Lorenz knot corresponding to any non-periodic word in x and y.

1) Write an ordered list of the cyclic permutations of the word, by successively transferring the first letter to the end of the word.

2) Label the words in the list according to their lexicographic order.

3) Place the same labels in numerical order on the singular line of the knot-holder.

4. Starting at the top of the list, connect p_i to p_j by x or y – whichever is the first letter of the word labelled i where words i and j are successive words on the list.

We have done exactly this algorithm to form the knot in Figure 16.3 corresponding to $(xy)^2y$. Note, as shown in this figure, that $(xy)^2y$ is indeed knotted. It is a trefoil knot.

Now we can apply the algorithm to the knot on the cover of the book. It is shown as a stereo pair in Figure 16.4.

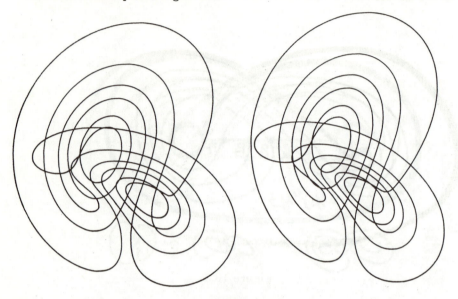

Figure 16.4

This knot is a computer generated Lorenz knot obtained by myself and Ivan Handler. Our method is accidental search. The knot is a periodic orbit **for our computer** at $r = 60$, $\sigma = 10$, $b = 8.3$, This is not a reproducible result! For a good discussion of the use of a generalization of Newton's method for the production of computer independent Lorenz knots, see [CUR]. Also, [JACK] has a good discussion of the generation of knots at high Reynolds number (say $r \approx 200$). In this range, one gets stable unknotted periodic orbits; and, as the Reynolds number is decreased, these bifurcate in a period-doubling cascade of stable - but hard to observe - knots.

Anyway, getting back to our knot in Figure 16.4, we see easily that it has the word x^3yxy^3xy.

Exercise 16.1. Apply the Lorenz knot algorithm to the word x^3yxy^3xy, and show that the corresponding knot is a torus knot of type $(3,4)$.

Exercise 16.2. Draw an infinite orbit on the knot-holder with word
$w = xyxyyxyyyx \ldots$. Figure 16.5 shows the first few turns.

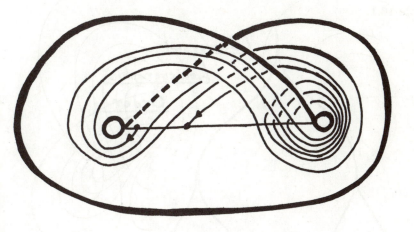

$$xyxy^2xy^3xy^4 \ldots$$

Figure 16.5

Remark. In regard to finite and infinite words, it is interesting to note that there is a correspondence between the lexicographic order of the partial trips on the knot-holder and the representations of numbers in the extended number system of John H. Conway [CON2]. To see this relationship, I use the representation of Conway numbers due to Martin Kruskal. For Kruskal, a Conway number is an ordinal sequence of x's and y's. Think of x as **left** and y as **right**. Then any word in x and y gives a sequence of "numbers" where

$$
\begin{array}{ccl}
0 & \leftrightarrow & \emptyset \text{ the empty word} \\
+1 & \leftrightarrow & y \\
+2 & \leftrightarrow & yy \\
-1 & \leftrightarrow & x \\
-2 & \leftrightarrow & xx \\
& \cdots & \\
-\tfrac{1}{2} & \leftrightarrow & xy
\end{array}
$$

508

In general, x is taken as an instruction to step back midway between the last two created numbers, and y as an instruction to step forward. We always have that $\omega < \tau$ on the Conway line if ω precedes τ lexicographically.

Example 16.1.

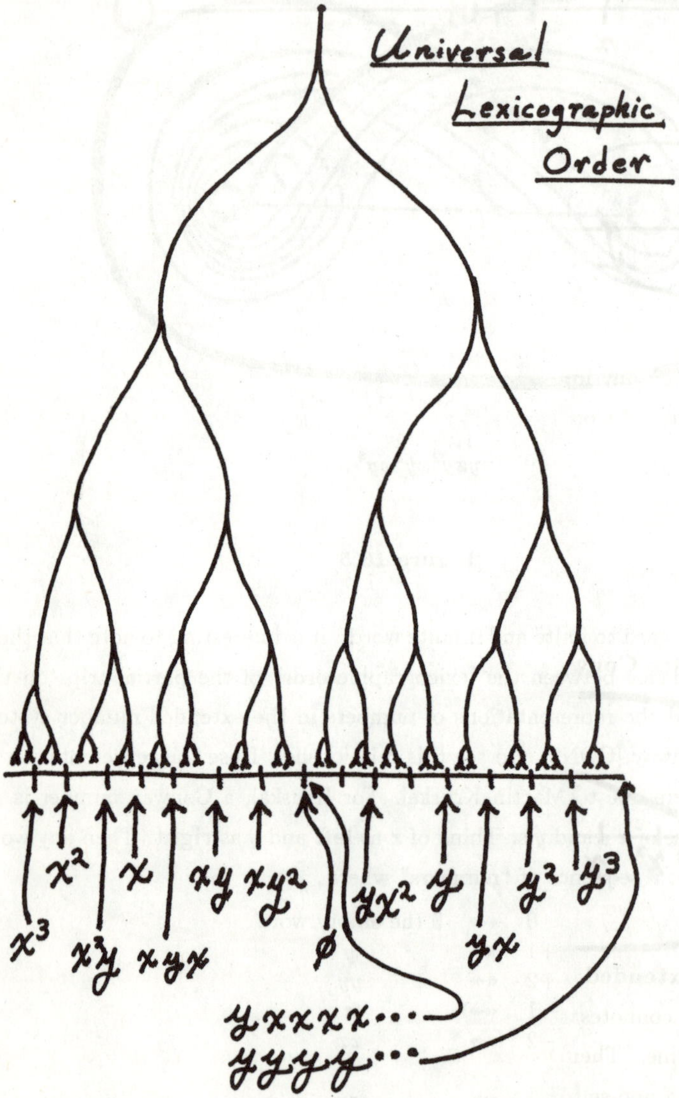

Universal Lexicographic Order

Example 16.2. $xyyyyy\ldots$ (This is a number infinitely close to zero.)

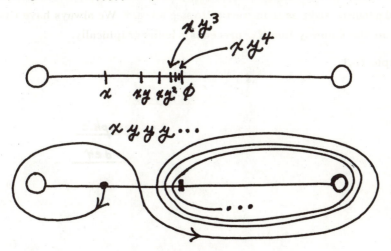

In general, given any infinite word $w = e_1 e_2 e_3 e_4 \ldots$ where $e_i = x$ or y we get an infinite sequence of Conway numbers:

$$w_1 = e_1 e_2 e_3 e_4 \ldots$$

$$w_2 = e_2 e_3 e_4 \ldots$$

$$w_3 = e_3 e_4 e_5 \ldots$$

$$\ldots$$

embedded in the **Conway Interval** $\mathcal{I}$

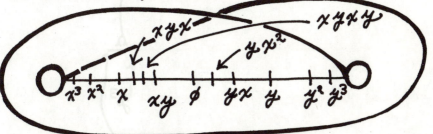

Define the **Extended Lorenz Holder** to be $\mathcal{S}^1 \times \mathcal{I} \sim$ where $\mathcal{S}^1$ is a Conway circle, and $\sim$ connotes the identification of the Conway annulus $\mathcal{S}^1 \times \mathcal{I}$ to form the singular line. Then for each w, the points $w_1, w_2, w_3, \ldots$ give the symbolic dynamics for a non-self intersecting orbit on the extended Lorenz Holder.

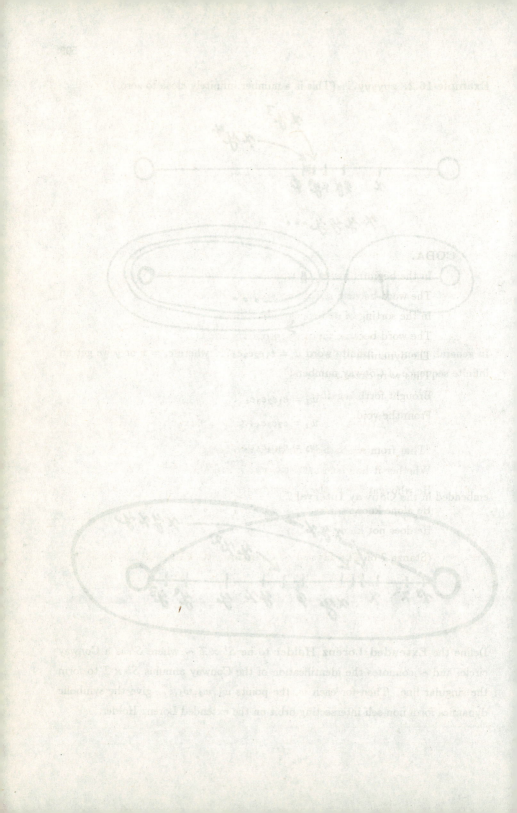

CODA.

In the beginning was the word.

The word became self-referential/periodic.

In the sorting of its lexicographic orders,

The word became topology, geometry and

The dynamics of forms;

Thus were chaos and order

Brought forth together

From the void.

"That from which creation came into being,

Whether it had held it together or it had not

He who watches in the highest heaven

He alone knows, unless. . .

He does not know."

(Stanza 7 of **The Hymn of Creation**, RG VEDA 10.129 [DN]).

REFERENCES

[AR] C. Agnes and M. Rasetti. Piecewise-linear category approach to entanglement. *Il. Nuovo Cimento*, vol. 3D, No. 1 (1984), pp. 121–136.

[AB] Y. Aharonov and D. Bohm. Significance of electromagnetic potentials in quantum theory. *Phys. Rev.*, 115(1959), pp. 485-491.

[AH] Y. Aharonov and L. Susskind. Observability of the Sign Change of Spinors under 2π Rotations. *Phys. Rev.*, vol. 158, No. 5. June 1967, pp. 1237.

[AW1] Y. Akutsu and M. Wadati. Knot invariants and critical statistical systems. *J. Phys. Soc. Japan*, 56(1987), 839-842.

[AW2] Y. Akutsu and M. Wadati. Exactly solvable models and new link polynomials. I. N-state vertex models. *J. Phys. Soc. Japan*, 56(1987), 3039-3051.

[AW3] Y. Akutsu and M. Wadati. Knots, links, braids and exactly solvable models in statistical mechanics. *Comm. Math. Phys.*, 117(1988), 243-259.

[AW4] Y. Akutsu, A. Kuniba and M. Wadati. Virasoro algebra, von Neumann algebra and critical eight-vertex SOS models. *J. Phys. Soc. Japan*, vol. 55, no. 10. October 1986, pp. 3285-3288.

[AW5] Y. Akutsu and M. Wadati. From solitons to knots and links. *Progress of Theoretical Physics Supplement*, no. 94 (1988), pp. 1-41.

[ADW1] Y. Akutsu, T. Deguchi and M. Wadati. Exactly solvable models and new link polynomials II. Link polynomials for closed 3-braids. *J. Phys. Soc. Japan*, 56(1987), 3464-3479.

[ADW2] Y. Akutsu, T. Deguchi and M. Wadati. Exactly solvable and new link polynomials III. Two-variable polynomial invariants. *J. Phys. Soc. Japan*, 57(1988), pp. 757-776.

[ALEX1] J. W. Alexander. A lemma on systems of knotted curves. *Proc. Nat. Acad. Sci., USA* 9(1923), pp. 93-95.

[ALEX2] J. W. Alexander. Topological invariants of knots and links. *Trans. Amer. Math. Soc.*, 20(1923), 275-306.

[ASH] A. Ashtekar. **New perspectives in canonical gravity**. Bibliopolis (1988).

[A1] M. F. Atiyah. K-**Theory**. W. A. Benjamin Inc. (1967).

[A2] M. F. Atiyah. **Geometry of Yang-Mills Fields**. Accademia Nazionale dei Lincei Scuola Normale Superiore - Lezioni Fermiane. Pisa (1979).

[A3] M. F. Atiyah. **The Geometry and Physics of Knots**. Cambridge University Press. (1990).

[BE] H. J. Bernstein and A. V. Phillips. Fiber bundles and quantum theory. *Scientific American*, vol. 245, no. 1. July 1981, pp. 122-137.

[BS] L. Baulieu and I. M. Singer. Topological Yang-Mills symmetry. *Nuclear Phys.* B (Proc. Suppl.) B(1988) pp. 12-19.

[BA1] R. J. Baxter. **Exactly Solved Models in Statistical Mechanics**. Academic Press (1982).

[BA2] R. J. Baxter. Chromatic polynomials of large triangular lattices. *J. Phys. A.: Math. Gen.*, **20**(1987), 5241-5261.

[BN] D. Bar-Natan. Perturbative Chern-Simons Theory. (Preprint 1990).

[BAT] E. P. Battey-Pratt and T. J. Racey. Geometric Model for Fundamental Particles. *Int. J. Theo. Phys.*, vol. 19, no. 6, (1980), pp. 437-475.

[BCW] W. R. Bauer, F. H. C. Crick and J. H. White. Supercoiled DNA. *Scientific American*, vol. 243. July 1980, pp. 118-133.

[BAY] B. F. Bayman. Theory of Hitches, *Amer. J. Physics*, vol. 45, no. 2, (1977), pp. 185-190.

[B1] J. S. Birman and H. Wenzl. **Braids, link polynomials and a new algebra**. *Trans. AMS*, **313** (1989), pp. 249–273.

[B2] J. S. Birman. **Braids, links and mapping class groups**. Annals of Math Study, no. 82. Princeton University Press (1974).

[B3] J. S. Birman and R. F. Williams. Knotted Periodic Orbits in Dynamical Systems - I. Lorenz's Equations. *Topology*, vol. 22, no. 1, (1983), pp. 47-82.

[B4] J. S. Birman. On the Jones polynomial of closed 3-braids. *Invent. Math.*, 81(1985), pp. 287-294.

[BO] H. Bondi. **Relativity and Common Sense**. Dover Pub. (1977).

[BR] R. D. Brandt, W. B. R. Lickorish, K. C. Millett. A polynomial invariant for unoriented knots and links. *Invent. Math.*, 84(1986), 563-573.

[BUC] G. Buck. Knots as dynamical systems. (Preprint 1990.)

[BZ] G. Burde and H. Zieschang. **Knots**. deGruyter (1986).

[BUR] P. N. Burgoyne. Remarks on the combinatorial approach to the Ising problem. *J. Math. Phys.*, vol. 4, no. 10, Oct. 1963.

[CLM] S. E. Cappell, R. Lee, E. Y. Miller. Invariants of 3-manifolds from conformal field theory. (Preprint 1990.)

[CG] Casson, A. and C. M. Gordon. On slice knots in dimension three. In **Geometric Topology**, R. J. Milgram ed., pp. 39-53, *Proc. Symp. Pure Math.*, XXXII, Amer. Math. Soc., Providence, R. I.

[CL] T. P. Cheng and L. F. Li. **Gauge Theory of Elementary Particle Physics**. Clarendon Press - Oxford (1988).

[CH] G. F. Chew and V. Poenaru. Single-surface basis for topological particle theory. *Phys. Rev. D.*, vol. 32, no. 10, pp. 2683-2697.

[CON] J. H. Conway. An enumeration of knots and links and some of their algebraic properties. **Computational Problems in Abstract Algebra.** Pergamon Press, New York (1970), 329-358.

[CON1] J. H. Conway. (Lecture at the University of Illinois at Chicago, Spring 1979.)

[CON2] J. H. Conway. **On Numbers and Games.** Academic Press (1976).

[COR] E. Corrigan, D. B. Fairlie, P. Fletcher and R. Sasaki. Some aspects of quantum groups and supergroups. (Preprint 1990.)

[COTT] P. Cotta-Ramusino, E. Guadagnini, M. Martellini and M. Mintchev. Quantum field theory and link invariants. (Preprint 1989).

[CSS] N. R. Cozzarelli, S. J. Spengler, A. Stasiak. The stereostructure of knots and catenanes produced by phase λ integrative recombination: implications for mechanism and DNA structure. *Cell* 42(1985), pp. 325-334.

[CR1] L. Crane. Topology of three manifolds and conformal field theories. (Preprint 1989).

[CR2] L. Crane. 2-D Physics and 3-D Topology. *Commun. Math. Phys.,* **135** (1991), 615–640.

[CR3] L. Crane. Conformal Field Theory, Spin Geometry and Quantum Gravity. (Preprint 1990).

[CF] R. H. Crowell, R. H. Fox. **Introduction to Knot Theory.** Blaisdell Pub. Co., (1963).

[CUR] J. H. Curry. An algorithm for finding closed orbits. *Lecture Notes in Mathematics,* no. 819, Springer-Verlag (1979) pp. 111-120.

[DN] A. T. de Nicolás. **Meditations through the Rg Veda.** Shambala Press (1978).

[DIR] P. A. M. Dirac. **The Principles of Quantum Mechanics.** Oxford University Press (1958).

[DRIN1] V. G. Drinfeld. Quantum Groups, *Proc. Intl. Congress Math.,* Berkeley, Calif. USA (1986), 789-820.

[DRIN2] V. G. Drinfeld. Hopf algebras and the quantum Yang-Baxter Equation. *Soviet Math. Dokl.,* **32**(1985), no. 1.

[ES] C. Ernst and D. W. Summers. A calculus for rational tangles: applications to DNA recombination. *Math. Proc. Camb. Phil. Soc.,* **108** (1990), pp. 489–515.

[EL] B. Ewing and K. Millett. A load balanced algorithm for the calculation of the polynomial knot and link invariants. (Preprint 1989.)

[FWu] C. Fan and F. Y. Wu. General lattice model of phase transitions. *Phys. Rev. B.,* vol. 2, no. 3, (1970), pp. 723-733.

516

[FRT] L. D. Faddeev, N. Yu Reshetikhin, L. A. Takhtajan. Quantization of Lie groups and Lie algebras. LOMI Preprint E-14-87, Steklov Mathematical Institute, Leningrad, USSR.

[FR] R. A. Fenn and C. P. Rourke. On Kirby's calculus of links. *Topology*, 18(1979), pp. 1-15.

[FE] R. P. Feynman. **Theory of Fundamental Processes**. W. A. Benjamin Pub., (1961).

[FE1] R. P. Feynman and S. Weinberg. **Elementary Particles and the Laws of Physics – The 1986 Dirac Memorial Lectures**. Cambridge University Press (1987).

[FI1] D. Finkelstein. Kinks. *J. Math. Phys.*, 7(1966), pp. 1218-1225.

[FI2] D. Finkelstein. Space-Time Code. *Phys. Rev.*, 184(1969), pp. 1261-1271.

[FR] D. Finkelstein and J. Rubenstein. Connection between spin, statistics and kinks. *J. Math. Phys.*, 9(1968), pp. 1762-1779.

[FOX1] R. H. Fox and J. W. Milnor. Singularities of 2-spheres in 4-space and cobordism of knots. *Osaka J. Math.*, 3, 257-267.

[FOX2] R. H. Fox. A quick trip through knot theory. In **Topology of 3-Manifolds**. Ed. by M. K. Fort Jr., Prentice-Hall (1962), pp. 120-167.

[FRA] G. K. Francis. **A Topologists Picturebook**. Springer-Verlag (1987).

[FW] J. Franks and R. Williams. Braids and the Jones polynomial. *Trans. Amer. Math. Soc.*, 303 (1987), pp. 97–108.

[FG] D. S. Freed and R. E. Gompf. Computer calculation of Witten's 3-Manifold Invariant. (Preprint 1990).

[F] P. Freyd, D. Yetter, J. Hoste, W. B. R. Lickorish, K. C. Millett and A. Ocneanu. A new polynomial invariant of knots and links. *Bull. Amer. Math. Soc.*, 12(1985), 239-246.

[FQS] D. Friedan, Z. Qui, S. Shenker. Conformal invariance, unitarity and two-dimensional critical exponents. In **Vertex operators in Mathematics and Physics**. Ed. by J. Lepowsky, S. Mandelstam, and I. M. Singer, pp. 419-450, (1985).

[FRO] J. Frohlich. Statistics of fields, the Yang-Baxter Equation and the theory of knots and links. (Preprint 1987).

[FROM] J. Frohlich and P. Marchetti. Quantum field theories of vortices and anyons. (Preprint 1988).

[FU] B. Fuller. Decomposition of the linking number of a closed ribbon: a problem from molecular biology. *Proc. Natl. Acad. Sci.* USA, vol. 75, no. 8. pp. 3557-3561 (1978).

[GA] K. Gawedski. Conformal Field Theory. Sem. Bourbaki. 41e annee (1988-89), 704, pp. 1-31.

[GE] M. L. Ge, N. Jing and Y. S. Wu. New quantum group associated with a "non-standard" braid group representation. (Preprint 1990.)

[GL] M. L. Glasser. Exact partition function for the two-dimensional Ising model. *Amer. J. Phys.*, vol. 38, no. 8, August 1970, pp. 1033-1036.

[G1] D. L. Goldsmith. The Theory of Motion Groups. *Michigan Math. J.*, 28(1981), pp. 3-17.

[G2] D. L. Goldsmith. Motion of Links in the Three Sphere, *Math. Scand.*, 50(1982), pp. 167-205.

[GR] B. Grossman. Topological quantum field theories: relations between knot theory and four manifold theory. (Preprint 1989.)

[GMM] E. Guadagnini, M. Martellini and M. Mintchev. Perturbative aspects of the Chern-Simons field theory. *Phys. Lett.*, B227 (1989), p. 111.

[HKC] P. de la Harpe, M. Kervaire and C. Weber. On the Jones polynomial. *L'Enseign Math.*, 32(1986), 271-335.

[HA] R. Hartley. Conway potential functions for links. *Comment. Math. Helv.*, 58(1983), pp. 365-378.

[HAS] B. Hasslacher and M. J. Perry. Spin networks are simplicial quantum gravity. *Phys. Lett.*, vol. 103B, no. 1, July (1981), pp. 21-24.

[HAY] M. V. Hayes. **A Unified Field Theory**. The Stinehour Press. (1964).

[HE1] M. A. Hennings. A polynomial invariant for oriented banded links. (Preprint 1989.)

[HE2] M. A. Hennings. A polynomial invariant for unoriented banded links. (Preprint 1989.)

[HE3] M. A. Hennings. Hopf algebras and regular isotopy invariants for link diagrams. (To appear in *Math. Proc. Cambridge Phil. Soc.*).

[HE4] M. A. Hennings. Invariants of links and 3-manifolds obtained from Hopf algebras. (To appear.)

[HO] C. F. Ho. A new polynomial invariant for knots and links - preliminary report. AMS Abstract, vol. 6, no. 4, Issue **39**(1985), 300.

[HOP] H. Hopf. Über die abbildungen der drei dimensionalen Sphäre die Kugelfläche. *Math. Ann.*, 104(1931), pp. 637-665.

[HOS1] J. Hoste. A polynomial invariant of knots and links. *Pacific J. Math.*, **124**(1986), 295-320.

[HOS2] J. Hoste and M. Kidwell. Dichromatic link invariants. *Trans. A.M.S.*, vol. 321, no. 1, (1990), pp. 197–229.

[HOS3] J. Hoste and J. H. Przytycki. Homotopy skein modules of orientable 3-manifolds. *Math. Proc. Camb. Phil. Soc.*, **108**(1990), pp. 475-488.

518

[HU] A. Hurwitz. "Über die Composition der quadratischen Formen von Beliebig vielen Variablen," **Nachrichten von der königlichen Gersellschaft der Wissenshaften in Göttingen** (1898), pp. 309-316; *Mathematische Werke,* vol. 11, pp. 565-571. Basel, 1932.

[HW] L. P. Hughston and R. S. Ward. **Advances in Twistor Theory.** Pitman (1979).

[IM1] T. D. Imbo, C. S. Imbo, R. S. Mahajan and E. C. G. Sudarshan. A User's Guide to Exotic Statistics. (Preprint 1990).

[IM2] T. D. Imbo and R. Brownstein. (private communication).

[IM3] T. D. Imbo, C. S. Imbo and E. C. G. Sudarshan. Identical particles, exotic statistics and braid groups. (Preprint 1990).

[JACI] R. Jackiw. Topological investigations of quantized gauge theories. In **Current Algebra and Anomalies.** Ed. by S. B. Treiman, R. Jackiw, B. Zumino, E. Witten. (1985), World Sci. Pub., pp. 211-340.

[JACK] E. A. Jackson. **Perspectives of Nonlinear Dynamics Vols. I and II.** Cambridge University Press (1990).

[JR] G. Jacussi and M. Rasetti. Exploring the use of artificial intelligence, logic programming, and computer-aided symbolic manipulation in computational physics I. 1. The mathematical structure of phase transition theory. *J. Phys. Chem.,* 91(1987), pp. 4970-4980.

[JA1] F. Jaeger. A combinatorial model for the Homfly polynomial. (Preprint 1988).

[JA2] F. Jaeger. On Tutte polynomials and cycles of plane graphs. *J. Comb. Theo. (B),* 44(1988), pp. 127-146.

[JA3] F. Jaeger. On transition polynomials of 4-regular graphs. (Preprint 1987).

[JA4] F. Jaeger. Composition products and models for the Homfly polynomial. *L'Enseignment Math.,* t35, (1989), pp. 323-361.

[JA5] F. Jaeger. Tutte polynomials and link polynomials. *Proc. Amer. Math. Soc.,* **103** (1988), pp. 647–654.

[JA6] F. Jaeger. On edge colorings of cubic graphs and a formula of Roger Penrose. *Ann. of Discrete Math.,* 41(1989), pp. 267-280.

[JA7] F. Jaeger, D. L. Vertigan and D. J. A. Welsh. On the computational complexity of the Jones and Tutte polynomials. *Math. Proc. Camb. Phil. Soc.,* 108(1990), pp. 35-53.

[JA8] F. Jaeger, L. H. Kauffman and H. Saleur. (In preparation.)

[JA9] F. Jaeger. (Announcement 1989.)

[JE] H. Jehle. Flux quantization and particle physics. *Phys. Rev. D.,* vol. 6, no. 2 (1972), pp. 441-457.

[JEN] H. Jenny. **Cymatics**, Basel: Basilius Presse, (1974).

[JI1] M. Jimbo. A q-difference analogue of $U(q)$ and the Yang-Baxter Equation. *Lect. in Math. Physics*, **10**(1985), 63-69.

[JI2] M. Jimbo. Quantum R-matrix for the generalized Toda system. *Comm. Math. Phys.*, **102**(1986), 537-547.

[JO1] V. F. R. Jones. A new knot polynomial and von Neumann algebras. *Notices of AMS*, **33**(1986), 219-225.

[JO2] V. F. R. Jones. A polynomial invariant for links via von Neumann algebras. *Bull. Amer. Math. Soc.*, **129**(1985), 103-112.

[JO3] V. F. R. Jones. Hecke algebra representations of braid groups and link polynomials. *Ann. of Math.*, **126**(1987), 335-388.

[JO4] V. F. R. Jones. On knot invariants related to some statistical mechanics models. *Pacific J. Math.*, vol. 137, no. 2, (1989), 311-334.

[JO5] V. F. R. Jones. On a certain value of the Kauffman polynomial. (To appear in *Comm. Math. Phys.*).

[JO6] V. F. R. Jones. Subfactors and related topics. In **Operator algebras and Applications**, vol. 2. Edited by D. E. Evans and M. Takesaki. Cambridge University Press (1988), pp. 103–118.

[JO7] V. F. R. Jones. Index for subfactors. *Invent. Math.*, **72**(1983), pp. 1-25.

[JO8] V. F. R. Jones. (Private communication.)

[JS] A. Joyal and R. Street. Braided monoidal categories. Macquarie reports 86008 (1986).

[JOY] D. Joyce. A classifying invariant of knots, the knot quandle. *J. Pure Appl. Alg.*, **23**, 37-65.

[KVA] L. H. Kauffman and F. J. Varela. Form dynamics. *J. Social and Bio. Struct.*, **3**(1980), pp. 171-206.

[LK] L. H. Kauffman. Link Manifolds. *Michigan Math. J.*, **21**(1974), pp. 33-44.

[LK1] L. H. Kauffman. The Conway polynomial. *Topology*, **20**(1980), 101-108.

[LK2] L. H. Kauffman. **Formal Knot Theory**. Princeton University Press, Mathematical Notes #30 (1983).

[LK3] L. H. Kauffman. **On Knots**. *Annals of Mathematics Studies* **115**, Princeton University Press (1987).

[LK4] L. H. Kauffman. State Models and the Jones Polynomial. *Topology*, **26**(1987), 395-407.

[LK5] L. H. Kauffman. Invariants of graphs in three-space. *Trans. Amer. Math. Soc.*, vol. 311, 2 (Feb. 1989), 697-710.

[LK6] L. H. Kauffman. New invariants in the theory of knots. (Lectures given in Rome, June 1986). – *Asterisque*, 163-164 (1988), pp. 137-219.

[LK7] L. H. Kauffman. New invariants in the theory of knots. *Amer. Math. Monthly*, vol. 95, no. 3, March 1988, pp. 195-242.

[LK8] L. H. Kauffman. An invariant of regular isotopy. *Trans. Amer. Math. Soc.*, vol. 318, no. 2, (1990), pp. 417-471.

[LK9] L. H. Kauffman. Statistical Mechanics and the Alexander polynomial. *Contemp. Math.*, 96(1989), Amer. Math. Soc. Pub., pp. 221-231.

[LK10] L. H. Kauffman. Statistical mechanics and the Jones polynomial. (Proceedings of the 1986 Santa Cruz conference on Artin's Braid Group. AMS Contemp. Math. Series (1989), vol. 78, pp. 263-297. Reprinted in **New Problems and Methods and Techniques in Quantum Field Theory and Statistical Mechanics.** pp. 175-222. Ed. by M. Rasetti. World Scientific Pub. (1990).

[LK11] L. H. Kauffman. A Tutte polynomial for signed graphs. *Discrete Applied Math.*, 25(1989), pp. 105-127.

[LK12] L. H. Kauffman. State models for knot polynomials - an introduction. (In the Proceedings of the meeting of the Brasilian Mathematical Society - July 1987.)

[LK13] L. H. Kauffman. State models for link polynomials. *l'Enseignment Mathematique*, t. 36 (1990), pp. 1-37.

[LK14] L. H. Kauffman. Knot polynomials and Yang-Baxter models. **IXth International Congress on Mathematical Physics − 17-27 July 1988 - Swansea, Wales,** Ed. Simon, Truman, Davies, Adam Hilger Pub. (1989), pp. 438-441.

[LK15] L. H. Kauffman. Polynomial Invariants in Knot Theory. **Braid Group, Knot Theory and Statistical Mechanics.** Ed. C. N. Yang and M. L. Ge. *World Sci. Pub. Advanced Series in Mathematical Physics*, 9(1989), 27-58.

[LK16] L. H. Kauffman. Spin networks and knot polynomials. *Intl. J. Mod. Phys. A.*, vol. 5, no. 1 (1990), pp. 93-115.

[LK17] L. H. Kauffman. Transformations in Special Relativity. *Int. J. Theo. Phys.*, vol. 24, no. 3, pp. 223-236, March 1985.

[LK18] L. H. Kauffman and H. Saleur. Free fermions and the Alexander-Conway polynomial. (To appear in *Comm. Math. Phys.*).

[LK19] L. H. Kauffman. Knots, Spin Networks and 3-Manifold Invariants. (To appear in the Proceedings of KNOTS 90 − Osaka Conference.)

[LK20] L. H. Kauffman and S. Lins. A 3-manifold invariant by state summation. (Preprint 1990).

[LK21] L. H. Kauffman and S. Lins. Computing Turaev-Viro invariants for 3-manifolds. (Preprint 1990).

[LK22] L. H. Kauffman and R. Baadhio. Classical knots, gravitational anomalies and exotic spheres. (To appear.)

[LK23] L. H. Kauffman. Knots, abstract tensors, and the Yang-Baxter Equation. In **Knots, Topology and Quantum Field Theories** - Proceedings of the Johns Hopkins Workshop on Current Problems in Particle Theory 13. Florence (1989). Ed. by L. Lussana. World Scientific Pub. (1989), pp. 179-334.

[LK24] L. H. Kauffman and P. Vogel. Link polynomials and a graphical calculus. (Announcement 1987. Preprint 1991.)

[LK25] L. H. Kauffman. Map coloring and the vector cross product. *J. Comb. Theor. B.*, vol. 48, no. 2. April 1990, pp. 145-154.

[LK26] L. H. Kauffman. From knots to quantum groups (and back). In **Proceedings of the CRM Workshop on Hamiltonian Systems, Transformation Groups and Spectral Transform Methods**. Ed. by J. Harnad and J. E. Marsden. Les Publications CRM (1990), pp. 161-176.

[LK27] L. H. Kauffman. **Map Reformulation**. Princelet Editions (1986).

[LK28] C. Anezeris, A. P. Balachandran, L. Kauffman and A. M. Srivastava. Novel statistics for strings and string 'Chern Simons' terms. (Preprint 1990.)

[LK29] L. H. Kauffman. An integral heuristic. *Intl. J. Mod. Phys. A.*, vol. 5, no. 7, (1990), pp. 1363-1367.

[LK30] L. H. Kauffman. Special relativity and a calculus of distinctions. In **Proceedings of the 9th Annual International Meeting of the Alternative Natural Philosophy Association - Cambridge, England (September 23, 1987)**. Published by ANPA West, Palo Alto, Calif., pp. 290-311.

[KA] T. Kanenobu. Infinitely many knots with the same polynomial. *Amer. Math. Soc.*, **97**(1986), pp. 158-161.

[KE] A. B. Kempe. On the geographical problem of four colors. *Amer. J. Math.*, **2**(1879), pp. 193-200.

[KV] M. A. Kervaire and J. W. Milnor. Groups of Homotopy Spheres: I. *Ann. of Math.*, vol. 77, no. 3, May 1963, pp. 504–537.

[K] M. Khovanov. New geometrical constructions in low dimensional topology. (Preprint 1990).

[KID] M. Kidwell. On the degree of the Brandt-Lickorish-Millett polynomial of a link. *Proc. Amer. Math. Soc.*, **100**(1987), pp. 755-762.

[KIRB] R. Kirby. A calculus for framed links in S^3. *Invent. Math.*, **45**(1978), 35–56.

[KM] R. Kirby and P. Melvin. On the 3-manifold invariants of Reshetikhin-Turaev for $sl(2, \mathbf{C})$. (Preprint 1990.)

[KI] T. P. Kirkman. The enumeration, description and construction of knots with fewer than 10 crossings. *Trans. Royal Soc. Edin.*, **32**(1865), 281-309.

[KR] A. N. Kirillov and N. Y. Reshetikhin. Representations of the algebra $U_q(sl\,2)$, q-orthogonal polynomials and invariants of links. In **Infinite dimensional Lie algebras and groups**, Ed. V. G. Kac, *Adv. Ser. in Math. Phys.*, **7**(1988), pp. 285-338.

[KO1] T. Kohno. Monodromy representations of braid groups and Yang-Baxter Equations. *Ann. Inst. Fourier, Grenoble,* **37**, 4 (1987), 139-160.

[KO2] T. Kohno. Topological invariants for 3-manifolds using representations of mapping class groups I. (Preprint 1990.)

[KON] M. Kontsevich. Rational conformal field theory and invariants of 3-dimensional manifolds. (Preprint 1988.)

[KU] P. P. Kulish, N. Yu, Reshetikhin and E. K. Skylyanin. Yang-Baxter Equation and representation theory: I. *Letters in Math. Physics,* **5**(1981), 393-403.

[KUS] P. P. Kulish and E. K. Sklyanin. Solutions of the Yang-Baxter Equation. *J. Soviet Math.,* **19**(1982), pp. 1596-1620. (Reprinted in **Yang-Baxter Equation in Integrable Systems.** Ed. by M. Jimbo. World Scientific Pub. (1990).)

[KAW] A. Kuniba, Y. Akutsu, M. Wadati. Virasoro algebra, von Neumann algebra and critical eight-vertex SOS models. *J. Phys. Soc. Japan,* **55**, no. 10 (1986), 3285-3288.

[KUP1] G. Kuperberg. Involutory Hopf algebras and 3-manifold invariants. (Preprint 1990.)

[KUP2] G. Kuperberg. The quantum G_2 link invariant. (Preprint 1990.)

[KMU] B. I. Kurpita and K. Murasugi. On a hierarchical Jones invariant. (Preprint 1990.)

[LAR] L. A. Lambe and D. E. Radford. Algebraic aspects of the quantum Yang-Baxter Equation. (Preprint 1990.)

[LAT] R. G. Larson and J. Towber. Braided Hopf algebras: a dual concept to quasitriangularity. (Preprint 1990.)

[LAW1] R. Lawrence. A universal link invariant using quantum groups. (Preprint 1988).

[LAW2] R. Lawrence. A functorial approach to the one-variable Jones polynomial. (Preprint 1990.)

[LEE1] H. C. Lee. Q-deformation of $sl(2, C) \times Z_N$ and link invariants. (Preprint 1988.)

[LEE2] H. C. Lee. Tangles, links and twisted quantum groups. (Preprint 1989.)

[LCS] H. C. Lee, M.Couture and N. C. Schmeling. Connected link polynomials (Preprint 1988).

[LM1] W. B. R. Lickorish and K. C. Millett. A polynomial invariant for oriented links. *Topology,* **26**(1987), 107-141.

[LM2] W. B. R. Lickorish and K. C. Millett. The reversing result for the Jones polynomial. *Pacific J. Math.*, **124**(1986), pp. 173-176.

[LICK1] W. B. R. Lickorish. A representation of orientable, combinatorial 3-manifolds. *Ann. Math.*, **76**(1962), pp. 531-540.

[LICK2] W. B. R. Lickorish. Polynomials for links. *Bull. London Math. Soc.*, 20(1988), 558-588.

[LICK3] W. B. R. Lickorish. 3-Manifolds and the Temperley-Lieb Algebra. (Preprint 1990.)

[LICK4] W. B. R. Lickorish. Calculations with the Temperley-Lieb algebra. (Preprint 1990.)

[LIP1] A. S. Lipson. An evaluation of a link polynomial. *Math. Proc. Camb. Phil. Soc.*, **100**(1986), 361-364.

[LIP2] A. S. Lipson. Some more states models for link invariants. *Pacific J. Math.*, (To appear).

[LIP3] A. S. Lipson. Link signatures, Goeritz matrices and polynomial invariants, *L'Enseignment Math.*, t.36, (1990), pp. 93-114.

[LIT] C. N. Little. Non-alternate + − knots. *Trans. Royal Soc. Edin.*, **35**(1889), 663-664.

[LOR] E. N. Lorenz. Nonperiodic flow. *J. Atmos. Sci.*, **20**(1963), pp. 130-141.

[M] S. Mac Lane. **Categories for the Working Mathematician.** Springer-Verlag (1971). GTM 5.

[MN] J. M. Maillet and F. W. Nijhoff. Multidimensional integrable lattices, quantum groups and the *D*-simplex equations. (Preprint – Institute for Non-Linear Studies – Clarkson University – INS #131 (1989).)

[MAJ] S. Majid. Quasitriangular Hopf algebras and Yang-Baxter equations. (Preprint 1988).

[MAN1] Yu. I. Manin. Quantum groups and non-commutative geometry. (Lecture Notes, Montreal, Canada, July 1988).

[MAN2] Yu. I. Manin. Some remarks on Kozul algebras and Quantum groups. *Ann. Inst. Fourier, Grenoble*, **37**, 4 (1987), 191-205.

[MAR] P. P. Martin. Analytic properties of the partition function for statistical mechanical models. *J. Phys. A: Math. Gen.*, **19**(1986), pp. 3267-3277.

[MAT] S. V. Matveev. Transformations of special spines and the Zeeman conjecture. *Math. USSR Izvestia* 31: 2 (1988), pp. 423-434.

[ME] D. A. Meyer. State models for link invariants from the classical Lie groups. (Preprint 1990.)

[MIE] E. W. Mielke. Knot wormholes in geometrodynamics. *General Relativity and Gravitation*, vol. 8, no. 3, (1977), pp. 175-196.

[MI] J. W. Milnor. **Singular Points of Complex Hypersurfaces.** *Annals Studies* **61**, Princeton University Press (1968).

[MD] J. Moody. A link not detected by the Burau matrix. (Preprint 1989.)

[MS] G. Moore and N. Seiberg. Classical and quantum conformal field theory. *Commun. Math. Phys.*, **123**(1989), pp. 177-254.

[MO1] H. R. Morton. The Jones polynomial for unoriented links. *Quart. J. Math. Oxford (2)*. **37**(1986), pp. 55-60.

[MO2] H. R. Morton. Seifert circles and knot polynomials. *Math. Proc. Camb. Phil. Soc.*, **99**(1986), pp. 107-109.

[MO3] H. R. Morton and H. B. Short. The two-variable polynomial for cable knots. *Math. Proc. Camb. Phil. Soc.*, **101**(1987), pp. 267-278.

[MO4] H. R. Morton and P. M. Strickland. Satellites and surgery invariants. (Preprint 1990.)

[MOU] J. P. Moussouris. **Quantum Models of Space-Time Based on Recoupling Theory.** (Thesis, Oxford Univ., 1983).

[MUK1] J. Murakami. The parallel version of link invariants. (Preprint Osaka University (1987).)

[MUK2] J. Murakami. A state model for the multi-variable Alexander polynomial. (Preprint 1990.)

[MUK3] J. Murakami. On local relations to determine the multi-variable Alexander polynomial of colored links. (Preprint 1990.)

[MUR1] K. Murasugi. The Jones polynomial and classical conjectures in knot theory. *Topology*, **26**(1987), 187-194.

[MUR2] K. Murasugi. Jones polynomials and classical conjectures in knot theory II. *Math. Proc. Camb. Phil. Soc.*, **102**(1987), 317-318.

[N] M. H. A. Newman. On a string problem of Dirac. *J. London Math. Soc.*, **17**(1942), pp. 173-177.

[O] E. Oshins. Quantum Psychology Notes. Vol. 1: A Personal Construct Notebook (1987). (Privately distributed.)

[PEN1] R. Penrose. Applications of negative dimensional tensors. **Combinatorial Mathematics and its Applications.** Edited by D. J. A. Welsh, Academic Press (1971).

[PEN2] R. Penrose. Theory of quantized directions. (Unpublished.)

[PEN3] R. Penrose. Angular momentum: an approach to Combinatorial Space-Time. In **Quantum Theory and Beyond.** Ed. T. A. Bastin. Cambridge University Press, (1969).

[PEN4] R. Penrose. Combinatorial quantum theory and quantized directions. **Advances in Twistor Theory.** Ed. by L. P. Hughston and R. S. Ward. Pitman (1979), pp. 301-307.

[PW] J. H. H. Perk and F. Y. Wu. Nonintersecting string model and graphical approach: equivalence with a Potts model. *J. Stat. Phys.*, vol. 42, nos. 5/6, (1986), pp. 727-742.

[PS] J. H. H. Perk and C. L. Schultz. New families of commuting transfer matrices in q-state vertex models. *Physics Letters*, vol. 84A, no. 8, August 1981, pp. 407-410.

[PO] A. M. Polyakov. Fermi-Bose transmutations induced by gauge fields. *Mod. Phys. Lett.*, A3 (1987), pp. 325-328.

[PT] J. H. Przytycki and P. Traczyk. Invariants of links of Conway type. *Kobe J. Math.*, 4(1987), pp. 115-139.

[Q] C. Quigg. **Gauge Theories of the Strong, Weak, and Electromagnetic Interactions**. Benjamin Pub. (1983).

[RAD] D. E. Radford. Solutions to the quantum Yang-Baxter Equation: the two-dimensional upper triangular case. (Preprint 1990.)

[RA] M. Rasetti and T. Regge. Vortices in He II, Current Algebras and Quantum Knots. *Physica* 80a (1975), pp. 217-233, Quantum vortices. In **Highlights of Condensed Matter Theory** (1985). Soc. Italiana di Fiscia - Bologna, Italy.

[REI] K. Reidemeister. **Knotentheorie**. Chelsea Publishing Co., New York (1948). Copyright 1932. Julius Springer, Berlin.

[RES] N. Y. Reshetikhin. Quantized universal enveloping algebras, the Yang-Baxter Equation and invariants of links, I and II. LOMI reprints E-4-87 and E-17-87, Steklov, Institute, Leningrad, USSR.

[RT1] N. Y. Reshetikhin and V. Turaev. Ribbon graphs and their invariants derived from quantum groups. *Comm. Math. Phys.*, **127**(1990), pp. 1-26.

[RT2] N. Y. Reshetikhin and V. Turaev. Invariants of three manifolds via link polynomials and quantum groups. *Invent. Math.*, **103** (1991), 547–597.

[ROB] G. D. Robertson. Torus knots are rigid string instantons. University of Durham - Centre for Particle Theory. (Preprint 1989.)

[ROLF] D. Rolfsen. **Knots and Links**. Publish or Perish Press (1976).

[ROSS1] M. Rosso. Groupes quantiques de Drinfeld et Woronowicz. (Preprint 1988).

[ROSS2] M. Rosso. Groupes quantiques et modeles a vertex de V. Jones en theorie des noeuds. *C. R. Scad. Sci. Paris*, t.307 (1988), pp. 207-210.

[RS] C. Rovelli and L. Smolin. Knot theory and quantum gravity. (Preprint 1988).

[RY] L. H. Ryder. **Quantum Field Theory**. Cambridge University Press (1985).

[SA] T. L. Saaty and P. C. Kainen. **The Four Color Problem**. Dover Pub. (1986).

[SAL1] H. Saleur. Virasoro and Temperley-Lieb algebras. In **Knots, Topology and Quantum Field theories** – Proceedings of the Johns Hopkins Workshop on Current Problems in Particle Theory 13. Florence (1989). Ed. by L. Lussana. World Scientific Pub. (1989), pp. 485-496.

[SAL2] H. Saleur. Zeroes of chromatic polynomials: a new approach to Beraha conjecture using quantum groups. *Commun. Math. Phys.*, **132**(1990), pp. 657-679.

[SAM] S. Samuel. The use of anticommuting variable integrals in statistical mechanics. I. The computation of partition functions. II. The computation of correlation functions. III. Unsolved models. *J. Math. Phys.*, **21(12)**(1980), pp. 2806-2814, 2815-2819, 2820-2833.

[SE] G. Segal. Two-dimensional conformal field theories and modular functors. **IXth International Congress on Mathematical Physics – 17-27 July-1988-Swansea, Wales.** Ed. Simon, Truman, Davies, Adam Hilger Pub. (1989), pp. 22-37.

[CS] C. Shultz. Solvable q-state models in lattice statistics and quantum field theory. *Phys. Rev. Letters.* **46** no. 10, March 1981.

[SI] L. Silberstein. **The Theory of Relativity.** MacMillan and Co. Ltd. (1914).

[SIM] J. Simon. Topological chirality of certain molecules. *Topology,* vol. 25, no. 2, (1986), pp. 229-235.

[SK] E. K. Skylanin. Some algebraic structures connected with the Yang-Baxter Equation. Representation of quantum algebras. *Funct. Anal. Appl.,* **17**(1983), pp. 273-284.

[LS1] L. Smolin. Link polynomials and critical points of the Chern-Simon path integrals. *Mod. Phys. Lett. A.,* vol. 4, no. 12, (1989), pp. 1091-1112.

[LS2] L. Smolin. Quantum gravity in the self-dual representation. *Contemp. Math.,* **28**(1988).

[SO1] R. Sorkin. Particle statistics in three dimensions. *Phys. Dev.,* D27, (1983), pp. 1787-1797.

[SO2] R. Sorkin. A general relation between kink exchange and kink rotation. *Commun. Math. Phys.,* **115**(1988), pp. 421-434.

[SB1] G. Spencer-Brown. **Cast and Formation Properties of Maps.** Privately distributed manuscript. Deposited in the Library of the Royal Society, London, March 1980.

[SB2] G. Spencer-Brown. **Laws of Form.** George Allen and Unwin Ltd., London (1969).

[STEW] R. L. Stewart. Pure field electromagnetism. (Unpublished - 1971).

[ST] J. Stillwell. **Classical Topology and Combinatorial Group Theory.** Springer-Verlag (1980).GTM 72.

[S] D. W. Sumners. Untangling DNA. *Math. Intelligencer.* vol. 12, no. 3, (1990), pp. 71-80.

[TAIT] P. G. Tait. On Knots I, II, III. Scientific Papers, vol. I, Cambridge University Press, London, 1898, 273-347.

[TEL] H. N. V. Temperley and E. H. Lieb. Relations between the 'percolation' and 'coloring' problem and other graph-theoretical problems associated with regular planar lattices: some exact results for the 'percolation' problem. *Proc. Roy. Soc. Lond. A.*, **322**(1971), pp. 251-280.

[TEU] R. D. Teuschner. Towards a topological spin-statistics theorem in quantum field theory, (Preprint 1989.)

[TH1] M. B. Thistlethwaite. Knot tabulations and related topics. **Aspects of Topology.** Ed. I. M. James and E. H. Kronheimer, Cambridge University Press (1985), 1-76.

[TH2] M. B. Thistlethwaite. A spanning tree expansion of the Jones polynomial. *Topology*, **26**(1987), pp. 297-309.

[TH3] M. B. Thistlethwaite. On the Kauffman polynomial of an adequate link. *Invent. Math.*, **93**(1988), pp. 285-296.

[TH4] M. B. Thistlethwaite. Kauffman's polynomial and alternating links. *Topology*, **27**(1988), pp. 311-318.

[TH5] M. B. Thistlethwaite. (Unpublished tables.)

[T] W. Thompson (Lord Kelvin). On vortex atoms. *Philosophical Magazine*, **34** July 1867, pp. 15-24. **Mathematical and Physical Papers**, vol. 4, Cambridge (1910).

[TR] B. Trace. On The Reidemeister moves of a classical knot. *Proc. Amer. Math. Soc.*, vol. 89, no. 4, (1983).

[TU1] V. G. Turaev. The Yang-Bater equation and invariants of links. LOMI preprint E-3-87, Steklov Institute, Leningrad, USSR, *Inventiones Math.*, **92**, Fasc. 3, 527–553.

[TU2] V. G. Turaev and O. Viro. State sum invariants of 3-manifolds and quantum $6j$ symbols. (Preprint 1990.)

[TU3] V. G. Turaev. Quantum invariants of links and 3-valent graphs in 3-manifolds. (Preprint 1990.)

[TU4] V. G. Turaev. Shadow links and face models of statistical mechanics. Publ. Inst. Recherche Math. Advance. Strasbourg (1990).

[TU5] V. G. Turaev. Quantum invariants of 3-manifolds and a glimpse of shadow topology. (Preprint 1990.)

[TU6] V. G. Turaev. The Conway and Kauffman modules of the solid torus with an appendix on the operator invariants of tangles. LOMI Preprint E-6-88.

[TU7] V. G. Turaev. Algebras of loops on surfaces, algebras of knots, and quantization. LOMI Preprint E-10-88. In **Braid Group, Knot Theory and Statistical Mechanics**. Ed. C. N. Yang and M. L. Ge. World Sci. Pub. *Advanced Series in Mathematical Physics*, 9(1989), pp. 59-96.

[TUT] W. T. Tutte. Graph Theory. **Encyclopedia of Mathematics and its Applications.** 21, Cambridge University Press, (1984).

[VMU] F. Varela, H. Maturana, R. Uribe. Autopoiesis: the organization of living systems, its characterization and a model. *Biosystems*, 5:187, (1974).

[V] P. Vogel. Representations of links by braids: A new algorithm. *Comment. Math. Helvetici*, **65**(1990), pp. 104-113.

[WE] H. Wenzl. Representations of braid groups and the quantum Yang-Baxter Equation, *Pacific J. Math.*, **145**(1990), pp. 153-180.

[WEYL] H. Weyl. Gravitation und Elektrizitat. *Sitzungsberichte der Königlich Preussischen Akademie der Wissenshäften*, 26(1918), pp. 465-480.

[W] J. H. White. Self-linking and the Gauss integral in higher dimensions. *Amer. J. Math.*, **91**(1969), pp. 693-728.

[WH1] H. Whitney. On regular closed curves in the plane. *Comp. Math.*, 4(1937), 276-284.

[WH2] H. Whitney. A logical expansion in mathematics. . *Bull. Amer. Math. Soc.*, **38**(1932), 572-579.

[WI] A. S. Wightman. (Private communication.)

[WIL] F. Wilczek. Gauge theory of deformable bodies. IASSNS-HEP-88/41 (Preprint 1988), in *Intl. Congress on Math. Phys.* Ed. by B. Simon, A. Truman and I. M. Davies. Adam Hilger (1989), pp. 220-233.

[WZ] F. Wilczek and A. Zee. Linking numbers, spin and statistics of solitons. *Phys. Rev. Lett.*, vol. 51, no. 25, (1983), pp. 2250-2252.

[WIT1] E. Witten. Physics and geometry. *Proc. Intl. Congress Math.*, Berkeley, Calif., USA (1986), 267-303.

[WIT2] E. Witten. Quantum field theory and the Jones polynomial. *Comm. Math. Phys.*, **121**(1989), 351-399.

[WIT3] E. Witten, Gauge theories vertex models and quantum groups. *Nucl. Phys. B.*, **330**(1990), pp. 225–246.

[WIT4] E. Witten. Global gravitational anomalies. *Comm. Math. Phys.*, **100**(1985), pp. 197-229.

[WIN] S. Winker. **Quandles, Knots Invariants and the N-fold Branched Cover**. Ph.D. Thesis, University of Illinois at Chicago, (1984).

[WU1] F. Y. Wu. The Potts Model. *Rev. Mod. Phys.*, vol. 54, no. 1, Jan. 1982, pp. 235-268.

[WU2] F. Y. Wu. Graph theory in statistical physics. In **Studies in Foundations and Combinatorics** - Advances in Mathematics Supplementary Studies. Vol. 1, (1978). Academic Press Inc., pp. 151-166.

[Y] S. Yamada. The minimum number of Seifert circles equals the braid index of a link. *Invent. Math.*, **89**(1987), pp. 346-356.

[YGW] K. Yamagishi, M. Ge, Y. Wu. New hierarchies of knot polynomials from topological Chern-Simons gauge theory. *Lett. Math. Phys.*, **19**(1990), pp. 15-24.

[YE] D. Yetter. Quantum groups and representations of monodial categories. *Math. Proc. Camb. Phil. Soc.*, **108**(1990), pp. 197-229.

[ZA] M. Zaganescu. Bosonic knotted strings and cobordism theory. *Europhys. Lett.*, **4(5)**(1987), pp. 521-525.

[Z1] A. B. Zamolodchikov. Factorized S matrices and lattice statistical systems. *Soviet Sci. Reviews.* Part A (1979-1980).

[Z2] A. B. Zamolodchikov. Tetrahedron equations and the relativistic S-matrix of straight strings in $2+1$ dimensions. *Comm. Math, Phys.*, **79**(1981), 489-505.

Index

538